Morphology Design Paradigms for Supercapacitors

Morphology Design Paradigms for Supercapacitors

Edited by
Inamuddin
Rajender Boddula
Mohammad Faraz Ahmer
Abdullah M. Asiri

CRC Press
Taylor & Francis Group
Boca Raton London New York

CRC Press is an imprint of the
Taylor & Francis Group, an **informa** business

CRC Press
Taylor & Francis Group
6000 Broken Sound Parkway NW, Suite 300
Boca Raton, FL 33487-2742

First issued in paperback 2021

© 2020 by Taylor & Francis Group, LLC
CRC Press is an imprint of Taylor & Francis Group, an Informa business

No claim to original U.S. Government works

ISBN 13: 978-1-03-223814-2 (pbk)
ISBN 13: 978-0-367-20754-0 (hbk)

Visit the Taylor & Francis Web site at
http://www.taylorandfrancis.com

and the CRC Press Web site at
http://www.crcpress.com

Contents

Preface

The fast growth of worldwide energy utilization has moved various research activities toward the development of clean and sustainable power sources (e.g., solar, wind, water splitting) during the most recent decade. As these energy systems are intermittent (e.g., solar and wind) or regionally constrained (e.g., water), there is a need to create advanced energy storage systems, for example, supercapacitors, for the productive storage of electrical energy. Supercapacitors have attracted significant attention during the last few decades because of their high specific power density, long cycle life, and capacity to connect the power/energy gap between customary capacitors and batteries/fuel cells. Even though the effectiveness of energy storage devices relies upon a variety of variables, their overall performance strongly depends on the structure and properties of the part materials. The current advancements in nanotechnology have opened up new frontiers by making new materials for effective energy storage, for example, fullerenes, carbon nanotubes, quantum dabs, graphene, carbon onions, hexagonal carbides, carbon curls, and carbonitrides-MXenes, and perovskites and structures, for example, multidimensional congregations, bioinspired structures, 0D nanoparticles, 1D nanowires, 2D nanosheets, and 3D nanofoams and networks, and so on. These nanostructured electrode materials have exhibited unrivaled electrochemical properties for creating elite supercapacitors.

This book depicts the current progress and advances made in nanostructured supercapacitor electrode materials of different **dimensions ranging** from zero to three, with microporous to mesoporous nature, and of subatomic to nanoscales. The potential applications, including customer electronics, wearable gadgets, hybrid electric vehicles, and stationary and industrial energy systems, are also highlighted. It also explores the impact of nanostructures on the properties of supercapacitors, including **specific capacitance**, **cycle stability**, **and rate capability**, which are deemed necessary for the upcoming generation of supercapacitor electrodes. It also discusses the recent research progress toward modification of the morphology of energy storage materials used in supercapacitor applications. The emphasis is given to provide in-depth knowledge on supercapacitor materials of various morphologies and structures including 0D and 1D carbon nanostructures, inorganic 1D nanomaterials, core–shell nanomaterials, hierarchical nanostructures, mesoporous and honeycomb nanostructures, 1D polymers, and metal–metal oxide systems. This book is a unique reference guide for postgraduates, researchers, and academicians of chemistry, physics, nanotechnology, advanced materials science, and other interdisciplinary fields of science and technology, more specially electrochemistry. This book is planned to provide the readers with a clear thought regarding the execution of materials of different morphologies and structures as an electrode in electrochemical supercapacitors.

Inamuddin
King Abdulaziz University, Saudi Arabia;
Aligarh Muslim University, Aligarh, India

Rajender Boddula
Chinese Academy of Sciences (CAS)
National Center for Nanoscience and Technology
Beijing, PR China

Mohammad Faraz Ahmer
Mewat College of Engineering and Technology,
Mewat, India

Abdullah M. Asiri
King Abdulaziz University, Jeddah, Saudi Arabia

Editors

Dr. Inamuddin is currently working as Assistant Professor in the Chemistry Department, Faculty of Science, King Abdulaziz University, Jeddah, Saudi Arabia. He is a permanent faculty member (Assistant Professor) at the Department of Applied Chemistry, Aligarh Muslim University, Aligarh, India. He obtained Master of Science degree in Organic Chemistry from Chaudhary Charan Singh (CCS) University, Meerut, India, in 2002. He received his Master of Philosophy and Doctor of Philosophy degrees in Applied Chemistry from Aligarh Muslim University (AMU), India, in 2004 and 2007, respectively. He has extensive research experience in multidisciplinary fields of Analytical Chemistry, Materials Chemistry, and Electrochemistry, and, more specifically, Renewable Energy and Environment. He has worked in different research projects as project fellow and senior research fellow funded by University Grants Commission (UGC), Government of India, and Council of Scientific and Industrial Research (CSIR), Government of India. He has received Fast Track Young Scientist Award from the Department of Science and Technology, India, to work in the area of bending actuators and artificial muscles. He has completed four major research projects sanctioned by UGC, Department of Science and Technology, CSIR, and Council of Science and Technology, India. He has published 147 research articles in international journals of repute and 18 book chapters in knowledge-based book editions published by renowned international publishers. He has published 60 edited books with Springer (UK), Elsevier, Nova Science Publishers, Inc. (USA), CRC Press Taylor & Francis Asia Pacific, Trans Tech Publications Ltd. (Switzerland), IntechOpen Limited (UK), and Materials Science Forum LLC (USA). He is a member of various journals' editorial boards. He is also serving as Associate Editor for journals (Environmental Chemistry Letter, Applied Water Science and Euro-Mediterranean Journal for Environmental Integration, Springer-Nature), Frontiers Section Editor (Current Analytical Chemistry, Bentham Science Publishers), Editorial Board Member (Scientific Reports-Nature), Editor (Eurasian Journal of Analytical Chemistry), and Review Editor (Frontiers in Chemistry, Frontiers, UK) He is also guest-editing various special thematic special issues to the journals of Elsevier, Bentham Science Publishers, and John Wiley & Sons, Inc. He has attended as well as chaired sessions in various international and national conferences. He has worked as a Postdoctoral Fellow, leading a research team at the Creative Research Initiative Center for Bio-Artificial Muscle, Hanyang University, South Korea, in the field of renewable energy, especially biofuel cells. He has also worked as a Postdoctoral Fellow at the Center of Research Excellence in Renewable Energy, King Fahd University of Petroleum and Minerals, Saudi Arabia, in the field of polymer electrolyte membrane fuel cells and computational fluid dynamics of polymer electrolyte membrane fuel cells. He is a life member of the Journal of the Indian Chemical Society. His research interests include ion-exchange materials, a sensor for heavy metal ions, biofuel cells, supercapacitors, and bending actuators.

Dr. Rajender Boddula is currently working for Chinese Academy of Sciences President's International Fellowship Initiative (CAS-PIFI) at National Center for Nanoscience and Technology (NCNST, Beijing). His academic honors includes University Grants Commission National Fellowship and many merit scholarships, and CAS-PIFI. He has published many scientific articles in international peer-reviewed journals and has authored twenty book chapters. He is also serving as an editorial board member and a referee for reputed international peer-reviewed journals. He has published edited books with Springer (UK), Elsevier, Materials Science Forum LLC (USA) and CRC Press Taylor & Francis Asia Pacific, Trans Tech Publications Ltd. (Switzerland). His specialized areas of research are energy conversion and storage, which include sustainable nanomaterials, graphene, polymer composites, heterogeneous catalysis for organic transformations, environmental remediation technologies, photoelectrochemical water-splitting devices, biofuel cells, batteries and supercapacitors.

Dr. Mohammad Faraz Ahmer is presently working as Assistant Professor in the Department of Electrical Engineering, Mewat Engineering College, Nuh Haryana, India, since 2012 after working as Guest Faculty in University Polytechnic, Aligarh Muslim University, Aligarh, India, during 2009–2011. He completed M.Tech. (2009) and Bachelor of Engineering (2007) degrees in Electrical Engineering in Aligarh Muslim University, Aligarh, in the first division. He obtained a Ph.D. degree in 2016 on his thesis entitled "Studies on Electrochemical Capacitor Electrodes." He has published six research papers in reputed scientific journals. He has edited two books with Materials Science Forum, USA. His scientific interests include electrospun nanocomposites and supercapacitors. He has presented his work at several conferences. He is actively engaged in searching of new methodologies involving the development of organic composite materials for energy storage systems.

Prof. Abdullah M. Asiri is the Head of the Chemistry Department at King Abdulaziz University since October 2009, and he is the Founder and the Director of the Center of Excellence for Advanced Materials Research (CEAMR) since 2010 till date. He is the Professor of Organic Photochemistry. He graduated from King Abdulaziz University (KAU) with B.Sc. in Chemistry in 1990 and obtained a Ph.D. from University of Wales, College of Cardiff, UK, in 1995. His research interests cover color chemistry, synthesis of novel photochromic and thermochromic systems, synthesis of novel coloring matters and dyeing of textiles, materials chemistry, nanochemistry and nanotechnology, polymers, and plastics. Prof. Asiri is the principal supervisor of more than 20 M.Sc. and six Ph.D. theses. He is the main author of ten books of different chemistry disciplines. Prof. Asiri is the Editor-in-Chief of King Abdulaziz University Journal of Science. A major achievement of Prof. Asiri is the discovery of tribochromic compounds, a new class of compounds that change from colorless or slightly colored to deep colored when subjected to small pressure or when ground. This research was introduced to the scientific community by a new terminology as published by International Union of Pure and Applied Chemistry (IUPAC) in 2000. This discovery was awarded a patent from European Patent Office and from the UK Patent Office. Prof. Asiri was involved in many committees at the

KAU level and on the national level. He took a major role in the advanced materials committee working for King Abdulaziz City for Science and Technology (KACST) to identify the national plan for science and technology in 2007. Prof. Asiri played a major role in advancing the chemistry education and research in KAU. He has been awarded the Best Researcher Award from KAU for the past five years. He was also awarded the Young Scientist Award from the Saudi Chemical Society in 2009 and received the first prize for the distinction in science from the Saudi Chemical Society in 2012. He also received a recognition certificate from the American Chemical Society (Gulf region chapter) for the advancement of chemical science in the Kingdom. He received a Scopus certificate as the most publishing scientist in Saudi Arabia in chemistry in 2008. He is also a member of the editorial board of various journals of international repute. He is the Vice-President of Saudi Chemical Society (Western Province Branch). He holds four US patents, more than one thousand publications in international journals, and several book chapters and edited books.

Contributors

Badekai Ramachandra Bhat
Catalysis and Materials Laboratory
Department of Chemistry
National Institute of Technology
Karnataka
Mangalore, India

Nivedhini Iswarya Chandrasekaran
Department of Chemical Engineering
National Institute of Technology
Tiruchirappalli, India

Huailin Fan
School of Material Science
and Engineering
University of Jinan
Jinan, PR China

Zhaoyang Fan
Department of Electrical and Computer
Engineering
Texas Tech University
Lubbock, Texas

Neena S. John
Centre for Nano and Soft Matter
Sciences
Jalahalli, India

Talam E. Kibona
Department of Physics, Mathematics
and Informatics
Mkwawa College, University of Dar es
Salaam P. O.
Iringa, Tanzania

Wenyue Li
Department of Electrical and Computer
Engineering
Texas Tech University
Lubbock, Texas

Manickam Matheswaran
Department of Chemical Engineering
National Institute of Technology
Tiruchirappalli, India

Xiangchao Meng
College of Chemistry and Chemical
Engineering
Ocean University of China
Qingdao, China

and

Department of Chemical and Biological
Engineering
University of Ottawa
Ottawa, Canada

Praveen Mishra
Catalysis and Materials Laboratory
Department of Chemistry
National Institute of Technology
Karnataka
Surathkal, India

Xuan Pan
Chinese Academy of Sciences
Institutes of Science and Development
Beijing, PR China

K. K. Purushothaman
Department of Physics
Arignar Anna Government Arts and
Science College
Karaikal, Pondicherry, India

Vivek Ramakrishnan
Centre for Nano and Soft Matter
Sciences
Jalahalli, India

M. Ramesh
Department of Mechanical Engineering
KIT-Kalaignarkarunanidhi Institute of
   Technology
Coimbatore, India

Arivumani Ravanan
Department of Mechanical Engineering
KIT-Kalaignarkarunanidhi Institute of
   Technology
Coimbatore, India

Manas Roy
Department of Chemistry
National Institute of Technology
Agartala, Tripura, India

Mitali Saha
Department of Chemistry
National Institute of Technology
Agartala, Tripura, India

B. Saravanakumar
Department of Physics
Dr. Mahalalingam College of
   Engineering and Technology
Pollachi, India

B. Sethuraman
Department of Physics
TRP Engineering College
Trichy, India

Godlisten N. Shao
Department of Chemistry
Mkwawa College
University of Dar es Salaam P. O.
Iringa, Tanzania

Liang Wang
College of Chemistry and Chemical
   Engineering
Ocean University of China
Qingdao, PR China

Mengqing Wang
Department of Chemical and Biological
   Engineering
University of Ottawa
Ottawa, Canada

Zisheng Zhang
Department of Chemical and Biological
   Engineering
University of Ottawa
Ottawa, Canada

1 Zero-Dimensional Carbon Nanostructures for Supercapacitors

Praveen Mishra and Badekai Ramachandra Bhat

CONTENTS

1.1 INTRODUCTION

The dependence of civilization on energy is rapidly increasing as the decades pass by. The world is now striving to find new sources to harness energy to meet this ever-increasing demand. This demand prompts research on energy generation and, more importantly, on energy storage devices. The latter has a high demand owing to the boom in the use of personal and portable electronic devices. Batteries have been catering to this requirement for quite some time now. Supercapacitors are the upcoming energy storing devices that can replace batteries due to their higher power output and fast recharge (Winter and Brodd 2004). A supercapacitor or an electrochemical capacitor is a simple device that can store charge at the interface of an electrode and an electrolyte (Kötz and Carlen 2000, Frackowiak and Béguin 2001). They are considered to be advantageous over batteries due to their very high specific capacitance, high power density, maintenance-free operations, no memory effect, and tapping into the massive void between the power-energy difference between capacitors and batteries (Winter and Brodd 2004, Poonam et al. 2019). A simplistic representation of this may be seen from a Ragone plot (Figure 1.1). As evident from

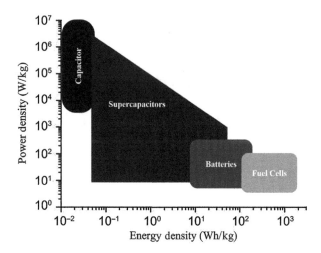

FIGURE 1.1   A Ragone plot depicting the relative positions of various electrochemical systems with respect to the energy density (E_d) and power density (P_d).

the figure, the supercapacitors fill the gap between capacitors and batteries in terms of power density (P_d) and energy density (E_d).

The history of the development of capacitors finds its roots in the development of the Leyden Jar in the year 1745–1746, which was essentially a glass vessel being the dielectric medium with metal foils acting as electrodes. The first electrolytic capacitor was reported in the 1920s. However, for supercapacitors, the first instance of the documented report surfaces in 1957, which is the electrostatic double-layer capacitor (EDLC) patented by General Electric and developed using activated charcoal, a form of highly porous carbon as plates. The supercapacitors made of carbonaceous materials usually result in an EDLC. EDLCs reflect their characteristic rectangular cyclic voltammetry curves and symmetrical galvanostatic charge-discharge profile. A typical EDLC is based on the principle of non-Faradaic charge transfer, in which the separation of charges is achieved by the formation of a few angstrom lengths, thick Helmholtz double layer between the surface of the conductive electrodes and the electrolyte (Figure 1.2). The reasons underlying the higher specific capacitance of the supercapacitor are as follows:

1. The capacitance is inversely proportional to the distance between the charged surfaces, which is very small for supercapacitor.
2. The specific charge per unit area, which is directly proportional to the capacitance, is a consequence of very high specific surface area of carbon electrodes, and the use of easily ionizable electrolyte results in the large deposit of electrostatic charge per unit surface area of the electrode.

Carbon materials are preferred for such applications due to their versatile existence, low cost, and environmental-friendly nature (Frackowiak 2007). Early carbon materials used to fabricate electrodes for supercapacitors have been amorphous carbons

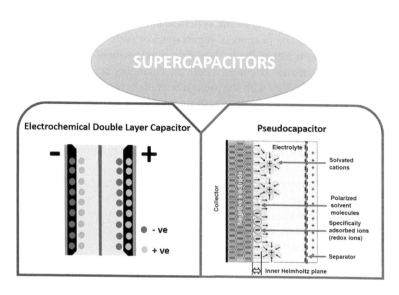

FIGURE 1.2 Types of supercapacitors.

(Pandolfo and Hollenkamp 2006). Carbon is available in various allotropic forms such as diamond, fullerenes, and graphite, which is a large family of nanostructures (0 to 3D) with microtextures (of varying orderedness) due to the degree of graphitization and polymorphism (e.g., aerogels, powders, fibers, foams, fabrics, and composites). The diversity in the structural morphology of carbonaceous materials makes them a promising material for an array of electrochemical applications with the storage of energy being an important one (Frackowiak and Béguin 2001, Georgakilas et al. 2015). The carbon-based electrodes can be easily polarized. The electrical conductivity of a carbon electrode is strictly dependent on the parameters such as thermal treatment, surface morphology, hybridization, and the nature of heteroatoms present in it. The amphoteric characteristic of carbon allows it to exhibit a wide range of electrochemical properties from its donor to acceptor state. Carbon nanomaterial, on the other hand, is further suitable as an electrode material due to their exceptionally high specific surface area. The extent of electrical conductivity of carbon nanomaterials is large enough and exceeds that of metals in some instances. The chemical and electrochemical stability and ability to synthesize in bulk economy are additional advantages of carbon nanomaterials. The shape, presence of surface defects, morphology, and surface functionality also affect the electrochemical performance of carbon nanomaterials (Frackowiak and Béguin 2001, Raymundo-Piñero et al. 2006, Poonam et al. 2019).

Despite the high specific capacitance of supercapacitors, the key challenge for them to be used as an energy storage device is to overcome its small E_d. The E_d of a capacitor is directly proportional to its capacitance and to the square of the voltage (V). Therefore, to obtain a higher E_d, one or both parameters need to be improved.

One common way to achieve this goal is by using electrodes with high specific capacitance and electrolytes, which are stable in large potential windows.

The optimization of the integrity of the system is another means to further improve it. The fabrication of the components of supercapacitor such as electrolyte and materials for the electrode is rather facile; however, making them work in harmony to achieve desired performance becomes a delicate balancing act between the optimization of surface morphology with the ionic size of electrolyte ions. Electrolytes are responsible for the P_d yield, operable temperature window, and the ionic conductivity of the supercapacitor. Apart from being able to operate in the wider potential window, the electrolyte should also possess a high ionic concentration, electrochemical stability, low equivalent circuit resistance, low volatility, viscosity, and toxicity (Wang, Zhang, and Zhang 2012).

Another method to improve the capacitance of a supercapacitor is the creation of pseudocapacitors (Figure 1.2). Pseudocapacitors work on the principle of Faradaic charge transfer, in which the components of electrode actively undergo redox process via intercalation mechanism, thereby generating faradic current (Conway 1991) (Figure 1.3). The Faradaic process is also responsible for the higher capacitance and E_d of pseudocapacitors than those of EDLCs. The ability of the electrodes to realize pseudocapacitance depends on the chemical affinity of the electrode materials toward the ions adsorbed on the electrode surface. The structure and morphology of the electrode surface also play a major role in this effect. Conway performed the pioneering work on the development of pseudocapacitors (Conway and Gileadi 1962, Conway 1991). Transition metal oxides such as IrO_2 (Liu et al. 2008), WO_3 (Ren et al. 2010), RuO_2 (Liu, Pell, and Conway 1997), Fe_3O_4 (Kulal et al. 2011), MnO_2 (Demarconnay, Raymundo-Piñero, and Béguin 2011), NiO (Kim et al. 2013), V_2O_5 (Saravanakumar, Purushothaman, and Muralidharan 2012), and Co_3O_4 (Meher and Rao 2011) exhibit excellent redox activity and are therefore used in pseudocapacitors. Transition metal sulfides such as Co_3S_4 (Patil, Kim, and Lee 2017), CuS (Xu, Liang et al. 2016), CdS (Xu, Liu et al. 2016), and $Co_{1.5}Ni_{1.5}S_4$ (Tang et al. 2015) are also excellent materials for pseudocapacitors. Additionally, conducting polymers such as polyaniline (PANI) (Gupta and Miura 2006, Eftekhari, Li, and Yang 2017), polythiophene (Laforgue et al. 1999), polypyrrole (PPy) (Fan and Maier 2006), polyvinylalcohol (PVA) (Liew, Ramesh, and Arof 2014), poly(3,4-ethylene dioxythiophene)

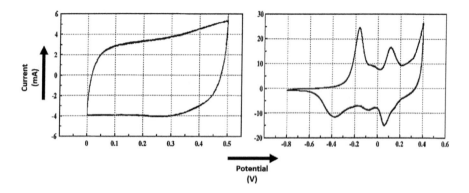

FIGURE 1.3   The typical shape of cyclic voltammogram of an EDLC and a pseudocapacitor.

(PEDOT) (Weng and Wu 2013), poly(4-styrene sulfonate) (Yuan et al. 2009), and poly(phenylene-vinylene) (Wee and Hong 2014) have also been reported to be used for the fabrication of pseudocapacitors.

Zero-dimensional (0D) nanomaterials are often used to produce a variety of electrochemical devices (Tiwari, Tiwari, and Kim 2012). 0D carbon nanostructure is a classification of carbon allotropes or derivatives, which came to light only approximately a decade ago (Sun, Zhou et al. 2006, Trauzettel et al. 2007). However, the oldest member of this family, fullerenes, were predicted to exist as early as 1965 (Schultz 1965) with its formal discovery made in 1985 (Kroto et al. 1985). These materials are highly functional and therefore possess the ability to assist the pseudocapacitive behavior in an otherwise conventionally favoring EDLC material. The following sections discuss the various methods used to synthesize 0D carbon nanostructures and their applications in the fabrication of supercapacitors.

1.2 SYNTHESIS OF ZERO-DIMENSIONAL CARBON NANOSTRUCTURES

0D carbon nanostructures are broadly classified as fullerenes, carbon dots (CD), and nanodiamonds (ND) (Figure 1.4). Fullerenes are principally a family of the molecule of carbon, which resembles a cage, with varying number of carbon atoms with C_{60} being the most commonly used among them (Kroto et al. 1985, Georgakilas et al. 2015). CD, on the other hand, are highly fluorescent materials that show excitonic

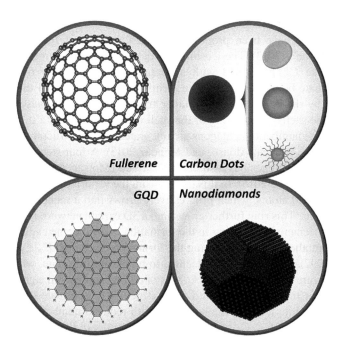

FIGURE 1.4 Types of zero-dimensional nanoparticles.

entrapment phenomenon in all the three spatial directions (Sun, Zhou et al. 2006). Hence, for this reason, they are called carbon quantum dots. A subclassification of CD is the graphene quantum dots (GQD), which are the fragments of graphene sheet with size so small that it affects the excitonic entrapment in all the three spatial dimensions. ND are 0D carbon nanomaterials with diamond core and amorphous carbon shell.

1.2.1 SYNTHESIS OF FULLERENES

Fullerenes are generally made from graphite by means of arc/plasma discharges (Krätschmer et al. 1990, Hare, Kroto, and Taylor 1991, Parker et al. 1991, Scrivens and Tour 1992) or laser irradiation (Kroto et al. 1985, Kroto 1988). Alternatively, naphthalene pyrolysis (Taylor et al. 1993) and combustion of hydrocarbons (Howard et al. 1991, Homann 1998) are also common methods for preparing fullerene. The large-scale production of fullerene is often realized by the combustion of hydrocarbons and therefore, is of commercial importance (Murayama et al. 2005). The product thus obtained contains a small fraction of fullerenes, among which C_{60} is the most commonly formed. Isolation of fullerene is achieved by solvent extraction in nonpolar solvents such as benzene or toluene. The separation of the various types of fullerenes present in the organic phase is performed by column or liquid chromatography (Lieber and Chen 1994). Despite using abundant precursor-like graphene and hydrocarbons, the preparation of fullerenes suffers from major limitations in terms of their low yield and requirement of sophisticated methods to isolate pure product, which greatly affects the overall cost of the materials (Anctil et al. 2011, Mojica, Alonso, and Méndez 2013).

Several models have been proposed for the formation of fullerenes (Mojica, Alonso, and Méndez 2013). The icospiral particle nucleation assumes that initially, corannulene-like C_{20} is formed with a pentagon surrounded by five hexagons. This intermediate being highly reactive forms an open spiral shell that resembles the Nautilus, by the aggregation of smaller fragments of carbon adsorbed on the surface of these shells. The closing of the shell occurs entirely on a statistical basis as an appropriate number of pentagons are integrated into the structure (Zhang et al. 1986, Kroto 1988, Mojica, Alonso, and Méndez 2013). A four-step mechanism for the formation of fullerenes has also been described (Mojica, Alonso, and Méndez 2013, Georgakilas et al. 2015). In this mechanism, the vaporized carbon atoms form a chain of up to 10 atoms, which progressively grows into a monocyclic ring with ~20 carbon atoms. This ring further grows into a 3D carbon network followed by the shell closing mechanism, resulting in the formation of a fullerene cage. Ring stacking model is another way of explaining the formation of C_{60} and C_{70} (Wakabayashi and Achiba 1992). Here, C_{10} ring is deemed to be the starting state, which deforms to form a structure comprising two hexagons of carbon atoms with eight free bonds. This structure is further stacked with a C_{18} structure to eliminate the preexisting dangling bonds while forming the new ones with reduced numbers. Similarly, the sequential stacking of C_{18}, C_{12}, and C_2 molecules results in the creation of the C_{60} molecule. The sequential stacking is in the order C_{10}, C_{18}, C_{20}, C_{16}, and $3C_2$ molecules to form C_{70}. The carbon clusters that are inclusive of chains and rings, bicyclic or tricyclic structures of 34–60 carbon atoms, or large random clusters may be annealed to produce fullerene according to the annealing model (Hunter, Fye, and Jarrold 1993, Mojica, Alonso, and Méndez 2013). The annealing of carbon clusters is often supplemented by the production of smaller clusters. Another model called the quantum chemical molecular dynamics is based on the shrinking hot gain road mechanism for the formation of fullerenes (Irle et al. 2006, Mojica, Alonso, and Méndez 2013, Georgakilas et al. 2015). This mechanism is a two-step process, in which the giant fullerenes are formed as hot vapors of carbon condense followed by C_{60} or C_{70} formation as giant fullerene shrinks irreversibly eliminating C_2 molecules. This irreversible elimination is caused due to the vibrational excitement in the giant fullerene because of the highly irregular and defective nature of its cage (Irle et al. 2006).

Fullerenes are also prepared with the bottom-up approach using various polycyclic hydrocarbon molecules (Georgakilas et al. 2015). Corannulene, which is the smallest unit of C_{60}, is considered to be a suitable precursor as it has a curved π surface (Scott 2004, Mojica, Alonso, and Méndez 2013). Hydrocarbons such as decacyclene, tribenzodecacyclene, and trinaphtodecacyclene are also good precursors to synthesize fullerene by the bottom-up approach (Mojica, Alonso, and Méndez 2013). The disadvantage of using these precursors is the highly complex, time-consuming multistep organic methodology that results in very small yield and therefore is not very practical (Scott 2004, Mojica, Alonso, and Méndez 2013). The *ab initio* calculations based on Density functional theory (DFT) supports the formation of fullerene from simple hydrocarbons (Viñes and Görling 2011). Metallic nanoparticles are used in such cases, both as a template and a catalyst. However, the procedure may be applicable only on strictly following the three conditions (Viñes and Görling 2011):

1. Complete dehydrogenation of the precursor, which is catalyzed by the metallic nanoparticles
2. Strong bonding of carbon to the surface of nanoparticles
3. The enthalpy of fullerene formation should be lower than that of the formation of other carbon nanoallotropes for a given set of reaction parameters.

Other miscellaneous routes to synthesize fullerene include the translation of precursors such as acetylene rods and benzene rings or reverse translation of CO_2 at high pressure and temperature (Chen and Lou 2009). These methods follow the intramolecular multimember ring formation reactions.

Onion-like carbons (OC) can be considered as fullerenes arranged concentrically to result in the formation of an onion made up of only carbon atoms. These structures are also regarded as multicore fullerene. OC were first prepared by intense irradiation of an electron beam on to carbon nanotubes (Ugarte 1992). A much higher amount of OC was prepared when carbon soot was heated at 2100°C–2250°C under vacuum (de Heer and Ugarte 1993). The OC thus prepared had 2–8 concentric nonspherical fullerenes with a diameter ranging between 3 and 10 nm. High quality and milligram scale synthesis of OC with an average diameter of 25–30 nm may also be achieved by producing a high voltage arc discharge in water between two submerged graphite electrodes (Sano et al. 2001, Alexandrou et al. 2004). Spherical OC in significant yield and diameters ranging 10–20 nm was successfully produced by embedding carbon ions into the silver matrix with an energy of 120 keV (Cabioc'h et al. 1997, 1998). An electron beam under high vacuum and high temperature of 1000°C–1500°C was used for annealing a highly dispersed diamond powder, which resulted in the gram-scale formation of OC (Kuznetsov et al. 1994). Similarly, OC of varying shapes and sizes can be produced by modulating the duration and temperature of annealing (Ugarte 1994, Barnard, Russo, and Snook 2003, Raty et al. 2003).

1.2.2 SYNTHESIS OF CARBON DOTS

CD were first discovered when the single-walled nanotubes prepared by the arc-discharge method was being purified by gel electrophoresis (Xu et al. 2004). Since then, a wide variety of methods and techniques have been perfected for the preparation of CD in the last decade (Li, Kang et al. 2012). As the most common precedence, CD are prepared by the top-down approach such as the laser ablation of suitably treated blends of cement and graphite. The as-synthesized CD are subsequently treated with acids to enrich the surface with reactive oxygen groups. These reactive oxygen sites act as centers where desired organic molecules and oligomers may be bound to passivate the surface of the CD. CD have a mean diameter of ~5 nm (Sun, Zhou et al. 2006). CD can be easily produced electrochemically using graphite electrodes dipped in pure water (Ming et al. 2012) or other electrolytes (Zhao et al. 2008, Zheng et al. 2009). The CD produced in such a way have a very narrow size distribution and contains both amorphous and graphitic components in its structure. Multiwalled nanotubes can also be used as a carbon source for the electrochemical synthesis of CD (Zhou et al. 2007).

Other methods of synthesizing CD include the carbonization of organic precursors in solvents with high boiling point (Wang, Pang et al. 2010, Dong, Wang et al. 2012, Lai et al. 2012, Wang et al. 2012), "water in oil" emulsions (Kwon, Do, and Rhee 2012, Kwon et al. 2014), or powders (Bourlinos, Stassinopoulos, Anglos, Zboril, Karakassides et al. 2008b, Bourlinos, Stassinopoulos, Anglos, Zboril, Georgakilas et al. 2008a, Wang, Pang et al. 2010, Deng et al. 2013, Xu et al. 2013, Jiang et al. 2014); pyrolysis of polymer-silica nanocomposites (Liu et al. 2009); ultrasonic carbonization (Li et al. 2011, Ma et al. 2012); and hydrothermal (Hsu and Chang 2012, Zhou, He, and Huang 2013, Hu et al. 2014, Liu et al. 2014, Yang et al. 2014), solvothermal (Zhang, Ma et al. 2012, Zhou et al. 2013), or microwave treatment (Chandra et al. 2011, Mitra et al. 2012, Puvvada et al. 2012, Niu et al. 2013) of a wide range of organic precursors. These hydrocarbon-based precursors may range from polymers, carbohydrates, organic macromolecules, and organosilanes to natural products. Among these methods of synthesis, the hydrothermal and microwave methods to synthesize CD are most prevalent (Yang, Li et al. 2013).

Like fullerene and OC, CD have been isolated from the carbon soot of combustion of candles or natural gases (Liu, Ye, and Mao 2007, Ray et al. 2009, Tian et al. 2009). The highly photoluminescent CD may be prepared by heating nitrogen- and oxygen-rich organic molecules such as cadaverine, ethylenediaminetetraacetic acid, glycine, and tris(hydroxymethyl)aminomethane in an autoclave at 300°C (Hsu and Chang 2012). CD obtained by this method are usually spherical, well dispersed, and 2.6–7.9 nm in diameter and are highly hydrophilic. X-ray diffraction patterns of CD have demonstrated them to be mostly amorphous with partly graphitic characteristic.

CD are often synthesized with atoms such as nitrogen as a dopant in a one-step method. A highly photoluminescent N-doped CD was reportedly made from carbon tetrachloride and $NaNH_2$ (used as N source) in an autoclave (Zhang, Ma et al. 2012). Similarly, a single-step thermal decomposition of substituted ammonium citrate salts also yielded N-doped CD with desired dispersibility (Bourlinos, Stassinopoulos, Anglos, Zboril, Karakassides et al. 2008b). In another instance, octadecylammonium citrate was used to obtain organophilic N-doped CD. A strongly hydrophilic N-doped CD was prepared by the decomposition of diethylene glycol ammonium citrate.

One of the most widely used carbon sources for the preparation of CD is citric acid because the carbonization of citric acid occurs at a relatively low temperature. Citric acid in octadecane carbonizes at 300°C under argon to produce CD (Wang, Pang et al. 2010). Branched poly(ethylenimine) and citric acid dissolved in water carbonize under 200°C (Dong, Wang et al. 2012). Glycerol is another interesting organic compound that acts as a precursor and a solvent for the synthesis of CD due to its high boiling point. CD are directly synthesized by refluxing glycerol at ~230°C (Xu et al. 2013).

The pyrolysis of organic compounds under microwave irradiation is a relatively new method to synthesize CD. The aqueous dextrin in the presence of sulfuric acid pyrolyzed under microwave irradiation results in highly photoluminescent CD, with the emission spectrum covering the entire visible region without the need of surface passivation (Puvvada et al. 2012). As-synthesized CD is highly dispersible in water. Similarly, the pyrolysis of citric acid in 1,2-ethylenediamine (EDA) under microwave irradiation results in well-dispersed, spherical, and amorphous EDA surface-passivated CD of 2–3 nm in diameter (Zhai et al. 2012).

Ultrasonic mixing of glucose and ammonium hydroxide results in amine-passivated CD with a diameter of ~10 nm and a lattice spacing of 3.43 Å, which is indicative of its graphitic nature (Ma et al. 2012). The solution phase reduction of carbon tetrachloride produces a highly photoluminescent CD in the reverse micelles of tetraoctylammonium bromide (Linehan and Doyle 2014). Additionally, the synthesis of water-soluble CD via single-step carbonization of ethylene glycol in concentrated H_2SO_4 at 80°C was also reported (Liu, Liu, and Zhang 2013).

CD can be purified by various methods such as column chromatography, dialysis, filtration, gel electrophoresis, ultracentrifugation, or a combination of these techniques (Wang, Cao et al. 2010, Anilkumar et al. 2011, Jaiswal, Ghosh, and Chattopadhyay 2012, Shen, Zhu, Yang, and Li 2012, Gong et al. 2014). The purification of CD eliminates the possibility of organic by-products formed during the synthesis of CD to interfere with the photoluminescence of CD.

Another challenge in the synthesis of CD is the control over size and dispersity of CD. A soft-template synthesis method was developed to control the size of the CD during the reaction process. Carbonation of a carbon-rich precursor in a water-in-oil emulsion at high temperature in one such way to produce size-controlled CD (Kwon et al. 2014). Hydrophilic carbon precursors such as citric acid are encapsulated by the water dispersed in a mixture of long-chain hydrocarbons such as octadecene and oleylamine. Heating this dispersion at 250°C causes the evaporation of water, which further leads to the formation of CD via carbonization of the precursor. The amount of oleylamine in the mixture affects the size of the formed CD (Kwon et al. 2014). Another template-based synthesis of CD is dependent on a soft–hard-template approach to control the size and other characteristics of the CD (Yang, Wu et al. 2013). Mesoporous silica, when used as a nanoreactor, result in CD with uniform size distribution (Zong et al. 2011).

1.2.3 SYNTHESIS OF GRAPHENE QUANTUM DOTS

As shown in the case of other two 0D carbon nanomaterials, GQD can be prepared on the basis of "top-down" and "bottom-up" approach (Shen, Zhu, Yang, and Li 2012, Li et al. 2013, Sun, Wu et al. 2013, Bacon, Bradley, and Nann 2014). The graphitic structures are cut from various sources to produce GQD in the "top-down" approach. Few precursors from which GQD can be prepared are carbon black (Dong, Chen et al. 2012), coal (Li et al. 2013, Ye et al. 2015), carbon fibers (Peng et al. 2012, Xie et al. 2013), graphene oxide (Pan et al. 2010, Li, Ji et al. 2012, Shen, Zhu, Yang, Zong et al. 2012, Yang et al. 2012), multiwalled nanotubes (Shinde and Pillai 2012), single-walled nanotubes (Dong et al. 2013), graphite (Markovic et al. 2012, Zhang, Bai et al. 2012, Sun, Wang et al. 2013), or graphene (Ponomarenko et al. 2008, Ananthanarayanan et al. 2014).

High-resolution electron lithography was one of the first methods for obtaining GQD from graphene, albeit with low yield (Ponomarenko et al. 2008). The most common method for preparing GQD is hydrothermal cutting of graphene oxide or reduced graphene oxide. GQD thus obtained are water-soluble and emit blue luminescence with diameters ranging from 5 to 13 nm. The cutting mechanism involves the dissociation of a series of epoxides and carbonyls on the surface of graphene oxide, thereby dividing them into GQD (Pan et al. 2010). Cutting graphene oxide sheets into GQD may also be enabled by an O_3 pre-oxidation treatment, enriching the graphene oxide surface with oxygen-rich functional groups before the hydrothermal treatment (Yang et al. 2012). The surface passivation of hydrothermally produced GQD can be performed in situ using commonly available pre-polymers such as polyethylene glycol (Shen, Zhu, Yang, Zong et al. 2012). Pre-polymers such as poly(ethylenimine) (PEI) assist in the hydrothermal cutting of graphene oxide by acting as a chemical scissor, producing virtually monodispersed GQD. PEI also acts as a stabilizing agent to avoid restacking of GQD (Xue et al. 2013).

Carbon black and carbon fibers are more unconventional sources for producing GQD. The process, in general, involves subjecting the carbon source to hot concentrated acid such as nitric acid to induce oxidative stress, leading to the stripping of smaller size GQD (Dong, Chen et al. 2012, Peng et al. 2012). The purification of GQD is performed by centrifugation and dialysis using a dialysis bag of desired molecular weight cutoff. Coal may also be used with the same methodology to yield GQD (Ye et al. 2015).

Electrochemical cutting of graphite is another popular method for the synthesis of GQD (Markovic et al. 2012, Zhang, Bai et al. 2012). A graphite rod is used as an anode in an alkaline electrolyte in a typical method. GQD produced from electrochemical cutting are usually having diameters of 5–10 nm. The method to purify the

GQD remains the same, i.e., centrifugation followed by dialysis. The synthesized GQD chiefly contains monolayer QD with the fluorescence quantum yield reaching 14% (Zhang, Bai et al. 2012). Other materials used for obtaining GQD via electrochemical cutting are multiwalled nanotubes or single-walled nanotubes (Shinde et al. 2011, John et al. 2014). Such electrochemical unzipping results in the formation of high-quality graphene nanoribbons, which are very similar to GQD. However, if the electrochemical unzipping is performed in the presence of an organic solvent, GQD are obtained. Electrochemical oxidation of nanotubes has reportedly yielded monodispersed GQD of diameters less than 10 nm (Shinde and Pillai 2012).

The broad size distribution of the produced GQD is a primary shortcoming of top-down methods. Few attempts have been made to achieve a much narrow size distribution of GQD from the top-down approach, as these methods are highly advantageous in producing GQD in higher yields. One such attempt was the synthesis of GQD by etching away the exposed graphene layer masked with self-assembled silica nanodots (Lee et al. 2012). The silica nanodots, having diameters within the range of 10–20 nm, were used as a mask for the intended GQD while etching the rest of the graphene layer using an oxygen plasma. Mono or bilayer GQD with narrow size distributions are obtained as a result of the masking effect. The diameter of obtained GQD depends on the dimension of silica nanodots (Lee et al. 2012). Other assorted precursors and methods for the preparation of GQD include oxidation of carbon nanofibers (Lee, Ryu, and Jang 2013), hydrothermal cutting of single-walled nanotubes (Dong et al. 2013), microwave-assisted acidic treatment of graphene oxide (Li, Ji et al. 2012), refluxing reduced graphene oxide in various solvents (Štengl et al. 2013), and acid-induced oxidation (Sun, Wang et al. 2013) or microwave irradiation of graphite (Shin et al. 2014); all these methods are summarily similar to the earlier discussed top-down methods.

On the other hand, the "bottom-up" approach for the synthesis of GQD involves assembling or fusing aromatic molecules to nanoscale using appropriate methods (Shen, Zhu, Yang, and Li 2012, Liu et al. 2011). The most prominent advantage of this method is greater control over the size distribution of GQD. This is achieved by carefully picking suitable precursor and optimizing reaction parameters. One such way is pyrolytic graphitization of hexa-peri-hexabenzocoronene (Liu et al. 2011), which results in artificial graphite that can undergo oxidative exfoliation like a top-down method to yield GQD. The obtained GQD were 60 nm wide and 2–3 nm thick. Another method is to synthesize GQD via a multistep organic reaction from the simplest of organic molecules such as 3-iodo-4-bromoaniline (Yan, Cui, and Li 2010). Similarly, GQD can be produced with yields of up to 63% by connecting pyrene derivatives (Wang, Wang et al. 2014). The method involves the pyrene nitration followed by the hydrothermal treatment of the pyrene derivative under basic conditions.

A high yield, large-scale bottom-up approach to synthesize GQD is using a thermal plasma jet of ethylene gas that results in carbon soot containing a large amount of GQD (Kim and Suh 2014). This can be simply visualized as a pyrolytic aggregation of carbon from ethylene into GQD. The procedure reportedly claims to produce 4 g/h of GQD dispersed in toluene with diameters between 10 and 20 nm.

GQD can also be synthesized using fullerene by means of cage-opening mechanism (Chua et al. 2015). As there is no loss of the number of carbon atoms from the precursor to the product, this may be thought of as a side-way method to synthesize GQD. Because cage opening is achieved by the oxidative cutting method, this process is much closer to the top-down method than to the bottom-up method.

Generally, GQD emit blue to deep blue fluorescence, whereas the oxides of GQD emit green fluorescence (Li, Ji et al. 2012). However, few amine- and hydrazine-functionalized GQD are reported to emit yellow and cyan fluorescence, respectively (Wang, Wang et al. 2014). Additionally, khaki luminescent nitrogen-doped GQD have been reported (Zhu et al. 2014)

1.2.4 Synthesis of Nanodiamonds

ND are possibly the most interesting carbon nanomaterials for research. The structure of ND consists of a diamond core (i.e., sp^3 hybridized carbon core) and amorphous carbon layers (sp^2 hybridized carbon shell). The physical characteristics of ND are determined by methods used for production. The most common method to produce ND is the detonation of a carbon-rich explosive blend. The blend of trinitrotoluene and hexogen (1,3,5-trinitro-1,3,5-triazinane) is one such combination (Dolmatov 2001, Danilenko 2004). Detonation method typically results in soot with 75% ND having size distribution between 4 and 5 nm. Aggregation of ND, which are nondispersible in solvents, is a disadvantage of this procedure. These aggregates are found enclosed within the layers of graphitic material, making it even harder to disperse and purify (Krueger 2008). Mechanical milling and ultra-sonication are few common methods to reduce such aggregates. Additionally, the ND become more hydrophilic and are therefore easily dispersible in polar solvents (Krüger et al. 2005, Ozawa et al. 2007). Alternatively, salt- or sugar-assisted milling to prevent the reaggregation of the ND is also available (Pentecost et al. 2010, Lai and Barnard 2011). The products of detonation are purified with ozone oxidation or by heating in concentrated acid to remove undesirable byproducts (Osswald et al. 2006, Shenderova et al. 2011).

ND are also produced at normal pressure and temperature by exposing the water-based suspension of carbon black to pulsed laser irradiation (Hu et al. 2009). The resulting products are boiled in perchloric acid to remove undesired carbon impurities. ND thus prepared can be easily dispersed in solvents and have a fine size distribution. Diamond microcrystals are also used as a precursor to produce ND by using a high-energy ball milling procedure (Boudou et al. 2009, Mahfouz et al. 2013). The procedure involves jet milling followed by nano-milling to produce spherical ND (Boudou et al. 2009).

The microplasma-induced dissociation of ethanol under an inert atmosphere has shown to produce high-purity ND (Kumar et al. 2013). The other carbonaceous byproducts are etched away using hydrogen gas. Pulsed irradiation exposure of graphite (Banhart and Ajayan 1996, Yang, Wang, and Liu 1998, Welz, Gogotsi, and McNallan 2003, Sun, Hu et al. 2006, Hu et al. 2008), electron irradiation of OC (Banhart and Ajayan 1996), and carbide chlorination (Welz, Gogotsi, and McNallan 2003) are few other methods used to produce ND.

1.3 ZERO-DIMENSIONAL CARBON NANOMATERIALS IN SUPERCAPACITORS

Conversely to the 1D, 2D, 3D, and amorphous carbon materials, the applicability of 0D carbon materials for supercapacitors has been limited. They are mostly used as a composite filler along with a higher-dimensional carbon nanomaterial for fabricating the electrodes for supercapacitors.

1.3.1 FULLERENE-BASED SUPERCAPACITORS

Fullerene may be the most used 0D carbon nanostructure to be incorporated in a supercapacitor electrode. A KOH-activated C_{70} microstructure has been reported to have a specific capacitance of 362 Fg^{-1} at the current density of 0.1 Ag^{-1} with over 92% capacitance retention even after 5000 deep cycles (Zheng, Ju, and Lu 2015). This is considered as excellent cyclic stability. Fullerene-based structures such as graphitic carbon microtubes obtained by high-temperature annealing (2000°C) of C_{70} have a specific capacitance of 212.2 Fg^{-1} at the current density of 0.5 Ag^{-1} (Bairi et al. 2016). Liquid–liquid interfacial precipitation of C_{60} results in the formation of a quasi 2D mesoporous microbelt of C_{60}. Further heat treatment at an elevated temperature of 900 C results in carbon microbelts. These structures derived from C_{60} have a specific capacitance of 360 Fg^{-1} due to the enhanced surface area and strong mesoporous framework (Tang et al. 2017).

Fullerenes may be used as a filler in conjunction with carbonaceous electrode for supercapacitors. The C_{60}-incorporated activated carbon electrode reportedly showed a higher performance than its unloaded counterpart (Okajima et al. 2005). The capacitance was affected by the amount of C_{60} in the activated carbon. However, a lower amount of C_{60} is preferred as it gave a better dispersion of fullerene in the activated carbon. The capacitance of the C_{60}-incorporated activated carbon electrode is also affected by the ultrasonication time. A typical 1 wt% C_{60} in activated carbon can yield a capacitance of 172 Fg^{-1} at a current density of 50 mA/cm^2. Incorporating of C_{60} into activated carbon also improves the cycle performance of supercapacitor for deep discharge cycles. Fullerene can also be incorporated into the single-walled nanotubes to fabricate electrodes with an exceptionally high specific surface area of 2200 m^2g^{-1} (Hiraoka et al. 2010). One such C_{60}-Pd single-walled nanotube composite yielded the highest specific capacitance of 994 Fg^{-1} (Grądzka et al. 2013). In the same report, the highest specific capacitance obtained for the C_{60}-Pd multi-walled nanotube composite was reported to be 758 Fg^{-1}. C_{60}-Pd natively has a specific capacitance of 239 Fg^{-1}. C_{60} graphene composite electrode for supercapacitor, on the other hand, reportedly has a specific capacitance of 135.36 Fg^{-1} at the current density of 1 Ag^{-1}, which is higher than that of graphene of 101.88 Fg^{-1} (Ma et al. 2015). This value is still significantly lower than the theoretical capacitance of graphene, which is 550 Fg^{-1} (Dey, Hjuler, and Chi 2015). This is due to the imperfection in the fabrication technique, which is still a major challenge in the synthesis of graphene. Fullerene-capped Au nanoparticle and graphene composite have shown a slight improvement in specific capacitance with 197 Fg^{-1} (Yong and Hahn 2013). However, this comparison is not much useful as there have been some variations in the reported capacitance with various methods used to obtain them. Doped fullerene and graphene composite have the potential to be an excellent supercapacitor material. N and Fe co-doped C_{60}-incorporated graphene electrode have resulted in a specific capacitance of 505.4 Fg^{-1} at a current density of 0.1 Ag^{-1} (Peng et al. 2019).

Incorporation of metal oxides into fullerene also opens excellent possibilities to improve the capacitance of the electrode material. For example, a multilayer fullerene and MnO_2 composite electrode showed specific capacitance as high as 1207 Fg^{-1} (Azhagan, Vaishampayan, and Shelke 2014). This capacitance is also significantly higher than that of conventional carbonaceous electrodes. An activated fullerene carbon soot and graphene-Co_3O_4 composite push the boundary of specific capacitance to even higher with a value of 1935 Fg^{-1} (Gao et al. 2018).

Fullerenes are also incorporated with conductive polymers such as PANI. C_{60} binds to PANI, resulting in a unique coral-like porous morphology (Xiong et al. 2012). The specific capacitance of this system is remarkably high at 776 Fg^{-1} at the current density of 1 mA/cm^2. The enhancement of the specific capacitance is due to the increased ionic conductivity of the C_{60}-PANI composite caused due to the low packing fraction caused by the porous morphology. Incorporation of fullerene also increases the electrical conductivity due to its strong electron-withdrawing nature when bonded to PANI. Fullerene whiskers with PANI have reportedly shown an even higher capacitance of 813 Fg^{-1} at a current density of 1 Ag^{-1} (Wang, Yan, and Piao 2017).

1.3.2 CARBON DOT-BASED SUPERCAPACITORS

CD find applications in supercapacitors in various capacities (Li et al. 2015). Natively, CD are not an ideal material for supercapacitor with a specific capacitance of 2.2 Fg^{-1} (Zhao et al. 2017). However, CD obtained from the cage opening of C_{60} by KOH activation followed by annealing resulted in a high-density electrode with a capacitance of 157.4 F/cm^3 (Chen et al. 2016). The annealed electrode with a density of 1.23 g/cm^3 was highly porous, and therefore, acted as an excellent platform for the adsorption of charged ions. A similarly high-porosity 3D CD structure noted as an aerogel was assembled in a multichannel arrangement using the sol–gel reaction of resorcinol and formaldehyde in the presence of CD, followed by pyrolysis (Lv et al. 2014). This aerogel had a specific capacitance of 294.7 Fg^{-1} at a current density of 0.5 Ag^{-1}. The cyclic stability noted for the electrode was over 1000 deep charge-discharge cycles.

CD are mostly used as a conductive filler to impart performance enhancement to the supercapacitors. CD-decorated RuO_2 electrode was possibly the first reported CD-based electrode being used in a supercapacitor (Zhu et al. 2013). The electrode could attain the specific capacitance of 460 Fg^{-1} at an exceptionally high current density of 50 Ag^{-1}. The CD-decorated electrode was reportedly highly stable even after 5000 ultrafast deep charge and discharge cycles with retention of ~97% of its capacitance. Similarly, a $CD/NiCo_2O_4$ composite electrode with porous spherical morphology exhibited specific capacitance of 856 Fg^{-1} with the current density of 1 Ag^{-1} (Zhu et al. 2015). The cyclic stability for this material was impressive with 10,000 cycles at 98.75% capacitance retention. In another report with a flower-shaped $CD/NiCO_2O_4$, a specific capacitance as high as 2168 Fg^{-1} with the current density of

1 Ag^{-1} was achieved (Wei, Ding et al. 2016). The highly graphitized cores and the abundant surface groups of CD are advantageous in enabling good wettability and low electrochemical impedance in the electrode. The above benefit, along with the reversible redox reactions of $NiCo_2O_4$ high-surface area of the flower-like morphology synergistically, works to obtain high capacitance. CD-NiO nanorods obtained by the complexation reaction of Ni^{2+} and $C_6H_5COO^-$ followed by a thermal treatment process are another wonderful material for pseudocapacitors (Xu, Xue et al. 2016). CD-NiO nanorods of 800 nm long and 30 nm wide have a specific capacitance of 1858 Fg^{-1} at the current density of 1 Ag^{-1}. CD with Ni-Al-layered double hydroxide composite also yield sufficiently high specific capacitance of 1794 Fg^{-1} owing to the redox behavior of the hydroxides and the electron transportability of CD (Wei, Zhang et al. 2016). CD are also used in conjunction with metal sulfides such as MoS_2 (Wu et al. 2016) and CuS (De, Balamurugan et al. 2017) for supercapacitor. CD, CuS, and graphene oxide hydrogel composite is another electrode with high specific capacitance (920 Fg^{-1} at the current density of 1 Ag^{-1}) electrode (De, Kuila et al. 2017). However, this composite-based supercapacitor is of asymmetric architecture with CuS@CD—graphene oxide hydrogel being the positive electrode and reduced graphene oxide being a negative electrode.

CD can also pair with conducting polymers such as polypyrrole (Jian, Li et al. 2017, Jian, Yang et al. 2017, Genc et al. 2017) and PANI (Zhao and Xie 2017). CD and PANI composite has higher specific capacitance (738 Fg^{-1}) among the two, mostly due to the better electrochemical activity of PANI (Zhao and Xie 2017). The specific capacitance (576 Fg^{-1}) of CD-polypyrrole system can be improved by forming a composite with graphene oxide (Zhang, Wang et al. 2017).

CD also enable the deposition of metal oxides such as MnO_2 over graphene due to its ability to homogenously disperse over graphene (Unnikrishnan et al. 2016). The oxide, therefore, is consistent with its pseudocapacitive behavior, resulting in good specific capacitance (~280 Fg^{-1}). However, CD dispersed directly on a reduced graphene oxide sheet natively has a high capacitance of 211 Fg^{-1} (Dang et al. 2016). Similarly, a 3D network of reduced graphene oxide with CD exhibits high specific capacitance of 308 Fg^{-1} (current density 0.5 Ag^{-1}) with cyclic stability up to 20,000 cycles (Zhao et al. 2017). CD can also be annealed to produce a highly porous 3D network of graphene, which itself is a useful material for supercapacitors (Strauss et al. 2018).

1.3.3　Graphene Quantum Dot-Based Supercapacitors

GQD is arguably the most investigated form of CD for the supercapacitor, and the same is true for other applications as well (Li, Kang et al. 2012, Bacon, Bradley, and Nann 2014, Georgakilas et al. 2015, Li et al. 2015, Bak, Kim, and Lee 2016). The maximum specific capacitance for GQD is reported to be 481 Fg^{-1} at a slow scan rate and 41 Fg^{-1} at a fast scan rate (Kumar et al. 2017). This remarkable reduction in the capacitance of GQD is attributed to the diffusion limits of electrolyte ions. As GQD share the properties with those of CD, the way these materials are used in supercapacitors is also similar. For example, direct KOH activation of GQD resulted in ultra-high specific surface area of 3000 m^2g^{-1} and specific capacitance of 236 Fg^{-1}, which is higher than that of bare GQD (108 Fg^{-1}) (Hassan et al. 2014). In another report, GQD have shown to have a native specific capacitance of 296 Fg^{-1} (Zhang, Sui et al. 2018). This difference in the observed capacitance is a compound result of the difference in the methods of measurement and the subtle variation in the synthesis parameters that affect the morphology of the GQD. The coal-derived GQD template is used for preparing porous carbon nanosheets to produce very high specific surface area electrodes (1332 m^2g^{-1}) (Zhang, Zhu et al. 2017). The electrode has been reported to have capacitance of 230 Fg^{-1} at 1 Ag^{-1} and excellent cyclic stability. Similarly, a GQD-based symmetrical micro-supercapacitor with comparable capacitance was reported elsewhere (Liu, Feng et al. 2013). An asymmetrical micro-supercapacitor with improved capacitance was also prepared with GQD and MnO_2 as negative and positive electrodes, respectively.

The GQD and MnO_2 heterostructure, on the other hand, takes the performance of the device in terms of capacitance (1170 Fg^{-1}) even further with active redox behavior of MnO_2 being emphasized by GQD's ability to transport electron (Jia et al. 2018). The GQD/MnO_2-based micro-supercapacitor with ionic liquid electrolyte is another such example (Shen et al. 2015). Ionic liquid provides the supercapacitor with the ability to operate over a wide potential window. Another example of coupling GQD with an oxide is the fabrication of GQD/Fe_2O_3 halloysite nanotubes that produce highly porous electrodes with GQD acting as an anchor to hold iron oxide on the nanotube (Ganganboina, Chowdhury, and Doong 2017a). These electrodes yield a capacitance of 418 Fg^{-1} in neutral electrolytes. Halloysite nanotubes are often preferred in conjunction with GQD for the fabrication of supercapacitors (Ganganboina, Dutta Chowdhury, and Doong 2017b).

Speaking of nanotubes, GQD with carbon nanotube hybrid array also formed viable supercapacitors (Hu et al. 2013). The water-assisted Chemical vapor deposition (CVD)-grown super long vertically aligned arrays of carbon nanotubes allow GQD to disperse evenly throughout the bulk, thereby enabling stronger retention of ions over the electrode surface in an ably formed supercapacitor. Graphene and graphene oxides are also the choices of carbon nanostructures to host GQD for improving the performance of supercapacitors (Chen et al. 2014, Lee et al. 2016, Muthurasu, Dhandapani, and Ganesh 2016). GQD grafted on carbon fiber allow the possibility of developing supercapacitor fabric, which could be the future of wearable electronics (Islam et al. 2017). The role of GQD in these composites are the same as that of carbon nanotubes.

Enriching GQD with suitable heteroatoms is another way to improve the performance of supercapacitors. The amine-functionalized GQD-derived carbon sheets can achieve a specific capacitance of 400–595 Fg^{-1}, which is closer to the theoretical prediction for pure graphene electrode (Dey, Hjuler, and Chi 2015, Li et al. 2016). Nitrogen and oxygen co-doped GQD-transformed hierarchal carbon network is another such example of high-performance supercapacitor (Li et al. 2017). Conversely, GQD-derived graphene structure enriched with heteroatoms such as nitrogen by photochemical method is another way to improve the capacitance of the electrode (Xu et al. 2017). Tryptophan-functionalized GQD, in conjunction with $NiCo_2S_4$, make up an excellent electrode for supercapacitors (1453.1 Fg^{-1}), which emphasizes the importance of functionalization of GQD (Wang et al. 2017).

As discussed for CD, GQD can be combined with conducting polymers such as polypyrrole (Wu et al. 2013), PVA (Syed Zainol Abidin et al. 2018), and PANI (Liu, Yan et al. 2013, Dinari, Momeni, and Goudarzirad 2015) to produce excellent electrodes for supercapacitors. GQD-PANI nanofibers, among these combinations, has been reported to have the highest specific capacitance of 1044 Fg^{-1} at the current density of 1 Ag^{-1} (Mondal, Rana, and Malik 2015). The electropolymerization of PEDOT on the PVA GQD-Co_3O_4 produces an electrode with the specific capacitance of 361.97 Fg^{-1}, low equivalent series resistance (ESR), and good cyclic stability (Syed Zainol Abidin et al. 2018).

One interesting use of GQD in supercapacitors was reported by Zhang et al., where they used GQD as an electrolyte for solid-state supercapacitor (Zhang et al. 2016). It was suggested that the ionic conductivity and ability to donate ions might be highly improved by neutralizing the acidic functional group of GQD using KOH, which, in turn, highly improves the capacitive performance of supercapacitors. This improvement is made possible due to the complete ionization of the weakly acidic oxygen functionality of GQD after the neutralization process.

1.3.4 NANODIAMOND-BASED SUPERCAPACITORS

Use of ND to fabricate supercapacitor electrode materials is possibly the least studied among the other 0D carbon nanomaterials. This is possibly due to the fact that the synthesis of ND requires stringent use of sophisticated instrumentation. However, due to exceptional mechanical strength and physicochemical stability, they are beneficial for such applications (Zhang, Rhee et al. 2018). The first studies exploring the suitability of ND for supercapacitors established that the more regular surface of ND allows for a smaller capacitance loss at high current density or higher operating frequency (Portet, Yushin, and Gogotsi 2007). It is also understood that the ND reduce the ESR of the supercapacitor by twofold due to the increase in electrical conductivity. It was also established that the ND-derived OC are a suitable filler to be used in composites for supercapacitors. The ND-derived OC are best prepared at 1700°C under Ar flow and yield a specific capacitance of 20 Fg^{-1} (Zeiger et al. 2015). The higher temperature results in a lower amount of heteroatoms in the ND-derived OC composite, which in turns improves its electrochemical performance. The ND and ND-derived OC composite with PANI have shown to be good materials for supercapacitors (Kovalenko, Bucknall, and Yushin 2010). The specific capacitance of this system was reported to be 640 Fg^{-1} with stability up to 10,000 cycles.

ND are used directly with reduced graphene oxide yielding a decent average specific capacitance of 184 Fg^{-1} (Wang, Plylahan et al. 2014). The ND-coated silicon wire prepared by the bottom-up approach has shown to improve the performance of a bare silicon wire in a supercapacitor (Gao et al. 2015). ND-derived foam, however, is the only available form of ND directly used as an electrode for supercapacitors to the best of our knowledge (Gao, Wolfer, and Nebel 2014). The specific capacitance of the foam was reported to be less than that of activated carbon. However, it was found to depend on the thickness of the foam and therefore, can be improved upon.

1.4 CONCLUSIONS

The 0D carbon nanostructures are, in various aspects, remarkably different from those of 1D or 2D carbon nanomaterials. However, these variations in properties make the family of carbon nanoallotropes quite complete in terms of the possibilities in diverse applications. The diverse synthetic methods used to produce these 0D nanomaterials have a direct relation to their functionality and morphology. These parameters, in turn, highly affect the performance of electrode synthesized using these materials. As discussed, there is substantial information available regarding how these 0D nanomaterials interact with other components in the composites used as electrodes. The careful selection of the material and its synthetic pathway, therefore, will result in availing desired device performance. There is still much scope for the study of the applicability of 0D carbon nanomaterials for supercapacitor, as the research in this field is still in its primary stage.

REFERENCES

Alexandrou, Ioannis, Hanguang Wang, Nobuyuki Sano, and Gehan A. J. Amaratunga. 2004. "Structure of carbon onions and nanotubes formed by arc in liquids." *The Journal of Chemical Physics* 120 (2):1055–1058. doi:10.1063/1.1629274.

Ananthanarayanan, Arundithi, Xuewan Wang, Parimal Routh, Barindra Sana, Sierin Lim, Dong-Hwan Kim, Kok-Hwa Lim, Jun Li, and Peng Chen. 2014. "Facile synthesis of graphene quantum dots from 3D graphene and their application for Fe^{3+} sensing." *Advanced Functional Materials* 24 (20):3021–3026. doi:10.1002/adfm.201303441.

Anctil, Annick, Callie W. Babbitt, Ryne P. Raffaelle, and Brian J. Landi. 2011. "Material and energy intensity of fullerene production." *Environmental Science & Technology* 45 (6):2353–2359. doi:10.1021/es103860a.

Anilkumar, Parambath, Xin Wang, Li Cao, Sushant Sahu, Jia-Hui Liu, Ping Wang, Katerina Korch, Kenneth N. Tackett Ii, Alexander Parenzan, and Ya-Ping Sun. 2011. "Toward quantitatively fluorescent carbon-based 'quantum' dots." *Nanoscale* 3 (5):2023–2027. doi:10.1039/C0NR00962H.

Azhagan, Muniraj Vedi Kuyil, Mukta V. Vaishampayan, and Manjusha V. Shelke. 2014. "Synthesis and electrochemistry of pseudocapacitive multilayer fullerenes and MnO_2 nanocomposites." *Journal of Materials Chemistry A* 2 (7):2152–2159. doi:10.1039/C3TA14076H.

Bacon, Mitchell, Siobhan J. Bradley, and Thomas Nann. 2014. "Graphene quantum dots." *Particle & Particle Systems Characterization* 31 (4):415–428. doi:10.1002/ppsc.201300252.

Bairi, Partha, Rekha Goswami Shrestha, Jonathan P. Hill, Toshiyuki Nishimura, Katsuhiko Ariga, and Lok Kumar Shrestha. 2016. "Mesoporous graphitic carbon microtubes derived from fullerene C_{70} tubes as a high performance electrode material for advanced supercapacitors." *Journal of Materials Chemistry A* 4 (36):13899–13906. doi:10.1039/C6TA04970B.

Bak, Sora, Doyoung Kim, and Hyoyoung Lee. 2016. "Graphene quantum dots and their possible energy applications: A review." *Current Applied Physics* 16 (9):1192–1201. doi:10.1016/j.cap.2016.03.026.

Banhart, Florian, and Pulickel M. Ajayan. 1996. "Carbon onions as nanoscopic pressure cells for diamond formation." *Nature* 382 (6590):433–435. doi:10.1038/382433a0.

Barnard, Amanda S., Salvy P. Russo, and Ian K. Snook. 2003. "Coexistence of bucky diamond with nanodiamond and fullerene carbon phases." *Physical Review B* 68 (7):073406. doi:10.1103/PhysRevB.68.073406.

Boudou, Jean-Paul, Patrick A. Curmi, Fedor Jelezko, Joerg Wrachtrup, Pascal Aubert, Mohamed Sennour, Gopalakrischnan Balasubramanian, Rolf Reuter, Alain Thorel, and Eric Gaffet. 2009. "High yield fabrication of fluorescent nanodiamonds." *Nanotechnology* 20 (23):235602. doi:10.1088/0957-4484/20/23/235602.

Bourlinos, Athanasios B., Andreas Stassinopoulos, Demetrios Anglos, Radek Zboril, Vasilios Georgakilas, and Emmanuel P. Giannelis. 2008a. "Photoluminescent carbogenic dots." *Chemistry of Materials* 20 (14):4539–4541. doi:10.1021/cm800506r.

Bourlinos, Athanasios B., Andreas Stassinopoulos, Demetrios Anglos, Radek Zboril, Michael Karakassides, and Emmanuel P. Giannelis. 2008b. "Surface functionalized carbogenic quantum dots." *Small* 4 (4):455–458. doi:10.1002/smll.200700578.

Cabioc'h, Thierry, Jean-Christophe Girard, Michel Jaouen, Marie Françoise Denanot, and Gilles Hug. 1997. "Carbon onions thin film formation and characterization." *Europhysics Letters (EPL)* 38 (6):471–476. doi:10.1209/epl/i1997-00270-x.

Cabioc'h, Thierry, Michel Jaouen, Marie Françoise Denanot, and P. Bechet. 1998. "Influence of the implantation parameters on the microstructure of carbon onions produced by carbon ion implantation." *Applied Physics Letters* 73 (21):3096–3098. doi:10.1063/1.122684.

Chandra, Sourov, Pradip Das, Sourav Bag, Dipranjan Laha, and Panchanan Pramanik. 2011. "Synthesis, functionalization and bioimaging applications of highly fluorescent carbon nanoparticles." *Nanoscale* 3 (4):1533–1540. doi:10.1039/C0NR00735H.

Chen, Changle, and Zhengsong Lou. 2009. "Formation of C_{60} by reduction of CO_2." *The Journal of Supercritical Fluids* 50 (1):42–45. doi:10.1016/j.supflu.2009.04.008.

Chen, Guanxiong, Shuilin Wu, Liwei Hui, Yuan Zhao, Jianglin Ye, Ziqi Tan, Wencong Zeng, Zhuchen Tao, Lihua Yang, and Yanwu Zhu. 2016. "Assembling carbon quantum dots to a layered carbon for high-density supercapacitor electrodes." *Scientific Reports* 6:19028. doi:10.1038/srep19028.

Chen, Qing, Yue Hu, Chuangang Hu, Huhu Cheng, Zhipan Zhang, Huibo Shao, and Liangti Qu. 2014. "Graphene quantum dots–three-dimensional graphene composites for high-performance supercapacitors." *Physical Chemistry Chemical Physics* 16 (36):19307–19313. doi:10.1039/C4CP02761B.

Chua, Chun Kiang, Zdeněk Sofer, Petr Šimek, Ondřej Jankovský, Kateřina Klímová, Snejana Bakardjieva, Štěpánka Hrdličková Kučková, and Martin Pumera. 2015. "Synthesis of strongly fluorescent graphene quantum dots by cage-opening buckminsterfullerene." *ACS Nano* 9 (3):2548–2555. doi:10.1021/nn505639q.

Conway, Brian Evans. 1991. "Transition from 'Supercapacitor' to 'Battery' behavior in electrochemical energy storage." *Journal of the Electrochemical Society* 138 (6):1539–1548.

Conway, Brian Evans, and Eliezer Gileadi. 1962. "Kinetic theory of pseudo-capacitance and electrode reactions at appreciable surface coverage." *Transactions of the Faraday Society* 58 (0):2493–2509. doi:10.1039/TF9625802493.

Dang, Yong-Qiang, Shao-Zhao Ren, Guoyang Liu, Jiangtao Cai, Yating Zhang, and Jieshan Qiu. 2016. "Electrochemical and capacitive properties of carbon dots/reduced graphene oxide supercapacitors." *Nanomaterials* 6 (11). doi:10.3390/nano6110212.

Danilenko, Viacheslav Vasilyovich. 2004. "On the history of the discovery of nanodiamond synthesis." *Physics of the Solid State* 46 (4):595–599. doi: 10.1134/1.1711431.

De, Bibekananda, Jayaraman Balamurugan, Nam Hoon Kim, and Joong Hee Lee. 2017. "Enhanced electrochemical and photocatalytic performance of core–shell CuS@carbon quantum dots@carbon hollow nanospheres." *ACS Applied Materials & Interfaces* 9 (3):2459–2468. doi:10.1021/acsami.6b13496.

De, Bibekananda, Tapas Kuila, Nam Hoon Kim, and Joong Hee Lee. 2017. "Carbon dot stabilized copper sulphide nanoparticles decorated graphene oxide hydrogel for high performance asymmetric supercapacitor." *Carbon* 122:247–257. doi:10.1016/j.carbon.2017.06.076.

de Heer, Walt A., and Daniel Ugarte. 1993. "Carbon onions produced by heat treatment of carbon soot and their relation to the 217.5 nm interstellar absorption feature." *Chemical Physics Letters* 207 (4):480–486. doi:10.1016/0009-2614(93)89033-E.

Demarconnay, Laurent, Encarnación Raymundo-Piñero, and Francandois Béguin. 2011. "Adjustment of electrodes potential window in an asymmetric carbon/ MnO_2 supercapacitor." *Journal of Power Sources* 196 (1):580–586. doi:10.1016/j. jpowsour.2010.06.013.

Deng, Yehao, Dongxu Zhao, Xing Chen, Fei Wang, Hang Song, and Dezhen Shen. 2013. "Long lifetime pure organic phosphorescence based on water soluble carbon dots." *Chemical Communications* 49 (51):5751–5753. doi:10.1039/C3CC42600A.

Dey, Ramendra Sundar, Hans Aage Hjuler, and Qijin Chi. 2015. "Approaching the theoretical capacitance of graphene through copper foam integrated three-dimensional graphene networks." *Journal of Materials Chemistry A* 3 (12):6324–6329. doi:10.1039/ C5TA01112D.

Dinari, Mohamad, Mohamad Mohsen Momeni, and Meysam Goudarzirad. 2015. "Nanocomposite films of polyaniline/graphene quantum dots and its supercapacitor properties." *Surface Engineering* 32 (7):535–540. doi:10.1080/02670844.2015. 1108047.

Dolmatov, Valerii Yu. 2001. "Detonation synthesis ultradispersed diamonds: Properties and applications." *Russian Chemical Reviews* 70 (7):607–626. doi:10.1070/ RC2001v070n07ABEH000665.

Dong, Yongqiang, Congqiang Chen, Xinting Zheng, Lili Gao, Zhiming Cui, Hongbin Yang, Chunxian Guo, Yuwu Chi, and Chang Ming Li. 2012. "One-step and high yield simultaneous preparation of single- and multi-layer graphene quantum dots from CX-72 carbon black." *Journal of Materials Chemistry* 22 (18):8764–8766. doi:10.1039/ C2JM30658A.

Dong, Yongqiang, Hongchang Pang, Shuyan Ren, Congqiang Chen, Yuwu Chi, and Ting Yu. 2013. "Etching single-wall carbon nanotubes into green and yellow single-layer graphene quantum dots." *Carbon* 64:245–251. doi:10.1016/j.carbon.2013.07.059.

Dong, Yongqiang, Ruixue Wang, Hao Li, Jingwei Shao, Yuwu Chi, Xiaomei Lin, and Guonan Chen. 2012. "Polyamine-functionalized carbon quantum dots for chemical sensing." *Carbon* 50 (8):2810–2815. doi:10.1016/j.carbon.2012.02.046.

Eftekhari, Ali, Lei Li, and Yang Yang. 2017. "Polyaniline supercapacitors." *Journal of Power Sources* 347:86–107. doi:10.1016/j.jpowsour.2017.02.054.

Fan, Li-Zhen, and Joachim Maier. 2006. "High-performance polypyrrole electrode materials for redox supercapacitors." *Electrochemistry Communications* 8 (6):937–940. doi:10.1016/j.elecom.2006.03.035.

Frackowiak, Elzbieta. 2007. "Carbon materials for supercapacitor application." *Physical Chemistry Chemical Physics* 9 (15):1774–1785. doi:10.1039/B618139M.

Frackowiak, Elzbieta, and François Béguin. 2001. "Carbon materials for the electrochemical storage of energy in capacitors." *Carbon* 39 (6):937–950. doi:10.1016/ S0008-6223(00)00183-4.

Ganganboina, Akhilesh Babu, Ankan Dutta Chowdhury, and Ruey-an Doong. 2017a. "Nano assembly of N-doped graphene quantum dots anchored Fe_3O_4/halloysite nanotubes for high performance supercapacitor." *Electrochimica Acta* 245:912–923. doi:10.1016/j. electacta.2017.06.002.

Ganganboina, Akhilesh Babu, Ankan Dutta Chowdhury, and Ruey-an Doong. 2017b. "New avenue for appendage of graphene quantum dots on halloysite nanotubes as anode materials for high performance supercapacitors." *ACS Sustainable Chemistry & Engineering* 5 (6):4930–4940. doi:10.1021/acssuschemeng.7b00329.

Gao, Fang, Georgia Lewes-Malandrakis, Marco T. Wolfer, Wolfgang Müller-Sebert, Pascal Gentile, David Aradilla, Thomas Schubert, and Christoph E. Nebel. 2015. "Diamond-coated silicon wires for supercapacitor applications in ionic liquids." *Diamond and Related Materials* 51:1–6. doi:10.1016/j.diamond.2014.10.009.

Gao, Fang, Marco T. Wolfer, and Christoph E. Nebel. 2014. "Highly porous diamond foam as a thin-film micro-supercapacitor material." *Carbon* 80:833–840. doi:10.1016/j. carbon.2014.09.007.

Gao, Zhiyong, Chen Chen, Jiuli Chang, Liming Chen, Dapeng Wu, Fang Xu, and Kai Jiang. 2018. "Balanced energy density and power density: Asymmetric supercapacitor based on activated fullerene carbon soot anode and graphene-Co_3O_4 composite cathode." *Electrochimica Acta* 260:932–943. doi:10.1016/j.electacta.2017.12.070.

Genc, Rukan, Melis Ozge Alas, Ersan Harputlu, Sergej Repp, Nora Kremer, Mike Castellano, Suleyman Gokhan Colak, Kasim Ocakoglu, and Emre Erdem. 2017. "High-capacitance hybrid supercapacitor based on multi-colored fluorescent carbon-dots." *Scientific Reports* 7 (1):11222. doi:10.1038/s41598-017-11347-1.

Georgakilas, Vasilios, Jason A. Perman, Jiri Tucek, and Radek Zboril. 2015. "Broad family of carbon nanoallotropes: Classification, chemistry, and applications of fullerenes, carbon dots, nanotubes, graphene, nanodiamonds, and combined superstructures." *Chemical Reviews* 115 (11):4744–4822. doi:10.1021/cr500304f.

Gong, Xiaojuan, Qin Hu, Man Chin Paau, Yan Zhang, Shaomin Shuang, Chuan Dong, and Martin M. F. Choi. 2014. "Red-green-blue fluorescent hollow carbon nanoparticles isolated from chromatographic fractions for cellular imaging." *Nanoscale* 6 (14):8162–8170. doi:10.1039/C4NR01453G.

Grądzka, Emilia, Krzysztof Winkler, Marta Borowska, Marta E. Plonska-Brzezinska, and Luis Echegoyen. 2013. "Comparison of the electrochemical properties of thin films of MWCNTs/C60-Pd, SWCNTs/C60-Pd and ox-CNOs/C60-E." *Electrochimica Acta* 96:274–284. doi:10.1016/j.electacta.2013.02.035.

Gupta, Vinay, and Norio Miura. 2006. "Polyaniline/single-wall carbon nanotube (PANI/ SWCNT) composites for high performance supercapacitors." *Electrochimica Acta* 52 (4):1721–1726. doi:10.1016/j.electacta.2006.01.074.

Hare, Jonathan P., Harold W. Kroto, and Roger Taylor. 1991. "Preparation and UV/visible spectra of fullerenes C_{60} and C_{70}." *Chemical Physics Letters* 177 (4–5):394–398. doi:10.1016/0009-2614(91)85072-5.

Hassan, Mahbub, Enamul Haque, Kakarla Raghava Reddy, Andrew I. Minett, Jun Chen, and Vincent G. Gomes. 2014. "Edge-enriched graphene quantum dots for enhanced photo-luminescence and supercapacitance." *Nanoscale* 6 (20):11988–11994. doi:10.1039/C4NR02365J.

Hiraoka, Tatsuki, Ali Izadi-Najafabadi, Takeo Yamada, Don N. Futaba, Satoshi Yasuda, Osamu Tanaike, Hiroaki Hatori, Motoo Yumura, Sumio Iijima, and Kenji Hata. 2010. "Compact and light supercapacitor electrodes from a surface-only solid by opened carbon nanotubes with 2 200 m^2g^{-1} surface area." *Advanced Functional Materials* 20 (3):422–428. doi:10.1002/adfm.200901927.

Homann, Klaus-Heinrich. 1998. "Fullerenes and soot formation—New pathways to large particles in flames." *Angewandte Chemie International Edition* 37 (18):2434–2451. doi:10.1002/(SICI)1521-3773(19981002)37:18 <2434::AID-ANIE2434>3.0.CO;2-L.

Howard, Jack B., J. Thomas McKinnon, Yakov Makarovsky, Arthur L. Lafleur, and M. Elaine Johnson. 1991. "Fullerenes C_{60} and C_{70} in flames." *Nature* 352 (6331):139–141. doi:10.1038/352139a0.

Hsu, Pin-Che, and Huan-Tsung Chang. 2012. "Synthesis of high-quality carbon nanodots from hydrophilic compounds: Role of functional groups." *Chemical Communications* 48 (33):3984–3986. doi:10.1039/C2CC30188A.

Hu, Liming, Yun Sun, Shengliang Li, Xiaoli Wang, Kelei Hu, Lirong Wang, Xing-jie Liang, and Yan Wu. 2014. "Multifunctional carbon dots with high quantum yield for imaging and gene delivery." *Carbon* 67:508–513. doi:10.1016/j.carbon.2013.10.023.

Hu, Shengliang, Jing Sun, Xiwen Du, Fei Tian, and Lei Jiang. 2008. "The formation of multiply twinning structure and photoluminescence of well-dispersed nanodiamonds produced by pulsed-laser irradiation." *Diamond and Related Materials* 17 (2):142–146. doi:10.1016/j.diamond.2007.11.009.

Hu, Shengliang, Fei Tian, Peikang Bai, Shirui Cao, Jing Sun, and Jing Yang. 2009. "Synthesis and luminescence of nanodiamonds from carbon black." *Materials Science and Engineering: B* 157 (1):11–14. doi:10.1016/j.mseb.2008.12.001.

Hu, Yue, Yang Zhao, Gewu Lu, Nan Chen, Zhipan Zhang, Hui Li, Huibo Shao, and Liangti Qu. 2013. "Graphene quantum dots–carbon nanotube hybrid arrays for supercapacitors." *Nanotechnology* 24 (19):195401. doi:10.1088/0957-4484/24/19/195401.

Hunter, Joanna, James Fye, and Martin F. Jarrold. 1993. "Annealing C60+: Synthesis of fullerenes and large carbon rings." *Science* 260 (5109):784. doi:10.1126/science.260.5109.784.

Irle, Stephan, Guishan Zheng, Zhi Wang, and Keiji Morokuma. 2006. "The C_{60} formation puzzle 'Solved': QM/MD simulations reveal the shrinking hot giant road of the dynamic fullerene self-assembly mechanism." *The Journal of Physical Chemistry B* 110 (30):14531–14545. doi:10.1021/jp061173z.

Islam, Mohammad S., Yan Deng, Liyong Tong, Anup K. Roy, Shaikh N. Faisal, Mahbub Hassan, Andrew I. Minett, and Vincent G. Gomes. 2017. "In-situ direct grafting of graphene quantum dots onto carbon fibre by low temperature chemical synthesis for high performance flexible fabric supercapacitor." *Materials Today Communications* 10:112–119. doi:10.1016/j.mtcomm.2016.11.002.

Jaiswal, Amit, Siddhartha Sankar Ghosh, and Arun Chattopadhyay. 2012. "One step synthesis of C-dots by microwave mediated caramelization of poly(ethylene glycol)." *Chemical Communications* 48 (3):407–409. doi:10.1039/C1CC15988G.

Jia, Henan, Yifei Cai, Jinghuang Lin, Haoyan Liang, Junlei Qi, Jian Cao, Jicai Feng, and Wei Dong Fei. 2018. "Heterostructural graphene quantum dot/MnO_2 nanosheets toward high-potential window electrodes for high-performance supercapacitors." *Advanced Science* 5 (5):1700887. doi:10.1002/advs.201700887.

Jian, Xuan, Jia-gang Li, Hui-min Yang, Le-le Cao, Er-hui Zhang, and Zhen-hai Liang. 2017a. "Carbon quantum dots reinforced polypyrrole nanowire via electrostatic self-assembly strategy for high-performance supercapacitors." *Carbon* 114:533–543. doi:10.1016/j.carbon.2016.12.033.

Jian, Xuan, Hui-min Yang, Jia-gang Li, Er-hui Zhang, Le-le Cao, and Zhen-hai Liang. 2017b. "Flexible all-solid-state high-performance supercapacitor based on electrochemically synthesized carbon quantum dots/polypyrrole composite electrode." *Electrochimica Acta* 228:483–493. doi:10.1016/j.electacta.2017.01.082.

Jiang, Zhiqiang, Andrew Nolan, Jeffrey G. A. Walton, Annamaria Lilienkampf, Rong Zhang, and Mark Bradley. 2014. "Photoluminescent carbon dots from 1,4-addition polymers." *Chemistry—A European Journal* 20 (35):10926–10931. doi:10.1002/chem.201403076.

John, Robin, Dhanraj B. Shinde, Lili Liu, Feng Ding, Zhiping Xu, Cherianath Vijayan, Vijayamohanan K. Pillai, and Thalappil Pradeep. 2014. "Sequential electrochemical unzipping of single-walled carbon nanotubes to graphene ribbons revealed by in situ Raman spectroscopy and imaging." *ACS Nano* 8 (1):234–242. doi:10.1021/nn403289g.

Kim, Juhan, and Jung Sang Suh. 2014. "Size-controllable and low-cost fabrication of graphene quantum dots using thermal plasma jet." *ACS Nano* 8 (5):4190–4196. doi:10.1021/nn404180w.

Kim, Sun-I., Jung-Soo Lee, Hyo-Jin Ahn, Hyun-Kon Song, and Ji-Hyun Jang. 2013. "Facile route to an efficient NiO supercapacitor with a three-dimensional nanonetwork morphology." *ACS Applied Materials & Interfaces* 5 (5):1596–1603. doi:10.1021/am3021894.

Kötz, Rüdiger, and Martin Carlen. 2000. "Principles and applications of electrochemical capacitors." *Electrochimica Acta* 45 (15):2483–2498. doi:10.1016/S0013-4686(00)00354-6.

Kovalenko, Igor, David G. Bucknall, and Gleb Yushin. 2010. "Detonation nanodiamond and onion-like-carbon-embedded polyaniline for supercapacitors." *Advanced Functional Materials* 20 (22):3979–3986. doi:10.1002/adfm.201000906.

Krätschmer, W., Lowell D. Lamb, K. Fostiropoulos, and Donald R. Huffman. 1990. "Solid C_{60}: A new form of carbon." *Nature* 347 (6291):354–358. doi:10.1038/347354a0.

Kroto, Harold W. 1988. "Space, stars, C_{60}, and soot." *Science* 242 (4882):1139. doi:10.1126/science.242.4882.1139.

Kroto, Harold W., James R. Heath, Sean C. O'Brien, Robert F. Curl, and Richard E. Smalley. 1985. "C_{60}: Buckminsterfullerene." *Nature* 318 (6042):162–163. doi:10.1038/318162a0.

Krueger, Anke. 2008. "Diamond nanoparticles: Jewels for chemistry and physics." *Advanced Materials* 20 (12):2445–2449. doi:10.1002/adma.200701856.

Krüger, Anke, Fumiaki Kataoka, Masaki Ozawa, T. Fujino, Y. Suzuki, Aleksandr E. Aleksenskii, Alexander Ya Vul', and Eiji Ōsawa. 2005. "Unusually tight aggregation in detonation nanodiamond: Identification and disintegration." *Carbon* 43 (8):1722–1730. doi:10.1016/j.carbon.2005.02.020.

Kulal, Prakash M., Deepak P. Dubal, Chandrakant D. Lokhande, and Vijay J. Fulari. 2011. "Chemical synthesis of Fe_2O_3 thin films for supercapacitor application." *Journal of Alloys and Compounds* 509 (5):2567–2571. doi:10.1016/j.jallcom.2010.11.091.

Kumar, Ajay, Pin Ann Lin, Albert Xue, Boyi Hao, Yoke Khin Yap, and R. Mohan Sankaran. 2013. "Formation of nanodiamonds at near-ambient conditions via microplasma dissociation of ethanol vapour." *Nature Communications* 4:2618. doi:10.1038/ncomms3618.

Kumar, Sumeet, Animesh K. Ojha, Bilal Ahmed, Ashok Kumar, Jayanta Das, and Arnulf Materny. 2017. "Tunable (violet to green) emission by high-yield graphene quantum dots and exploiting its unique properties towards sun-light-driven photocatalysis and supercapacitor electrode materials." *Materials Today Communications* 11:76–86. doi:10.1016/j.mtcomm.2017.02.009.

Kuznetsov, Vladimir L., Andrey L. Chuvilin, Yuri V. Butenko, Igor Yu Mal'kov, and Vladimir M. Titov. 1994. "Onion-like carbon from ultra-disperse diamond." *Chemical Physics Letters* 222 (4):343–348. doi:10.1016/0009-2614(94)87072-1.

Kwon, Woosung, Sungan Do, and Shi-Woo Rhee. 2012. "Formation of highly luminescent nearly monodisperse carbon quantum dots via emulsion-templated carbonization of carbohydrates." *RSC Advances* 2 (30):11223–11226. doi:10.1039/C2RA22186A.

Kwon, Woosung, Gyeongjin Lee, Sungan Do, Taiha Joo, and Shi-Woo Rhee. 2014. "Size-controlled soft-template synthesis of carbon nanodots toward versatile photoactive materials." *Small* 10 (3):506–513. doi:10.1002/smll.201301770.

Laforgue, Alexis, Patrice Simon, Christian Sarrazin, and Jean-François Fauvarque. 1999. "Polythiophene-based supercapacitors." *Journal of Power Sources* 80 (1):142–148. doi:10.1016/S0378-7753(98)00258-4.

Lai, Chih-Wei, Yi-Hsuan Hsiao, Yung-Kang Peng, and Pi-Tai Chou. 2012. "Facile synthesis of highly emissive carbon dots from pyrolysis of glycerol; gram scale production of carbon dots/mSiO_2 for cell imaging and drug release." *Journal of Materials Chemistry* 22 (29):14403–14409. doi:10.1039/C2JM32206D.

Lai, Lin, and Amanda S. Barnard. 2011. "Modeling the thermostability of surface functionalisation by oxygen, hydroxyl, and water on nanodiamonds." *Nanoscale* 3 (6):2566–2575. doi:10.1039/C1NR10108K.

Lee, Eunwoo, Jaehoon Ryu, and Jyongsik Jang. 2013. "Fabrication of graphene quantum dots via size-selective precipitation and their application in upconversion-based DSSCs." *Chemical Communications* 49 (85):9995–9997. doi:10.1039/C3CC45588B.

Lee, Jinsup, Kyungho Kim, Woon Ik Park, Bo-Hyun Kim, Jong Hyun Park, Tae-Heon Kim, Sungyool Bong et al. 2012. "Uniform graphene quantum dots patterned from self-assembled silica nanodots." *Nano Letters* 12 (12):6078–6083. doi:10.1021/nl302520m.

Lee, Keunsik, Hanleem Lee, Yonghun Shin, Yeoheung Yoon, Doyoung Kim, and Hyoyoung Lee. 2016. "Highly transparent and flexible supercapacitors using graphene-graphene quantum dots chelate." *Nano Energy* 26:746–754. doi:10.1016/j.nanoen.2016.06.030.

Li, Haitao, Xiaodie He, Yang Liu, Hui Huang, Suoyuan Lian, Shuit-Tong Lee, and Zhenhui Kang. 2011. "One-step ultrasonic synthesis of water-soluble carbon nanoparticles with excellent photoluminescent properties." *Carbon* 49 (2):605–609. doi:10.1016/j.carbon.2010.10.004.

Li, Haitao, Zhenhui Kang, Yang Liu, and Shuit-Tong Lee. 2012. "Carbon nanodots: Synthesis, properties and applications." *Journal of Materials Chemistry* 22 (46):24230–24253. doi:10.1039/C2JM34690G.

Li, Ling-Ling, Jing Ji, Rong Fei, Chong-Zhi Wang, Qian Lu, Jian-Rong Zhang, Li-Ping Jiang, and Jun-Jie Zhu. 2012. "A facile microwave avenue to electrochemiluminescent two-color graphene quantum dots." *Advanced Functional Materials* 22 (14):2971–2979. doi:10.1002/adfm.201200166.

Li, Lingling, Gehui Wu, Guohai Yang, Juan Peng, Jianwei Zhao, and Jun-Jie Zhu. 2013. "Focusing on luminescent graphene quantum dots: Current status and future perspectives." *Nanoscale* 5 (10):4015–4039. doi:10.1039/C3NR33849E.

Li, Xiaoming, Muchen Rui, Jizhong Song, Zihan Shen, and Haibo Zeng. 2015. "Carbon and graphene quantum dots for optoelectronic and energy devices: A review." *Advanced Functional Materials* 25 (31):4929–4947. doi:10.1002/adfm.201501250.

Li, Zhen, Yanfeng Li, Liang Wang, Ling Cao, Xiang Liu, Zhiwen Chen, Dengyu Pan, and Minghong Wu. 2017. "Assembling nitrogen and oxygen co-doped graphene quantum dots onto hierarchical carbon networks for all-solid-state flexible supercapacitors." *Electrochimica Acta* 235:561–569. doi:10.1016/j.electacta.2017.03.147.

Li, Zhen, Ping Qin, Liang Wang, Chengshuai Yang, Yanfeng Li, Zhiwen Chen, Dengyu Pan, and Minghong Wu. 2016. "Amine-enriched graphene quantum dots for high-pseudocapacitance supercapacitors." *Electrochimica Acta* 208:260–266. doi:10.1016/j.electacta.2016.05.030.

Lieber, Charles M., and Chia-Chun Chen. 1994. "Preparation of fullerenes and fullerene-based materials." In *Solid State Physics*, edited by Henry Ehrenreich and Frans Spaepen, 109–148. San Diego: Academic Press.

Liew, Chiam-Wen, Subramaniam T. Ramesh, and Abdul K. Arof. "Good prospect of ionic liquid based-poly(vinyl alcohol) polymer electrolytes for supercapacitors with excellent electrical, electrochemical and thermal properties." *International Journal of Hydrogen Energy* 39 (6):2953–2963. doi:10.1016/j.ijhydene.2013.06.061.

Linehan, Keith, and Hugh Doyle. 2014. "Efficient one-pot synthesis of highly monodisperse carbon quantum dots." *RSC Advances* 4 (1):18–21. doi:10.1039/C3RA45083J.

Liu, Dong-Qiang, Sung-Hun Yu, Se-Wan Son, and Seung-Ki Joo. 2008. "Electrochemical performance of iridium oxide thin film for supercapacitor prepared by radio frequency magnetron sputtering method." *ECS Transactions* 16 (1):103–109.

Liu, Haipeng, Tao Ye, and Chengde Mao. 2007. "Fluorescent carbon nanoparticles derived from candle soot." *Angewandte Chemie International Edition* 119 (34):6593–6595. doi:10.1002/ange.200701271.

Liu, Jing, Xinling Liu, Hongjie Luo, and Yanfeng Gao. 2014. "One-step preparation of nitrogen-doped and surface-passivated carbon quantum dots with high quantum yield and excellent optical properties." *RSC Advances* 4 (15):7648–7654. doi:10.1039/C3RA47577H.

Liu, Ruili, Dongqing Wu, Xinliang Feng, and Klaus Müllen. 2011. "Bottom-up fabrication of photoluminescent graphene quantum dots with uniform morphology." *Journal of the American Chemical Society* 133 (39):15221–15223. doi:10.1021/ja204953k.

Liu, Ruili, Dongqing Wu, Shuhua Liu, Kaloian Koynov, Wolfgang Knoll, and Qin Li. 2009. "An aqueous route to multicolor photoluminescent carbon dots using silica spheres as carriers." *Angewandte Chemie International Edition* 48 (25):4598–4601. doi:10.1002/anie.200900652.

Liu, Tongchang, Wendy G. Pell, and Brian Evans Conway. 1997. "Self-discharge and potential recovery phenomena at thermally and electrochemically prepared RuO_2 supercapacitor electrodes." *Electrochimica Acta* 42 (23):3541–3552. doi:10.1016/S0013-4686(97)81190-5.

Liu, Wen-Wen, Ya-Qiang Feng, Xing-Bin Yan, Jiang-Tao Chen, and Qun-Ji Xue. 2013. "Superior micro-supercapacitors based on graphene quantum dots." *Advanced Functional Materials* 23 (33):4111–4122. doi:10.1002/adfm.201203771.

Liu, Wenwen, Xingbin Yan, Jiangtao Chen, Yaqiang Feng, and Qunji Xue. 2013. "Novel and high-performance asymmetric micro-supercapacitors based on graphene quantum dots and polyaniline nanofibers." *Nanoscale* 5 (13):6053–6062. doi:10.1039/C3NR01139A.

Liu, Yun, Chun-yan Liu, and Zhi-ying Zhang. 2013. "Graphitized carbon dots emitting strong green photoluminescence." *Journal of Materials Chemistry C* 1 (32):4902–4907. doi:10.1039/C3TC30670D.

Lv, Lingxiao, Yueqiong Fan, Qing Chen, Yang Zhao, Yue Hu, Zhipan Zhang, Nan Chen, and Liangti Qu. 2014. "Three-dimensional multichannel aerogel of carbon quantum dots for high-performance supercapacitors." *Nanotechnology* 25 (23):235401. doi:10.1088/0957-4484/25/23/235401.

Ma, Jia, Qi Guo, Hai-Ling Gao, and Xue Qin. 2015. "Synthesis of C_{60}/graphene composite as electrode in supercapacitors." *Fullerenes, Nanotubes and Carbon Nanostructures* 23 (6):477–482. doi:10.1080/1536383X.2013.865604.

Ma, Zheng, Hai Ming, Hui Huang, Yang Liu, and Zhenhui Kang. 2012. "One-step ultrasonic synthesis of fluorescent N-doped carbon dots from glucose and their visible-light sensitive photocatalytic ability." *New Journal of Chemistry* 36 (4):861–864. doi:10.1039/C2NJ20942J.

Mahfouz, Remi, Daniel L. Floyd, Wei Peng, Jennifer T. Choy, Marko Loncar, and Osman M. Bakr. 2013. "Size-controlled fluorescent nanodiamonds: A facile method of fabrication and color-center counting." *Nanoscale* 5 (23):11776–11782. doi:10.1039/C3NR03320A.

Markovic, Zoran M., Biljana Z. Ristic, Katarina M. Arsikin, Djordje G. Klisic, Ljubica M. Harhaji-Trajkovic, Biljana M. Todorovic-Markovic, Dejan P. Kepic et al. 2012. "Graphene quantum dots as autophagy-inducing photodynamic agents." *Biomaterials* 33 (29):7084–7092. doi:10.1016/j.biomaterials.2012.06.060.

Meher, Sumanta Kumar, and G. Ranga Rao. 2011. "Ultralayered Co_3O_4 for high-performance supercapacitor applications." *The Journal of Physical Chemistry C* 115 (31):15646–15654. doi:10.1021/jp201200e.

Ming, Hai, Zheng Ma, Yang Liu, Keming Pan, Hang Yu, Fang Wang, and Zhenhui Kang. 2012. "Large scale electrochemical synthesis of high quality carbon nanodots and their photocatalytic property." *Dalton Transactions* 41 (31):9526–9531. doi:10.1039/C2DT30985H.

Mitra, Shouvik, Sourov Chandra, Tanay Kundu, Rahul Banerjee, Panchanan Pramanik, and Arunava Goswami. 2012. "Rapid microwave synthesis of fluorescent hydrophobic carbon dots." *RSC Advances* 2 (32):12129–12131. doi:10.1039/C2RA21048G.

Mojica, Martha, Julio A. Alonso, and Francisco Méndez. 2013. "Synthesis of fullerenes." *Journal of Physical Organic Chemistry* 26 (7):526–539. doi:10.1002/poc.3121.

Mondal, Sanjoy, Utpal Rana, and Sudip Malik. 2015. "Graphene quantum dot-doped polyaniline nanofiber as high performance supercapacitor electrode materials." *Chemical Communications* 51 (62):12365–12368. doi:10.1039/C5CC03981A.

Murayama, Hideki, Shigeki Tomonoh, J. Michael Alford, and Michael E. Karpuk. 2005. "Fullerene production in tons and more: From science to industry." *Fullerenes, Nanotubes and Carbon Nanostructures* 12 (1–2):1–9. doi:10.1081/FST-120027125.

Muthurasu, Alagan, Perumal Dhandapani, and Vattikondala Ganesh. 2016. "Facile and simultaneous synthesis of graphene quantum dots and reduced graphene oxide for bio-imaging and supercapacitor applications." *New Journal of Chemistry* 40 (11):9111–9124. doi:10.1039/C6NJ00586A.

Niu, Qingyuan, Kezheng Gao, Zhihui Lin, and Wenhui Wu. 2013. "Amine-capped carbon dots as a nanosensor for sensitive and selective detection of picric acid in aqueous solution via electrostatic interaction." *Analytical Methods* 5 (21):6228–6233. doi:10.1039/C3AY41275J.

Okajima, Keiichi, Atsushi Ikeda, Kazunori Kamoshita, and Masao Sudoh. 2005. "High rate performance of highly dispersed C_{60} on activated carbon capacitor." *Electrochimica Acta* 51 (5):972–977. doi:10.1016/j.electacta.2005.04.055.

Osswald, Sebastian, Gleb Yushin, Vadym Mochalin, Sergei O. Kucheyev, and Yury Gogotsi. 2006. "Control of sp^2/sp^3 carbon ratio and surface chemistry of nanodiamond powders by selective oxidation in air." *Journal of the American Chemical Society* 128 (35):11635–11642. doi:10.1021/ja063303n.

Ozawa, M., M. Inaguma, M. Takahashi, F. Kataoka, A. Krüger, and E. Ōsawa. 2007. "Preparation and behavior of brownish, clear nanodiamond colloids." *Advanced Materials* 19 (9):1201–1206. doi:10.1002/adma.200601452.

Pan, Dengyu, Jingchun Zhang, Zhen Li, and Minghong Wu. 2010. "Hydrothermal route for cutting graphene sheets into blue-luminescent graphene quantum dots." *Advanced Materials* 22 (6):734–738. doi:10.1002/adma.200902825.

Pandolfo, Anthony G., and Anthony F. Hollenkamp. 2006. "Carbon properties and their role in supercapacitors." *Journal of Power Sources* 157 (1):11–27. doi:10.1016/j.jpowsour.2006.02.065.

Parker, Deborah Holmes, Peter Wurz, Kuntal Chatterjee, Keith R. Lykke, Jerry E. Hunt, Michael J. Pellin, John C. Hemminger, Dieter M. Gruen, and Leon M. Stock. 1991. "High-yield synthesis, separation, and mass-spectrometric characterization of fullerenes C_{60} to C_{266}." *Journal of the American Chemical Society* 113 (20):7499–7503. doi:10.1021/ja00020a008.

Patil, Shrinivas J., Jinhyeok Kim, and Dong Weon Lee. 2017. "Graphene-nanosheet wrapped cobalt sulphide as a binder free hybrid electrode for asymmetric solid-state supercapacitor." *Journal of Power Sources* 342:652–665. doi: https://doi.org/10.1016/j.jpowsour.2016.12.096.

Peng, Juan, Wei Gao, Bipin Kumar Gupta, Zheng Liu, Rebeca Romero-Aburto, Liehui Ge, Li Song et al. 2012. "Graphene quantum dots derived from carbon fibers." *Nano Letters* 12 (2):844–849. doi:10.1021/nl2038979.

Peng, Zhiyao, Yajing Hu, Jingjing Wang, Sijie Liu, Chenxi Li, Qinglong Jiang, Jun Lu, Xiaoqiao Zeng, Ping Peng, and Fang-Fang Li. 2019. "Fullerene-based in situ doping of N and Fe into a 3D cross-like hierarchical carbon composite for high-performance supercapacitors." *Advanced Energy Materials* 9 (11):1802928. doi:10.1002/aenm.201802928.

Pentecost, Amanda, Shruti Gour, Vadym Mochalin, Isabel Knoke, and Yury Gogotsi. 2010. "Deaggregation of nanodiamond powders using salt- and sugar-assisted milling." *ACS Applied Materials & Interfaces* 2 (11):3289–3294. doi:10.1021/am100720n.

Ponomarenko, Leonid A., Fredrik Schedin, Mikhail I. Katsnelson, Ruyan Yang, Ernie W. Hill, Kostya S. Novoselov, and Andre K. Geim. 2008. "Chaotic Dirac billiard in graphene quantum dots." *Science* 320 (5874):356. doi:10.1126/science.1154663.

Poonam, Kriti Sharma, Anmol Arora, and Surya Kant Tripathi. 2019. "Review of supercapacitors: Materials and devices." *Journal of Energy Storage* 21:801–825. doi:10.1016/j.est.2019.01.010.

Portet, Cristelle, Gleb N. Yushin, and Yuri Gogotsi. 2007. "Electrochemical performance of carbon onions, nanodiamonds, carbon black and multiwalled nanotubes in electrical double layer capacitors." *Carbon* 45 (13):2511–2518. doi:10.1016/j.carbon.2007.08.024.

Puvvada, Nagaprasad, B. N. Prashanth Kumar, Suraj Konar, Himani Kalita, Mahitosh Mandal, and Amita Pathak. 2012. "Synthesis of biocompatible multicolor luminescent carbon dots for bioimaging applications." *Science and Technology of Advanced Materials* 13 (4):045008. doi:10.1088/1468-6996/13/4/045008.

Raty, Jean-Yves, Giulia Galli, C. Bostedt, Tony W. van Buuren, and Louis J. Terminello. 2003. "Quantum confinement and fullerenelike surface reconstructions in nanodiamonds." *Physical Review Letters* 90 (3):037401. doi:10.1103/PhysRevLett.90.037401.

Ray, S. C., Arindam Saha, Nikhil R. Jana, and Rupa Sarkar. 2009. "Fluorescent carbon nanoparticles: Synthesis, characterization, and bioimaging application." *The Journal of Physical Chemistry C* 113 (43):18546–18551. doi:10.1021/jp905912n.

Raymundo-Piñero, E., K. Kierzek, J. Machnikowski, and F. Béguin. 2006. "Relationship between the nanoporous texture of activated carbons and their capacitance properties in different electrolytes." *Carbon* 44 (12):2498–2507. doi:10.1016/j.carbon.2006.05.022.

Ren, Xiu-Bin, Hai-Yan Lu, Hai-Bo Lin, Ya-Nan Liu, and Yan Xing. 2010. "Preparation and characterization of the Ti/IrO$_2$/WO$_3$ as supercapacitor electrode materials." *Russian Journal of Electrochemistry* 46 (1):77–80. doi:10.1134/S102319351001009X.

Sano, Noboru, Houqi Wang, Manish Chhowalla, Ioannis Alexandrou, and Gehan A. J. Amaratunga. 2001. "Synthesis of carbon 'onions' in water." *Nature* 414:506. doi:10.1038/35107141.

Saravanakumar, Balasubramaniam, Kamatchi K. Purushothaman, and Gopalan Muralidharan. 2012. "Interconnected V$_2$O$_5$ nanoporous network for high-performance supercapacitors." *ACS Applied Materials & Interfaces* 4 (9):4484–4490. doi:10.1021/am301162p.

Schultz, Harry P. 1965. "Topological organic chemistry. Polyhedranes and prismanes." *The Journal of Organic Chemistry* 30 (5):1361–1364. doi:10.1021/jo01016a005.

Scott, Lawrence T. 2004. "Methods for the chemical synthesis of fullerenes." *Angewandte Chemie International Edition* 43 (38):4994–5007. doi:10.1002/anie.200400661.

Scrivens, Walter A., and James M. Tour. 1992. "Synthesis of gram quantities of C$_{60}$ by plasma discharge in a modified round-bottomed flask. Key parameters for yield optimization and purification." *The Journal of Organic Chemistry* 57 (25):6932–6936. doi:10.1021/jo00051a047.

Shen, Baoshou, Junwei Lang, Ruisheng Guo, Xu Zhang, and Xingbin Yan. 2015. "Engineering the electrochemical capacitive properties of microsupercapacitors based on graphene quantum dots/MnO$_2$ using ionic liquid gel electrolytes." *ACS Applied Materials & Interfaces* 7 (45):25378–25389. doi:10.1021/acsami.5b07909.

Shen, Jianhua, Yihua Zhu, Xiaoling Yang, and Chunzhong Li. 2012. "Graphene quantum dots: Emergent nanolights for bioimaging, sensors, catalysis and photovoltaic devices." *Chemical Communications* 48 (31):3686–3699. doi:10.1039/C2CC00110A.

Shen, Jianhua, Yihua Zhu, Xiaoling Yang, Jie Zong, Jianmei Zhang, and Chunzhong Li. 2012. "One-pot hydrothermal synthesis of graphene quantum dots surface-passivated by polyethylene glycol and their photoelectric conversion under near-infrared light." *New Journal of Chemistry* 36 (1):97–101. doi:10.1039/C1NJ20658C.

Shenderova, Olga, Alexey P. Koscheev, N. Zaripov, Ivan Petrov, Yuriy Alekseevich Skryabin, P. Detkov, Stuart Turner, and Gustaaf Van Tendeloo. 2011. "Surface chemistry and properties of ozone-purified detonation nanodiamonds." *The Journal of Physical Chemistry C* 115 (20):9827–9837. doi:10.1021/jp1102466.

Shin, Yonghun, Junghyun Lee, Junghee Yang, Jintaek Park, Keunsik Lee, Sungjin Kim, Younghun Park, and Hyoyoung Lee. 2014. "Mass production of graphene quantum dots by one-pot synthesis directly from graphite in high yield." *Small* 10 (5):866–870. doi:10.1002/smll.201302286.

Shinde, Dhanraj B., Joyashish Debgupta, Ajay Kushwaha, Mohammed Aslam, and Vijayamohanan K. Pillai. 2011. "Electrochemical unzipping of multi-walled carbon nanotubes for facile synthesis of high-quality graphene nanoribbons." *Journal of the American Chemical Society* 133 (12):4168–4171. doi:10.1021/ja1101739.

Shinde, Dhanraj B., and Vijayamohanan K. Pillai. 2012. "Electrochemical preparation of luminescent graphene quantum dots from multiwalled carbon nanotubes." *Chemistry—A European Journal* 18 (39):12522–12528. doi:10.1002/chem.201201043.

Štengl, Václav, Snejana Bakardjieva, Jiří Henych, Kamil Lang, and Martin Kormunda. 2013. "Blue and green luminescence of reduced graphene oxide quantum dots." *Carbon* 63:537–546. doi:10.1016/j.carbon.2013.07.031.

Strauss, Volker, Kris Marsh, Matthew D. Kowal, Maher El-Kady, and Richard B. Kaner. 2018. "A simple route to porous graphene from carbon nanodots for supercapacitor applications." *Advanced Materials* 30 (8):1704449. doi:10.1002/adma.201704449.

Sun, Hanjun, Li Wu, Weili Wei, and Xiaogang Qu. 2013. "Recent advances in graphene quantum dots for sensing." *Materials Today* 16 (11):433–442. doi:10.1016/j.mattod.2013.10.020.

Sun, Jing, Sheng-Liang Hu, Xi-Wen Du, Yi-Wen Lei, and Lei Jiang. 2006. "Ultrafine diamond synthesized by long-pulse-width laser." *Applied Physics Letters* 89 (18):183115. doi:10.1063/1.2385210.

Sun, Ya-Ping, Bing Zhou, Yi Lin, Wei Wang, K. A. Shiral Fernando, Pankaj Pathak, Mohammed Jaouad Meziani et al. 2006. "Quantum-sized carbon dots for bright and colorful photoluminescence." *Journal of the American Chemical Society* 128 (24):7756–7757. doi:10.1021/ja062677d.

Sun, Yiqing, Shiqi Wang, Chun Li, Peihui Luo, Lei Tao, Yen Wei, and Gaoquan Shi. 2013. "Large scale preparation of graphene quantum dots from graphite with tunable fluorescence properties." *Physical Chemistry Chemical Physics* 15 (24):9907–9913. doi:10.1039/C3CP50691F.

Syed Zainol Abidin, Shariffah Nur Jannah, Md Shuhazlly Mamat, Suraya Abdul Rasyid, Zulkarnain Zainal, and Yusran Sulaiman. 2018. "Electropolymerization of poly(3,4-ethylenedioxythiophene) onto polyvinyl alcohol-graphene quantum dot-cobalt oxide nanofiber composite for high-performance supercapacitor." *Electrochimica Acta* 261:548–556. doi:10.1016/j.electacta.2017.12.168.

Tang, Qin, Partha Bairi, Rekha Goswami Shrestha, Jonathan P. Hill, Katsuhiko Ariga, Haibo Zeng, Qingmin Ji, and Lok Kumar Shrestha. 2017. "Quasi 2D mesoporous carbon microbelts derived from fullerene crystals as an electrode material for electrochemical supercapacitors." *ACS Applied Materials & Interfaces* 9 (51):44458–44465. doi:10.1021/acsami.7b13277.

Tang, Yongfu, Teng Chen, Shengxue Yu, Yuqing Qiao, Shichun Mu, Shaohua Zhang, Yufeng Zhao, Li Hou, Weiwei Huang, and Faming Gao. 2015. "A highly electronic conductive cobalt nickel sulphide dendrite/quasi-spherical nanocomposite for a supercapacitor electrode with ultrahigh areal specific capacitance." *Journal of Power Sources* 295:314–322. doi:10.1016/j.jpowsour.2015.07.035.

Taylor, Roger, G. John Langley, Harold W. Kroto, and David R. M. Walton. 1993. "Formation of C_{60} by pyrolysis of naphthalene." *Nature* 366 (6457):728–731. doi:10.1038/366728a0.

Tian, Lei, Debraj Ghosh, Wei Chen, Sulolit Pradhan, Xijun Chang, and Shaowei Chen. 2009. "Nanosized carbon particles from natural gas soot." *Chemistry of Materials* 21 (13):2803–2809. doi:10.1021/cm900709w.

Tiwari, Jitendra N., Rajanish N. Tiwari, and Kwang S. Kim. 2012. "Zero-dimensional, one-dimensional, two-dimensional and three-dimensional nanostructured materials for advanced electrochemical energy devices." *Progress in Materials Science* 57 (4):724–803. doi:10.1016/j.pmatsci.2011.08.003.

Trauzettel, Björn, Denis V. Bulaev, Daniel Loss, and Guido Burkard. 2007. "Spin qubits in graphene quantum dots." *Nature Physics* 3:192. doi:10.1038/nphys544.

Ugarte, Daniel. 1992. "Curling and closure of graphitic networks under electron-beam irradiation." *Nature* 359 (6397):707–709. doi:10.1038/359707a0.

Ugarte, Daniel. 1994. "High-temperature behaviour of 'fullerene black'." *Carbon* 32 (7):1245–1248. doi:10.1016/0008-6223(94)90108-2.

Unnikrishnan, Binesh, Chien-Wei Wu, I. Wen Peter Chen, Huan-Tsung Chang, Chia-Hua Lin, and Chih-Ching Huang. 2016. "Carbon dot-mediated synthesis of manganese oxide decorated graphene nanosheets for supercapacitor application." *ACS Sustainable Chemistry & Engineering* 4 (6):3008–3016. doi:10.1021/acssuschemeng.5b01700.

Viñes, Francesc, and Andreas Görling. 2011. "Template-assisted formation of fullerenes from short-chain hydrocarbons by supported platinum nanoparticles." *Angewandte Chemie International Edition* 50 (20):4611–4614. doi:10.1002/anie.201006588.

Wakabayashi, Tomonari, and Yohji Achiba. 1992. "A model for the C_{60} and C_{70} growth mechanism." *Chemical Physics Letters* 190 (5):465–468. doi:10.1016/0009-2614(92)85174-9.

Wang, Chuanfu, Xia Wu, Xiangping Li, Wentai Wang, Lianzhou Wang, Min Gu, and Qin Li. 2012. "Upconversion fluorescent carbon nanodots enriched with nitrogen for light harvesting." *Journal of Materials Chemistry* 22 (31):15522–15525. doi:10.1039/C2JM30935A.

Wang, Fu, Shuping Pang, Long Wang, Qin Li, Maximilian Kreiter, and Chun-Yan Liu. 2010. "One-step synthesis of highly luminescent carbon dots in noncoordinating solvents." *Chemistry of Materials* 22 (16):4528–4530. doi:10.1021/cm101350u.

Wang, Guoping, Lei Zhang, and Jiujun Zhang. 2012. "A review of electrode materials for electrochemical supercapacitors." *Chemical Society Reviews* 41 (2):797–828. doi:10.1039/C1CS15060J.

Wang, Haibin, Xiaomei Yan, and Guangzhe Piao. 2017. "A high-performance supercapacitor based on fullerene C_{60} whisker and polyaniline emeraldine base composite." *Electrochimica Acta* 231:264–271. doi:10.1016/j.electacta.2017.02.057.

Wang, Huiying, Yongqiang Yang, Xiaoyan Zhou, Ruiyi Li, and Zaijun Li. 2017. "NiCo2S4/tryptophan-functionalized graphene quantum dot nanohybrids for high-performance supercapacitors." *New Journal of Chemistry* 41 (3):1110–1118. doi:10.1039/C6NJ03443H.

Wang, Liang, Yanli Wang, Tao Xu, Haobo Liao, Chenjie Yao, Yuan Liu, Zhen Li et al. 2014. "Gram-scale synthesis of single-crystalline graphene quantum dots with superior optical properties." *Nature Communications* 5:5357. doi:10.1038/ncomms6357.

Wang, Qi, Nareerat Plylahan, Manjusha V. Shelke, Rami Reddy Devarapalli, Musen Li, Palaniappan Subramanian, Thierry Djenizian, Rabah Boukherroub, and Sabine Szunerits. 2014. "Nanodiamond particles/reduced graphene oxide composites as efficient supercapacitor electrodes." *Carbon* 68:175–184. doi:10.1016/j.carbon.2013.10.077.

Wang, Xin, Li Cao, Sheng-Tao Yang, Fushen Lu, Mohammed J. Meziani, Leilei Tian, Katherine W. Sun, Mathew A. Bloodgood, and Ya-Ping Sun. 2010. "Bandgap-like strong fluorescence in functionalized carbon nanoparticles." *Angewandte Chemie International Edition* 49 (31):5310–5314. doi:10.1002/anie.201000982.

Wee, Boon-Hong, and Jong-Dal Hong. 2014. "Multilayered poly(p-phenylenevinylene)/reduced graphene oxide film: An efficient organic current collector in an all-plastic supercapacitor." *Langmuir* 30 (18):5267–5275. doi:10.1021/la500636m.

Wei, Ji-Shi, Hui Ding, Peng Zhang, Yan-Fang Song, Jie Chen, Yong-Gang Wang, and Huan-Ming Xiong. 2016. "Carbon dots/$NiCo_2O_4$ nanocomposites with various morphologies for high performance supercapacitors." *Small* 12 (43):5927–5934. doi:10.1002/smll.201602164.

Wei, Ying, Xuelei Zhang, Xuanyu Wu, Di Tang, Kedi Cai, and Qingguo Zhang. 2016. "Carbon quantum dots/Ni–Al layered double hydroxide composite for high-performance supercapacitors." *RSC Advances* 6 (45):39317–39322. doi:10.1039/C6RA02730J.

Welz, Sascha, Yury Gogotsi, and Michael J. McNallan. 2003. "Nucleation, growth, and graphitization of diamond nanocrystals during chlorination of carbides." *Journal of Applied Physics* 93 (7):4207–4214. doi:10.1063/1.1558227.

Weng, Yu-Ting, and Nae-Lih Wu. 2013. "High-performance poly(3,4-ethylene-dioxythiophene): Polystyrenesulfonate conducting-polymer supercapacitor containing hetero-dimensional carbon additives." *Journal of Power Sources* 238:69–73. doi:10.1016/j.jpowsour.2013.03.070.

Winter, Martin, and Ralph J. Brodd. 2004. "What are batteries, fuel cells, and supercapacitors?" *Chemical Reviews* 104 (10):4245–4270. doi:10.1021/cr020730k.

Wu, Jinzhu, Jun Dai, Yanbin Shao, Meiqi Cao, and Xiaohong Wu. 2016. "Carbon dot-assisted hydrothermal synthesis of flower-like MoS_2 nanospheres constructed by few-layered multiphase MoS_2 nanosheets for supercapacitors." *RSC Advances* 6 (81):77999–78007. doi:10.1039/C6RA15074H.

Wu, Kun, Si-zhe Xu, Xue-jiao Zhou, and Hai-xia Wu. 2013. "Graphene quantum dots enhanced electrochemical performance of polypyrrole as supercapacitor electrode." *Journal of Electrochemistry* 4:13.

Xie, Minmin, Yanjie Su, Xiaonan Lu, Yaozhong Zhang, Zhi Yang, and Yafei Zhang. 2013. "Blue and green photoluminescence graphene quantum dots synthesized from carbon fibers." *Materials Letters* 93:161–164. doi:10.1016/j.matlet.2012.11.029.

Xiong, Shanxin, Fan Yang, Hao Jiang, Jan Ma, and Xuehong Lu. 2012. "Covalently bonded polyaniline/fullerene hybrids with coral-like morphology for high-performance supercapacitor." *Electrochimica Acta* 85:235–242. doi:10.1016/j.electacta.2012.08.056.

Xu, Juan, Yufei Xue, Jianyu Cao, Guoxin Wang, Yang Li, Wenchang Wang, and Zhidong Chen. 2016. "Carbon quantum dots/nickel oxide (CQDs/NiO) nanorods with high capacitance for supercapacitors." *RSC Advances* 6 (7):5541–5546. doi:10.1039/C5RA24192H.

Xu, Panpan, Jijun Liu, Peng Yan, Chenxu Miao, Ke Ye, Kui Cheng, Jinling Yin, Dianxue Cao, Kaifeng Li, and Guiling Wang. 2016. "Preparation of porous cadmium sulphide on nickel foam: a novel electrode material with excellent supercapacitor performance." *Journal of Materials Chemistry A* 4 (13):4920–4928. doi:10.1039/C5TA09740A.

Xu, Wence, Yanqin Liang, Yungao Su, Shengli Zhu, Zhenduo Cui, Xianjin Yang, Akihisa Inoue, Qiang Wei, and Chunyong Liang. 2016. "Synthesis and properties of morphology controllable copper sulphide nanosheets for supercapacitor application." *Electrochimica Acta* 211:891–899. doi:10.1016/j.electacta.2016.06.118.

Xu, Xiaoyou, Robert Ray, Yunlong Gu, Harry J. Ploehn, Latha Gearheart, Kyle Raker, and Walter A. Scrivens. 2004. "Electrophoretic analysis and purification of fluorescent single-walled carbon nanotube fragments." *Journal of the American Chemical Society* 126 (40):12736–12737. doi:10.1021/ja040082h.

Xu, Yang, Ming Wu, Yang Liu, Xi-Zeng Feng, Xue-Bo Yin, Xi-Wen He, and Yu-Kui Zhang. 2013. "Nitrogen-doped carbon dots: A facile and general preparation method, photoluminescence investigation, and imaging applications." *Chemistry—A European Journal* 19 (7):2276–2283. doi:10.1002/chem.201203641.

Xu, Yongjie, Xinyu Li, Guanghui Hu, Ting Wu, Yi Luo, Lang Sun, Tao Tang, Jianfeng Wen, Heng Wang, and Ming Li. 2017. "Graphene oxide quantum dot-derived nitrogen-enriched hybrid graphene nanosheets by simple photochemical doping for high-performance supercapacitors." *Applied Surface Science* 422:847–855. doi:10.1016/j.apsusc.2017.05.189.

Xue, Qi, He Huang, Liang Wang, Zhiwen Chen, Minghong Wu, Zhen Li, and Dengyu Pan. 2013. "Nearly monodisperse graphene quantum dots fabricated by amine-assisted cutting and ultrafiltration." *Nanoscale* 5 (24):12098–12103. doi:10.1039/C3NR03623E.

Yan, Xin, Xiao Cui, and Liang-shi Li. 2010. "Synthesis of large, stable colloidal graphene quantum dots with tunable size." *Journal of the American Chemical Society* 132 (17):5944–5945. doi:10.1021/ja1009376.

Yang, Feng, Meilian Zhao, Baozhan Zheng, Dan Xiao, Li Wu, and Yong Guo. 2012. "Influence of pH on the fluorescence properties of graphene quantum dots using ozonation pre-oxide hydrothermal synthesis." *Journal of Materials Chemistry* 22 (48):25471–25479. doi:10.1039/C2JM35471C.

Yang, Guo-Wei, Jin-Bin Wang, and Qui-Xiang Liu. 1998. "Preparation of nano-crystalline diamonds using pulsed laser induced reactive quenching." *Journal of Physics: Condensed Matter* 10 (35):7923–7927. doi:10.1088/0953-8984/10/35/024.

Yang, Yuxing, Dongqing Wu, Sheng Han, Pengfei Hu, and Ruili Liu. 2013. "Bottom-up fabrication of photoluminescent carbon dots with uniform morphology via a soft–hard template approach." *Chemical Communications* 49 (43):4920–4922. doi:10.1039/C3CC38815H.

Yang, Zhi, Zhaohui Li, Minghan Xu, Yujie Ma, Jing Zhang, Yanjie Su, Feng Gao, Hao Wei, and Liying Zhang. 2013. "Controllable synthesis of fluorescent carbon dots and their detection application as nanoprobes." *Nano-Micro Letters* 5 (4):247–259. doi:10.1007/BF03353756.

Yang, Zhi, Minghan Xu, Yun Liu, Fengjiao He, Feng Gao, Yanjie Su, Hao Wei, and Yafei Zhang. 2014. "Nitrogen-doped, carbon-rich, highly photoluminescent carbon dots from ammonium citrate." *Nanoscale* 6 (3):1890–1895. doi:10.1039/C3NR05380F.

Ye, Ruquan, Zhiwei Peng, Andrew Metzger, Jian Lin, Jason A. Mann, Kewei Huang, Changsheng Xiang et al. 2015. "Bandgap engineering of coal-derived graphene quantum dots." *ACS Applied Materials & Interfaces* 7 (12):7041–7048. doi:10.1021/acsami.5b01419.

Yong, Virginia, and H. Thomas Hahn. 2013. "Synergistic effect of fullerene-capped gold nanoparticles on graphene electrochemical supercapacitors." *Scientific Research* 2 (1):1–5.

Yuan, Changzhou, Li Chen, Bo Gao, Linghao Su, and Xiaogang Zhang. 2009. "Synthesis and utilization of $RuO_2 \cdot xH_2O$ nanodots well dispersed on poly(sodium 4-styrene sulfonate) functionalized multi-walled carbon nanotubes for supercapacitors." *Journal of Materials Chemistry* 19 (2):246–252. doi:10.1039/B811548F.

Zeiger, Marco, Nicolas Jäckel, Daniel Weingarth, and Volker Presser. 2015. "Vacuum or flowing argon: What is the best synthesis atmosphere for nanodiamond-derived carbon onions for supercapacitor electrodes?" *Carbon* 94:507–517. doi:10.1016/j.carbon.2015.07.028.

Zhai, Xinyun, Peng Zhang, Changjun Liu, Tao Bai, Wenchen Li, Liming Dai, and Wenguang Liu. 2012. "Highly luminescent carbon nanodots by microwave-assisted pyrolysis." *Chemical Communications* 48 (64):7955–7957. doi:10.1039/C2CC33869F.

Zhang, Mo, Linling Bai, Weihu Shang, Wenjing Xie, Hong Ma, Yingyi Fu, Decai Fang et al. 2012. "Facile synthesis of water-soluble, highly fluorescent graphene quantum dots as a robust biological label for stem cells." *Journal of Materials Chemistry* 22 (15):7461–7467. doi:10.1039/C2JM16835A.

Zhang, Q. L., Sean C. O'Brien, James R. Heath, Yichang Liu, Robert F. Curl, Harold W. Kroto, and Richard E. Smalley. 1986. "Reactivity of large carbon clusters: Spheroidal carbon shells and their possible relevance to the formation and morphology of soot." *The Journal of Physical Chemistry* 90 (4):525–528. doi:10.1021/j100276a001.

Zhang, Shuo, Lina Sui, Hongzhou Dong, Wenbo He, Lifeng Dong, and Liyan Yu. 2018. "High-performance supercapacitor of graphene quantum dots with uniform sizes." *ACS Applied Materials & Interfaces* 10 (15):12983–12991. doi:10.1021/acsami.8b00323.

Zhang, Su, Yutong Li, Huaihe Song, Xiaohong Chen, Jisheng Zhou, Song Hong, and Minglu Huang. 2016. "Graphene quantum dots as the electrolyte for solid state supercapacitors." *Scientific Reports* 6:19292. doi:10.1038/srep19292.

Zhang, Su, Jiayao Zhu, Yan Qing, Chengwei Fan, Luxiang Wang, Yudai Huang, Rui Sheng et al. 2017. "Construction of hierarchical porous carbon nanosheets from template-assisted assembly of coal-based graphene quantum dots for high performance supercapacitor electrodes." *Materials Today Energy* 6:36–45. doi:10.1016/j.mtener.2017.08.003.

Zhang, Xiang, Jingmin Wang, Jian Liu, Ji Wu, Hao Chen, and Hong Bi. 2017. "Design and preparation of a ternary composite of graphene oxide/carbon dots/polypyrrole for supercapacitor application: Importance and unique role of carbon dots." *Carbon* 115:134–146. doi:10.1016/j.carbon.2017.01.005.

Zhang, Yan-Qing, De-Kun Ma, Yan Zhuang, Xi Zhang, Wei Chen, Li-Li Hong, Qing-Xian Yan, Kang Yu, and Shao-Ming Huang. 2012. "One-pot synthesis of N-doped carbon dots with tunable luminescence properties." *Journal of Materials Chemistry* 22 (33):16714–16718. doi:10.1039/C2JM32973E.

Zhang, Yinhang, Kyong Yop Rhee, David Hui, and Soo-Jin Park. 2018. "A critical review of nanodiamond based nanocomposites: Synthesis, properties and applications." *Composites Part B: Engineering* 143:19–27. doi:10.1016/j.compositesb.2018.01.028.

Zhao, Qiao-Ling, Zhi-Ling Zhang, Bi-Hai Huang, Jun Peng, Min Zhang, and Dai-Wen Pang. 2008. "Facile preparation of low cytotoxicity fluorescent carbon nanocrystals by electro-oxidation of graphite." *Chemical Communications* (41):5116–5118. doi:10.1039/B812420E.

Zhao, Xiao, Ming Li, Hanwu Dong, Yingliang Liu, Hang Hu, Yijin Cai, Yeru Liang, Yong Xiao, and Mingtao Zheng. 2017. "Interconnected 3D network of graphene-oxide nanosheets decorated with carbon dots for high-performance supercapacitors." *ChemSusChem* 10 (12):2626–2634. doi:10.1002/cssc.201700474.

Zhao, Zhichao, and Yibing Xie. 2017. "Enhanced electrochemical performance of carbon quantum dots-polyaniline hybrid." *Journal of Power Sources* 337:54–64. doi:10.1016/j.jpowsour.2016.10.110.

Zheng, Liyan, Yuwu Chi, Yongqing Dong, Jianpeng Lin, and Binbin Wang. 2009. "Electrochemiluminescence of water-soluble carbon nanocrystals released electrochemically from graphite." *Journal of the American Chemical Society* 131 (13):4564–4565. doi:10.1021/ja809073f.

Zheng, Shushu, Hui Ju, and Xing Lu. 2015. "A high-performance supercapacitor based on KOH activated 1D C_{70} microstructures." *Advanced Energy Materials* 5 (22):1500871. doi:10.1002/aenm.201500871.

Zhou, Jigang, Christina Booker, Ruying Li, Xingtai Zhou, Tsun-Kong Sham, Xueliang Sun, and Zhifeng Ding. 2007. "An electrochemical avenue to blue luminescent nanocrystals from multiwalled carbon nanotubes (MWCNTs)." *Journal of the American Chemical Society* 129 (4):744–745. doi:10.1021/ja0669070.

Zhou, Jin, Pei Lin, Juanjuan Ma, Xiaoyue Shan, Hui Feng, Congcong Chen, Jianrong Chen, and Zhaosheng Qian. 2013. "Facile synthesis of halogenated carbon quantum dots as an important intermediate for surface modification." *RSC Advances* 3 (25):9625–9628. doi:10.1039/C3RA41243A.

Zhou, Li, Benzhao He, and Jiachang Huang. 2013. "Amphibious fluorescent carbon dots: One-step green synthesis and application for light-emitting polymer nanocomposites." *Chemical Communications* 49 (73):8078–8080. doi:10.1039/C3CC43295E.

Zhu, Xiaohua, Xiaoxi Zuo, Ruiping Hu, Xin Xiao, Yong Liang, and Junmin Nan. 2014. "Hydrothermal synthesis of two photoluminescent nitrogen-doped graphene quantum dots emitted green and khaki luminescence." *Materials Chemistry and Physics* 147 (3):963–967. doi:10.1016/j.matchemphys.2014.06.043.

Zhu, Yirong, Xiaobo Ji, Chenchi Pan, Qingqing Sun, Weixin Song, Laibing Fang, Qiyuan Chen, and Craig E. Banks. 2013. "A carbon quantum dot decorated RuO_2 network: Outstanding supercapacitances under ultrafast charge and discharge." *Energy & Environmental Science* 6 (12):3665–3675. doi:10.1039/C3EE41776J.

Zhu, Yirong, Zhibin Wu, Mingjun Jing, Hongshuai Hou, Yingchang Yang, Yan Zhang, Xuming Yang, Weixin Song, Xinnan Jia, and Xiaobo Ji. 2015. "Porous $NiCo_2O_4$ spheres tuned through carbon quantum dots utilised as advanced materials for an asymmetric supercapacitor." *Journal of Materials Chemistry A* 3 (2):866–877. doi:10.1039/C4TA05507A.

Zong, Jie, Yihua Zhu, Xiaoling Yang, Jianhua Shen, and Chunzhong Li. 2011. "Synthesis of photoluminescent carbogenic dots using mesoporous silica spheres as nanoreactors." *Chemical Communications* 47 (2):764–766. doi:10.1039/C0CC03092A.

2 One-Dimensional Nanomaterials for Supercapacitors

M. Ramesh and Arivumani Ravanan

CONTENTS

2.1 INTRODUCTION

This chapter reports one-dimensional (1D) inorganic nanomaterials for supercapacitors and their performance in terms of capacitive behavior, power density, energy density, specific capacitance, retention efficiency, etc. Additionally, improvements in design behavior for the huge categories of demands, which include wearable devices, stationary supercapacitors, manufacturing-oriented industrial systems, consumer electronics, hybrid electric vehicles, etc., are discussed. Here, the role of 1D inorganic nanostructured materials is discussed, which includes the shapes of nanowires (NWs), nanotubes, nanosponges, and nanoneedles. These are the most promising building blocks for nanoscale devices. A typical supercapacitor is also called as electrochemical capacitor. It comprises three significant elements termed as electrolytes, separator, and electrodes. Physical characteristics and chemical attributes of electrolyte solutions and electrode materials determine the overall behavior of supercapacitors. The electrode is one of the most essential elements for storing charge and delivering the same. It also plays a key role in determining the power density and the energy density of supercapacitors (Figure 2.1).

Novel electrodes with excellent physical attributes are required to replace the present "activated carbon electrode materials." Enhanced and newly designed structures are highly desirable for supercapacitors, for example, all-solid-state supercapacitors and fiber-shaped supercapacitors, which are optically transparent and mechanically flexible and stretchable. Nanotechnology enables to use advanced approaches and unique techniques to study material efficiency for energy storage. Generally, 1D inorganic nanomaterials are used as the electrode for enhancing the performance of the supercapacitor (Figure 2.2).

Various carbon materials are used in the fabrication of electrodes. Activated carbon, carbon fiber-cloth, carbide-derived carbon, carbon aerogel, graphitec (graphene), and carbon nanotubes (CNTs) are some of the major reported carbon materials used in the field of nanostructured electrode materials for supercapacitors. Carbon NWs, carbon nanorods (NRs), carbon nanosponges, hexagonal carbides, and carbonitrides—MXenes, and perovskites are some of these novel materials. Structures such as zero-dimensional (0D) nanoparticles, 1D NWs, two-dimensional (2D) nanosheets, and three-dimensional (3D) nanofoams are developed to enhance

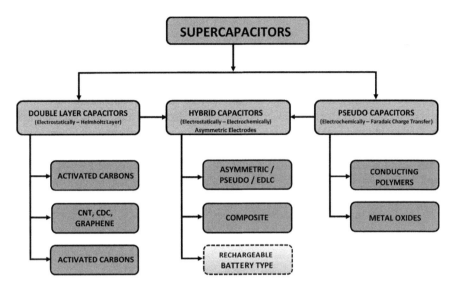

FIGURE 2.1   Types of supercapacitors based on materials.

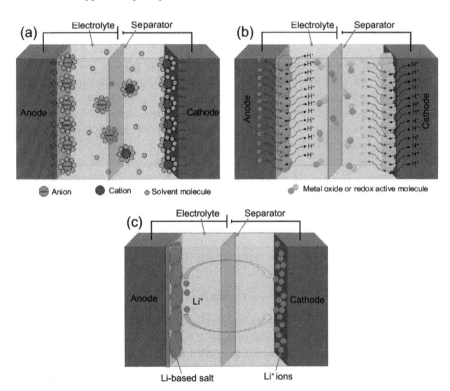

FIGURE 2.2   Schematic representation of (a) Electrode Double layer Capacitor (EDLC), (b) Pseudocapacitor (PC), and (c) Hybrid supercapacitor (HSC). (From Chen, X. et al., *Natl. Sci. Rev.*, 4, 37, 2017.)

and enrich the quality of the energy storage system. Double-layer supercapacitors and pseudocapacitors (redox supercapacitors) are the two major individual types of supercapacitors. Double-layer supercapacitors are prepared from carbon powders with large surface area. The charge separation in the assemblage of solid electrode/ electrolyte provides the capacitance.

Charge is mainly stored in electrostatic nature [2]. Pseudocapacitors are fabricated using metal oxides or conducting polymers, also known as redox supercapacitors. Reversible and rapid Faradaic charge transfer occurs at the region of electrode/ electrolyte assemblage in metal oxides and conducting polymers. Polymers deliver higher specific capacitance in the conducting polymers due to the participation of whole bulk polymers. Hence, when comparing with noble metal oxides, conducting polymers are cheaper [3].

2.2   INORGANIC NANOMATERIALS FOR SUPERCAPACITORS

2.2.1   GRAPHENE

Supercapacitor is a promising energy storage system, which is also referred to as electrochemical double-layer capacitor. Attributes such as high conductivity and large surface area and the potential use of CNT-based supercapacitors make this energy storage system a highly exciting one. A suitable data collection mechanism is required for the energy storage systems; batteries, supercapacitors, flywheels, and fuel cells are a few of such energy storage systems (Figure 2.3).

The study on the conductivity of superconductor-graphene-superconductor (SGS) junctions by using nonequilibrium techniques of green's function revealed that the SGS junction plays a significant role in the measurement of supercurrent (I_c). The on-site potential (U) in the center zigzag graphene nanoribbon at the SGS junction is suddenly transformed to an "ON" state from an "OFF" state, when the effective Fermi energy (μ_{eff}) shifts from negative to positive. Supercurrent (I_c) will continue to increase when the Fermi energy (μ_{eff}) further increases [5]. This increase is contributed by the ON/OFF action of the SGS junction. Lithium-ion battery stores approximately twenty times that of the energy stored in the supercapacitors for a given origination [6] with the support of the zigzag graphite nanoribbon valley-isospin valve. In addition, densities at different stages possess an impact on the repression of the supercurrent (I_c).

2.2.2   GRAPHENE-BASED PHOTONIC CRYSTALS

Band theory describes the electromagnetic wave propagation of equally doped graphene monolayers in a periodic arrangement. It has been found the enrichment of the photonic response in this arrangement, while the level of doping is consecutively modulated acquiring harmonic profiles, quasi-periodic besides the super-lattice axis. It is due to the dielectric environment in the backdrop, which assists the homogeneous layers of graphene. The graphene layers are superimposed by the doping modulation, which generates a specific photonic band structure. Fixed partitions between the layers produce a predominant and consistent structural bandgap, and it can be

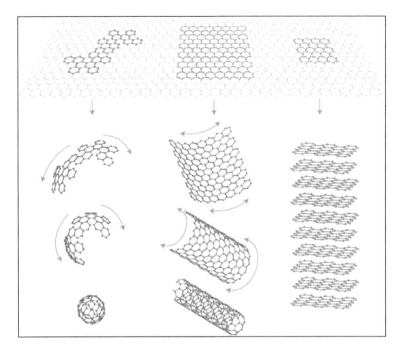

FIGURE 2.3 The father of all graphitic structures. Graphene is a basic 2D unit from which alternative dimensionalities can be developed. 0D bulky balls can be availed by wrapping it up, 1D nanotubes by rolling it, and 3D graphite sheets by stacking it. (From Novoselov, K.S. and Geim, A.K., *Nat. Mat.*, 6, 6, 2007.)

tuned by gating. In addition, the calculations for this experiment indicate that the doping levels between 0.2 and 1.2 eV of μ correspond to chemical potentials [7]. Under the low-temperature limit, the model of graphene conductivity is an effective one in the THz spectrum.

2.2.3 1D Nanoribbons in Supercapacitors

Graphene is considered as the primary element of the graphitic materials in all measurable extents. Graphene is a 2D monolayer of carbon atoms in carbon nanostructures firmly packed into a 2D honeycomb lattice [8]. In 2004, an experimental analysis of graphene showed that these materials could exist on noncrystalline substrates and in liquid suspensions as well [9]. Among graphene-based carbon nanostructures, graphene nanoribbons (GNRs) have shown a great promise in terms of width in nanometers, which can be achieved by altering the dimensions of the bandgap (Figure 2.4).

GNRs are known as the elongated stripes of graphene. By slitting the graphene sheet in a particular direction, GNRs can be obtained. Zigzag edge and armchair edge are the two types of GNRs, which are obtained by cutting the graphene honeycomb structure with two different edge shapes. Here, the 2D structure is refined to

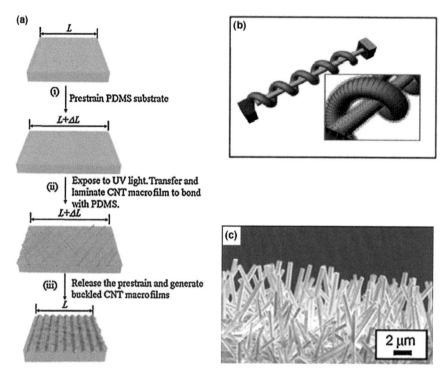

FIGURE 2.4 Illustration of a novel structure of a supercapacitor. (a) Schematic diagram of fabrication stages of an Single-walled carbon nanotube (SWNT) macrofilm on polydimethylsiloxane (PDMS) substrate. (From Dai, L. and Chen, T., *Mat. Today*, 16, 9, 2013.) (b) Schematic illustration of a fiber-based supercapacitor. (c) Scanning electron microscopy (SEM) image of arrays of nanosized plastic wires. (From Bae, J. et al., *Angew. Chem. Int. Ed.*, 50, 7, 2011.)

1D by reducing size of any one of the layers during thermal activation series of small waves of a honeycomb lattice of graphene. Strongest corrugations are rendered by the narrow nanoribbons rather than the square graphene sheets, which results in the reduction of 2D to 1D as reported by Costamagna et al. [12]. It causes a high impact on the electron transport properties of freestanding graphene systems. Barone and Scuseria prepared carbon nanoribbons by unrolling infinite periodic CNTs and slitting/elongating them into the required width and removing dangling bonds through hydrogenation. Figure 2.5 shows two types of CNTs, which are unfolded and slit to obtain carbon nanoribbons. Moreover, their crystallographic direction and width are also a significant factor in deciding the electronic properties of them [13].

2.2.4 1D CARBON NANOTUBES IN SUPERCAPACITORS

When discussing 1D materials for supercapacitors, it becomes necessary to study CNTs with a different combination of electrode materials. Supercapacitors have been investigated and used in many fields as energy-storing devices, in which high

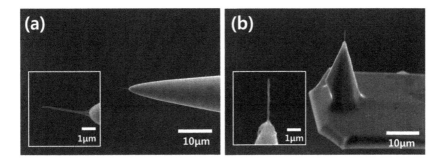

FIGURE 2.5 SEM images of a carbon nanotube nanoneedle with different magnifications: (a) tip of a tungsten material and (b) tip of an Atomic Force Microscopy (AFM) tip. Scale: 10 µm. (Insets show a magnified view—scale bar: 1 µm). (From An, T. et al., *Nanoscale Res. Lett.*, 6, 6, 2011.)

capacitance, low resistance, and stability are the three major fundamental requirements to be satisfied by the electrode material of the supercapacitor. CNT-based electrode materials include CNTs, CNT/oxide composite, and CNT/polymer composite. These materials have been broadly studied in recent decades and received interest for their application in supercapacitor because they fulfill the requirement.

As an electric double-layer capacitor, the behavior of CNT-based supercapacitor is nearly related to the physical properties of CNTs, such as specific surface area and electrochemical conductivity. The behavior depends on the post-treatment methods and synthesis of CNTs [15]. It is obvious that the specific surface area is not only the dominant factor related to the behavior. The capacitance of CNTs is affected by different factors, including pore size, specific surface area, distribution of pore size, and electrical conductance. Only by optimizing these factors, the behavior can be enhanced. Different methods have been suggested for the synthesis of CNTs, such as doping, functionalization, and oxidization. In addition, it can provide high capacitance by the enhancement in conductivity, addition of pseudocapacitance, and quick ion diffusivity.

Other methods have also been proposed to change/incorporate the structure of CNTs, such as tubes-in-tube CNTs and densely packed and ordered CNTs. They result in the enhancement of the conductivity and capacitance without introducing any functional groups. These modifications lead to the achievement of higher stability. The capacitance of CNT-based supercapacitor is still lower, even though the capacitance of the pure CNTs was improved and its stability was enhanced when compared with porous and amorphous carbons [15]. However, uniform distribution of pore size, optimization of pore size, and surface area are the parameters that can improve these qualities.

By applying the mixture of the CNTs with oxide, polymer, or both, as the electrode, a higher capacitance can also be achieved. There are some requirements for preparing a hybrid capacitor, which can be either achieved by attaching the oxide nanoparticles to the walls of CNTs or the uniform covering of the CNTs by the polymers with precisely controlled thickness. The enhancement of the capacitance becomes critical due to the ratio between oxide/polymer and CNTs. At present, there are no common answers to the ratio in any studies reported in the literature, due to the CNTs, starting materials, polymer, and oxide. In addition, processing methods vary from one research study to another. Some systematic investigations are required to solve such issues. Nevertheless, the stability of the hybrid supercapacitor is still a topic under research.

2.2.5 1D Nanoribbons (Graphene) Embedded with Zinc Manganite

The preparation of a flexible type of supercapacitor demands extensive efforts in the fabrication of electrodes including good electrical and mechanical attributes. Ahujaand et al. have designed a well-performed flexible supercapacitor by using $ZnMn_2O_4$/GNRs (zinc manganite/graphene nanoribbons) as the electrode. A transparent and independent gel electrolyte was used. That host polymer was designed to dissolve in acetone, followed by mixing with ionic liquid and stirring process [16]. The ionic liquid contains 1-butyl-3-methylimidazolium tetra fluoroborate. The available graphene nanoribbons cause the constant dispersion of $ZnMn_2O_4$ nanospheres (7 nanometers), resulting in improved migration of ions of electrolyte. The prepared flexible supercapacitor, $ZnMn_2O_4$/GNRkZnMn$_2$O$_4$/GNR showed a superior electrochemical response with an excellent specific capacitance of approximately 2.7 V. It exhibits an excellent energy density and power density of 37 Wh kg^{-1} and 30 kW kg^{-1}, respectively, at 1.25 Ag^{-1} and delivers extremely good cycling firmness, and retention efficiency of 91% after over 4000 cycles [17]. The superior flexibility of the solid-state flexible supercapacitor under severe situations shows its huge potentiality in the elementary research on flexible energy-storing systems.

2.2.6 1D Nanoneedle Arrays: Ni-Co Oxide and Sulfide Nanoarray/Carbon Aerogel Hybrid Nanostructures

Hao et al. performed an experiment on hybrid supercapacitor electrode materials. The natural characteristics of a carbon aerogel, such as electrical conductivity, own a hierarchical porous formation, and higher specific surface areas are aided to transition metal oxides, which lead to achieving extremely good cycling stability and superior energy density electrodes. $NiCo_2O_4$ nanoneedle array/carbon aerogel and $NiCo_2S_4$ nanotube array/carbon aerogel were synthesized, in which, nickel-cobalt precursor needle arrays are deposited over the top face of the walls by following the higher order ranking of porous carbon aerogels [18]. 1D nanostructures develop tightly and vertically over this surface and play a role in the enhancement of electrochemical behavior and extremely high energy density (Figure 2.6).

To achieve various electrochemical behaviors, two major aspects can be applied to nanostructured materials. The first aspect is about the outstanding electrical conductivity of $NiCo_2S_4$ with the minimal of twice the greater magnitude than that of $NiCo_2O_{4.35}$. The second aspect is the hollow structure of $NiCo_2S_4$ nanotubes that afford greater contact area across the electrolyte and the active material. This attribute leads to good electron transportation, thus enriching electrochemical behavior

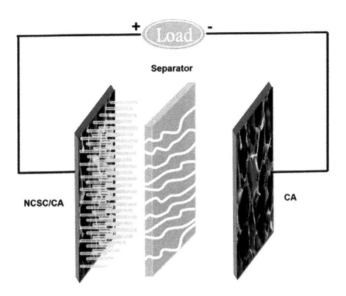

FIGURE 2.6 Asymmetric supercapacitor: Schematic diagram of composed Ni-Co Supercapacitance (NCSC)/Carbon Aerogels (CA) and CA electrodes. (From Hao, P. et al., *Nanoscale*, 8, 10, 2016.)

of the $NiCo_2S_4$ nanotube array/carbon aerogel. The furthermore identification of the electrochemical performance these two nanostructure materials, set of asymmetric supercapacitors were fabricated where CA as the negative electrode, and the samples of $NiCo_2O_4$.

Nanoneedle array/carbon aerogel or $NiCo_2S_4$ nanotube array/carbon aerogel is used as the positive electrode. Between the electrolyte and active material, a 1D hollow structure of $NiCo_2S_4$ nanotubes provides greater contact areas. Hence, it affords more efficient electron and ion transportation, which leads to the enriched electrochemical performance. The cycling stability is relatively excellent. First, support is the need for the transition of metal oxides to multiply the energy density; hence, the skeleton of carbon aerogel is used. When comparing with mono-metal oxides, mixed transition metal compounds show enhanced electrical conductivity, reversible capacity, and structural stability [18].

A pair of electrons is involved in the redox reaction of M–O/M–O–OH for the hybrid nanostructures, where M represents Ni or Co, which contributes to enhanced energy storage. In addition, the direct, robust, and vertical contacts between the carbon aerogels and 1D array could confirm rapid electron migration between pseudocapacitor nanostructures and EDLC. The relatively higher electrical conductivity of $NiCo_2S_4$ setup shows greater electrochemical behavior of $NiCo_2O_4$ than that of $NiCo_2S_4$. $NiCo_2O_4$ nanoneedle array/carbon aerogel and $NiCo_2S_4$ nanotube array/ carbon aerogel hybrid asymmetric supercapacitors achieved the energy densities up to 47.5 and 55.3 Wh kg^{-1}, respectively, and the power density of 400 W kg^{-1}. Moreover, the capacitance retention efficiency of these asymmetric devices is approximately 92% and 96.6% over 5000 cycles [18].

2.2.7 Mesoporous Strontium-Doped Lanthanum Manganite Perovskite

Lang et al. reported that strontium-doped lanthanum manganite perovskite is a potential composition for preparing electrode materials. By using an improved sol-gel method, lanthanum manganite ($La_{1-x}Sr_xMnO_3$, LSM) perovskites were prepared; among these LSM samples, a sample of $x = 0.15$ showed excellent electrochemical performance and provided lower intrinsic resistance in 1 M KOH. In supercapacitors, a mesoporous $La_{1-x}Sr_xMnO_3$ perovskite was introduced as anion-intercalation-type electrode [19]. This material was synthesized by a green process, in which the ratio of strontium doping was tuned. Embedding of OH− into the lattice will increase the electrochemical performance. This LSM (lanthanum-strontium-manganite) showed the pseudocapacitance effect due to the OH− ions present in the electrolyte. Manganese and strontium elements in the electrolyte separated after a few discharge and charge cycles, which led to poor cycle stability and damage of the perovskite structure. The results indicated that superior capacitive behavior and good charge storage performance are achieved by LSM15 through a mechanism known as anion-intercalation [19].

2.2.8 1D Nanowires of Double Perovskite Y_2NiMnO_6

The remarkable characteristics of Mn, Ni, and Y type oxides have attracted researchers to investigate about these double perovskite Y_2NiMnO_6 complex oxide, a novel emerging multiferroic material. The notable supercapacitance attributes of manganese, nickel and Y type oxides have moved the attraction to analyze the complicated oxide, paired perovskite Y_2NiMnO_6 [20]. It is an emerging multiferroic material. Furthermore, supercapacitance properties are improved by enhancing the charge transport and surface area at the nanoscale (Figure 2.7).

Hence, a simplistic, hydrothermal method was developed to synthesize Y_2NiMnO_6 NWs at a lower temperature. In terms of electrochemical properties, the double

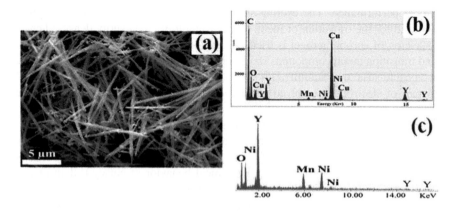

FIGURE 2.7 (a) Energy Dispersive X-ray (EDX) spectrum of the Y_2NiMnO_6 NWs and (b) bulk. (c) Specifying the presence of yttrium, nickel, manganese, and oxygen. (From Alam, M. et al., *RSC Adv.*, 6, 5, 2016.)

perovskite Y_2NiMnO_6 showed energy density, power density, and specific capacitance of 0.98, 19.27 W kg^{-1}, and 77.76 Fg^{-1}, respectively. In addition, it showed a retention efficiency of approximately 70.17% after 1800 cycles. This study revealed the electrochemical characteristics of a recently established and prominent multi-ferroic candidate, double perovskite Y_2NiMnO_6. Hence, this energetic material is identified as a positive electrode component for electrochemical supercapacitors [21].

2.2.9 1D NANORODS OF Mn_3O_4

One-dimensional Mn_3O renders unquestionably steady attributes and shows great electrochemical behavior in neutral aqueous solution. The experimental study of Zhang et al. indicated that a set of nanostructures and Mn_3O_4 NRs were prepared with different micro-morphologies, which indicate that 107.5 nm magnesium oxide (Mn_3O_4) long-rod electrodes show a current density of approximately 0.1 Ag^{-1} and high specific capacitance of approximately 136.5 Fg^{-1}. An asymmetric supercapacitor device of superior power density with a maximum work voltage of 1.8 V was prepared using reduced graphene oxide as a negative electrode and Mn_3O_4 long rods as a positive electrode. Furthermore, a sodium sulfate (Na_2SO_4) solution with an initial concentration of 500 mgL^{-1} was registered by this asymmetric capacitor for capacitive deionization, which can be considered an exceptional electro-adsorption performance. Moreover, waste energy during the desorption processes was efficiently reused [22].

2.2.10 MIXED VALENCE NANOSTRUCTURED Mn_3O_4

By using the same material Mn_3O_4, Bose and Biju conducted an experiment and reported the following result (Figure 2.8). A nanostructured Mn_3O_4 was synthesized with the particle size of approximately 19 nm by using the microwave-assisted

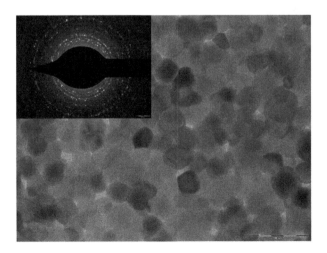

FIGURE 2.8 Transmission Electron Microscopy (TEM) image of the nanostructured Mn_3O_4 together with the selected area electron diffraction pattern. (From Bose, V.C. et al., *B. Mater. Sci.*, 38, 9, 2015.)

chemical method. Ratio of lattice constants (C/A) of the specimen related to the lattice of single-crystalline magnesium oxide (Mn_3O_4), where observed the conversion in the proportion of lattice constants to the existence of Mn^{4+} ions in octahedral sites. X-ray photoelectron spectroscopy results reveal the presence of Mn^{4+} ions. The value of specific capacitance achieved for symmetric supercapacitor with Mn_3O_4 active material is approximately 665.08 Fg^{-1} at slow scan rate. In addition, it yields an energy density and a high power density values of 4.36 $\times10^{-2}$ Wh kg^{-1} and 4.27 kW kg^{-1}, respectively, gained for discharge-charge cycles about 12.74 mA cm^{-2} at a constant current [23].

2.2.11 Hybrid Nanostructure of $MnCo_2O_{4.5}$ Nanoneedle/Carbon Aerogel

Hao et al. conducted an experiment on hybrid supercapacitor electrode materials. The natural characteristics of a carbon aerogel, such as electrical conductivity, possess the porous arrangement in order of rank and superior specific surface area are aided to transition metal oxides, which lead to achieving extremely good cycling stability and high energy density electrodes [24] (Figure 2.9).

The porous $MnCo_2O_{4.5}$ nanoneedle/carbon aerogel nanoneedle component was merged over the top face of the carbon aerogels with the order of rank of the porous structure. Over this surface, the 1D nanostructures developed tightly and vertically and lead to the enhancement of electrochemical performance along with ultrahigh energy density. This hybrid nanostructure was synthesized in two steps during the synthetic process, where the first step was the hydrothermal process, through which magnesium-cobalt precursors were grown on carbon aerogel. The second step was the transformation of the precursor as $MnCo_2O_{4.5}$ nanoneedles through the calcination process [24].

The carbon aerogel showed a superior specific surface area, porous structure arrangement, and higher electrical conductivity. When the aerogel was combined with the nanoneedles, (porous $MnCo_2O_{4.5}$), it achieved enhancement in electrochemical behavior of hybrid nanostructure. These hybrid nanostructures exhibited an excellent high specific surface area of approximately 888.6 m^2g^{-1} in order of ranks of porous structures. By using the neutral sodium sulfate electrolyte (Na_2SO_4), these

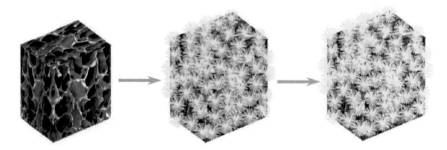

FIGURE 2.9 Schematic drawings of the formation mechanism of $MnCo_2O_{4.5}$/carbon aerogel hybrid nanostructures. (From Hao, P. et al., *Nanoscale*, 7, 7, 2015.)

hybrid nanostructures obtained a higher stable potential window (1.5 V) for the symmetric supercapacitor, whereas for other electrolytes such as potassium hydroxide (KOH) and sulfuric acid (H_2SO_4), they obtained a lesser amount [24].

2.2.12   1D NICKEL VANADIUM OXIDE (NiV_2O_6)

In an experimental study, by using hydrothermal method, which is a facile and economic synthetic process, nickel vanadium oxide was prepared and electrochemically characterized with lithium chloride as the electrolyte. It achieved a maximal specific capacitance of 412 Fg^{-1} at 2 mV/s and a high energy density of 165 Wh kg^{-1} at the power density of 3015 W kg^{-1} along with the capacitance retention efficiency of approximately 89.5% over 500 successive cyclic voltammetry (CV) cycles, which indicates the high utility of the fabricated nickel vanadium oxide as a supercapacitor electrode [25].

2.2.13   NANOCHAINS OF VANADIUM PENTOXIDE (V_2O_5)

In an experimental study, by using the simple hydrothermal method, a distinct, 1D in order of rank of nanochain of V_2O_5 was synthesized, where cetyltrimethylammonium bromide (CTAB) was used as the soft template. Galvanostatic charge-discharge and electrochemical impedance spectroscopy techniques were used to evaluate the electrochemical performance of the resulting V_2O_5 (vanadium pentoxide) electrode materials. These V_2O_5 nanochains (V_2O_5-CTAB) showed approximately a maximum specific capacitance of 631 Fg^{-1} at a regular current density of 0.5 Ag^{-1}. Even at a higher current density of approximately 15 Ag^{-1}, these nanochains retained approximately 300 Fg^{-1}. Furthermore, a capacitance retention of 75% was maintained, which indicates good cyclic stability. When the specific capacitance of vanadium pentoxide was compared with that of its different forms, the commercial bulk V_2O_5 showed a specific capacitance of 160 Fg^{-1} and the agglomerated V_2O_5 particles showed 395 Fg^{-1}, whereas V_2O_5 nanochains showed a maximum of approximately 631 Fg^{-1}. This resulted in V_2O_5-interconnected section that resembles a nanochain-like morphology [26] (Figure 2.10).

This nanochain-like design structure and higher specific surface area are the major considerations in contributing to the excellent electrochemical behavior of

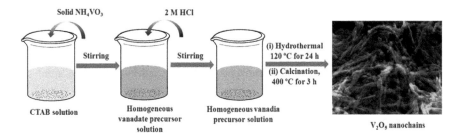

FIGURE 2.10   Nanochains: The evolution of V_2O_5 (V_2O_5-CTAB) nanochains under hydrothermal conditions in the presence of CTAB surfactant. (From Umeshbabu, E. and Ranga Rao, G., *J. Colloid Interface Sci.*, 472, 10, 2016.)

V_2O_5 chains at the nanolevel. During repeated charge-discharge process, these two factors enhance the rapid migration of Li^+ ions. Electrode prepared with these V_2O_5 nanochains has the greatest amount of supercapacitance, without introducing any conducting support materials such as CNTs or graphene oxide. The enhanced electrochemical behavior of V_2O_5 nanochains is the result of its mesoporous structure and higher surface area [26].

2.2.14   NANOWIRES OF V_2O_5 ASSISTED BY DISODIUM CITRATE

In an experimental study, using the facile hydrothermal process, uniform morphological orthorhombic V_2O_5 NWs were fabricated. Followed by the disodium citrate dosage on morphology, electrochemical characteristics and crystallinity of the resultant product were examined. The results showed that when the disodium citrate dose is 0.236 g at 180°C for 24 h, approximately 20 nm diameter V_2O_5 NWs with high crystallinity can be obtained. These NWs yielded 528.2 Fg^{-1} specific peak capacitance at 0.5 Ag^{-1} and a retention efficiency of approximately 85% [27]. Warburg slope in the mid-frequency portion was vertical to V_2. This indicated that diffusion and higher ion mobility are favorable for capability rate and cycling stability [28]. This study also reported that K^+ ion was reversibly deintercalated/intercalated in the lattice of prepared orthorhombic V_2O_5 NWs.

2.2.15   NANOSPONGES OF VO_2/TiO_2

An experimental study fabricated this new type of hybrid energy storage system, in which the prepared nanoarchitectures were synthesized with VO_2/TiO_2 nanosponges through electrostatic spray deposition. The resultant binder-free VO_2/TiO_2 electrodes displayed a high capacitance of 548 Fg^{-1} along with the appreciable cycle stability of 84.3% retention efficiency after 1000 cycles. This beneficial result was due to the synergistic integration of capable VO_2 and steady TiO_2 and distinct interconnected matrix pore of VO_2/TiO_2 components. Hence, this study provided an additive-free composite with high potential electrodes [29]. This excellent supercapacitive behavior is attributed to the highly porous distinctive arrangement, which enhances electrolyte infiltration and null strain of TiO_2. In addition, a low-cost electrostatic spray deposition (ESD) with a simplistic way was expanded for preparing other forms of practical nanostructured thin films to design better energy storing systems.

2.2.16   NANORODS OF ZnO WITH HETEROSTRUCTURE OF $ZnCo_2O_4$

Because of the direct electron/ion transport pathways and large surface area, 1D nanostructures such as NRs and NWs are used as supercapacitor electrode materials. Moreover, to achieve the heterostructures, these materials serve as unique substrates for the growth or coating of other functional materials [30,31] (Figure 2.11).

Furthermore, the direct growth of 1D nanostructure on current collectors induces an optimum ion/charge transport to decrease the junction resistance. It controls the capacitive drop and the rapid voltage during the electrolytic process [32]. In addition, most of the high-performance supercapacitor electrode materials are limited only

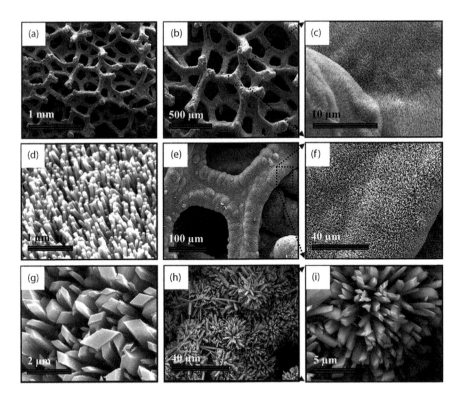

FIGURE 2.11 Samples for microstructures of the as-grown nanostructures on 3D NiF: (a–d) Lower and higher magnifying images of the as-grown H:ZnO NRs on 3D NiF, respectively. The as-grown ZnCo$_2$O$_4$ NRs on NiF at (e) low, (f) lower, and (g) higher magnifications. (h and i) The lower and higher magnifying images of ZnCo$_2$O$_4$ NRs developed on H:ZnO NRs. (From Boruah, B.D. et al., *Nanoscale*, 9, 10, 2017.)

to energy storage applications due to their effectiveness only on the electrochemical process. Hence, for meeting both optical energy conversion process and energy storage applications, even though it is still a challenge for the researcher's community, focus on the fabrication of hybrid novel heterostructure materials is needed (Figure 2.12).

In a research study by Boruah et al., a novel heterostructure was prepared by combining both photoactive and electrochemical materials. A group of ZnCo$_2$O$_4$ NRs was grown on the 3D matrix of H:ZnO nanorods (ZnCo$_2$O$_4$/H:ZnO NRs) on 3D NiF. These NRs have the synergistic integrated advantages during the electrolytic process by providing large electrochemically energetic surface, steady capacitive response, and the most appropriate ion/charge migration passage. A high power density is provided by the prepared asymmetric solid-state supercapacitor (ZnCo$_2$O$_4$/H:ZnO NRs/activated carbon (AC)). Moreover, the ZnCo$_2$O$_4$/H:ZnO NRs electrode material exhibits a superior photosensitive response under visible and ultraviolet illumination. Furthermore, the prepared solid-state asymmetric supercapacitor can provide the desired power during the process of the Ag-ZnCo$_2$O$_4$/H:ZnO NRs-Ag

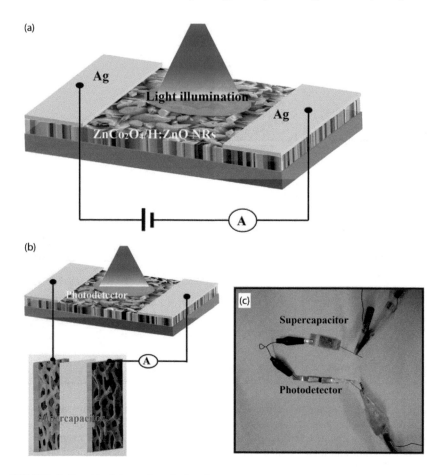

FIGURE 2.12 (a) Schematic of the fabricated ZnCo₂O₄/H:ZnO NRs photodetector. (b) Schematic of the supercapacitor connected in series with the photodetector during the operation. (c) Image of the respective configuration. (From Boruah, B.D. et al., *Nanoscale*, 9, 10, 2017.)

photodetector. Therefore, the heterostructure material $ZnCo_2O_4$/H:ZnO NRs has a great potential for storage applications and energy conversion [30].

2.2.17   1D C70 Fullerenes Activated by KOH

Because of the higher electronic affinity and other aspects about its structure and electron movement property in 3D, fullerenes are considered as a significant family of carbon materials. Fullerenes have been the subject of research for a very long time. Nevertheless, the use of fullerenes in supercapacitors has been under-evolved for a long time. In a previous study, potassium hydroxide activation of C70 microtubes was reported at elevated temperatures. It produced activated samples that deliver the greatest capacitive features.

The enhanced capacitive behavior is attributed to three aspects: the introduction of oxygen functionalities, generation of micropores and macropores, and the development of graphitic carbons. Result of a study indicates that both micropores and macropores originated at the time of activation process. Ellipsoidal fullerenes provide the graphitic carbons. They enable the incorporation of the redox-active functional group. This incorporation facilitates the conversion of fullerene molecules as graphitic arrangement. The optimum initialized condition is reached at 600°C, where the system shows the good continued possession efficiency, i.e., with the minimal loss of 7.5% over 5000 cycles at 1 Ag^{-1}. It mainly exhibits high electrochemical behavior with a gravimetric capacitance of 362.0 Fg^{-1} at 0.1 Ag^{-1} [33].

2.2.18 MESOPOROUS GRAPHITIC CARBON MICROTUBES (DERIVED FROM FULLERENE C70)

In a study, the direct conversion of mesoporous crystalline fullerene C70 was performed in which the microtubes were converted as the mesoporous graphitic carbon microtubes (Figure 2.13). It was achieved by conducting the heat treatment process at a high temperature of around 2000°C, where the 1D tubular material structure is registered. The outer face of the processed graphitic microtubes is comprised and combined in the order of sp^2 carbon with a strong mesoporous structure. The chronopotentiometry measurements and cyclic voltammetry revealed that the resultant new carbon material shows a higher specific capacitance of approximately 184.6 Fg^{-1}. Moreover, this material exhibited a high rate of performance [34] (Figure 2.14).

The outcome of this research study was that p-electron carbon origin (of fullerene C70) provided the optimistic electrode component of mesoporous graphitic carbon microtubes. The highlight was the significance of the combined graphitic carbon material arrangement of mesoporous carbons in the enhancement of general electrochemical behavior.

2.2.19 TUBULAR NANOSTRUCTURE OF CuCo₂S₄ WITH MoO₂ @ NITROGEN-DOPED CARBON

Quasi-solid-state supercapacitors have attracted researchers' curiosity to satisfy the requirement of wearable and portable electronic equipment. Nevertheless, these supercapacitors face several challenges in terms of electrode materials. An

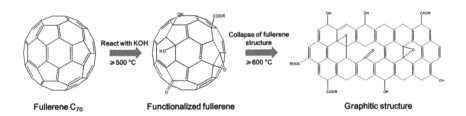

FIGURE 2.13 Schematic illustration of fullerene C70 transformed into graphitic carbons. (From Bairi, P. et al., *J. Mat. Chem. A*, 4, 8, 2016.)

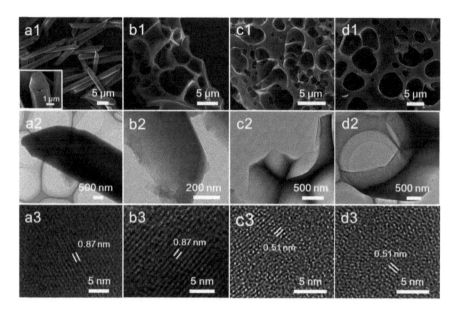

FIGURE 2.14 SEM images (a1, b1, c1, d1), TEM images (a2, b2, c2, d2), and high-resolution TEM images (a3, b3, c3, d3). SEM images of (a) C70 microtube, (b) fullerene and graphitic carbon composite processed at 500°C, (c) fullerene and graphitic carbon processed at 600°C, and (d) fullerene and graphitic carbon processed at 700°C. (From Bairi, P. et al., *J. Mat. Chem. A*, 4, 8, 2016.)

experimental study reported well-aligned MoO_2@NC and $CuCo_2S_4$ nanolevel tubular structures developed using carbon fibers. Molybdenum dioxide and nitrogen-doped carbon and cobalt sulfide nanostructures are integrated on conductive substrates. Hence, they deliver excellent electrochemical performance such as efficient 1D electron transport, chemically stable interface, and affluent active sites. It enables good electrochemical performance and ensures superior mechanical flexibility with the capacitance retention of approximately 92.2% after 2000 bending cycles [35].

2.2.20 Nanotube Coaxial Arrays of Titanium Nitride with MnO_2

In an experimental study, a nanostructured composite material was fabricated to produce a highly methodical storage arrangement with an efficient, fast charge separation network, in which MnO_2 and titanium nitride nanotube coaxial arrays were the materials used for better performances. By using the electrodeposition method, mesoporous magnesium oxide was deposited over titanium nitrates nanotube arrays. This was achieved by anodizing the titanium foil substrate, followed by nitridation using ammonia annealing. This material exhibited a specific capacitance of approximately 681.0 Fg^{-1} at 2 Ag^{-1} with the retention efficiency of 55% achieved from 2 to 2000 mV/s with the life cycle stability of 97% after 1000 cycles [7]. These results reveal the coaxial composite nanostructure and its effective role in the preparation of high-performance supercapacitors.

2.2.21 1D Nanowires of MnO_2 at 1D Nanoflakes of $Ni_{1-x}Mn_xO_y$

In an experimental study, as an initial step, a hydrothermal method was applied to synthesize the ultra-level lengthy α-MnO_2 NW materials. As the next step, to produce the core-shell nanostructures, $Ni_{1-x}Mn_xO_y$ 1D nanoflakes were developed on the α-MnO_2 1D NWs through chemical bath deposition; subsequently, thermal annealing was carried out. These methods are highly suitable for mass production. After testing in supercapacitors applications, the results showed that this hierarchical nanostructured core-shell setup (ultra-lengthy 1D α-MnO_2 NWs) provided consistent and effective, broad support for charge transport. These advantageous attributes offer a reliable, stable, and high-capacitance electrode component [36].

2.2.22 1D Hollow Nanofibers of MnO_2

One-dimensional hollow-structured components are considered as one of the attractive nanomaterials in the application of supercapacitors. In an experimental study, the electrospun carbon nanofibers were used as the base material, which produced hollow MnO_2 nanofibers. These resultant hollow nanofibers comprised extremely thin ultra-level MnO_2 nanosheets, which produced electrochemical energetic spots and enabled good electrochemical reactions. Though at higher current density, these ultrathin MnO_2 nanosheets increased the use of active material, because these nanosheets carried the open and free interspaces between them. These hollow MnO_2 nanostructured electrode materials exhibited, in energy storage devices, superior electrochemical behavior such as specific capacitance (291 Fg^{-1} at 1 Ag^{-1},), higher capability rate (73%), and efficiency of 90.9% after 5000 cycles [37].

2.2.23 1D MnO_2 Nanorods/Graphene Composites

In an experimental study, MnO_2 nanorods/graphene composite materials were synthesized by a simple hydrothermal method. The study results revealed that the proportions of MnO_2 nanorods in graphene play a very important role in the electrochemical behavior of the composite electrode. In addition, the ratio of the most desirable composite displayed higher energy density (16 Wh kg^{-1}) and excellent retention efficiency of 94% after 1000 cycles [38].

2.2.24 1D Ternary Hybrid Nanotubes of Cobalt Oxide-Manganese Dioxide-Nickel Oxide

In an experimental study, to obtain active supercapacitor electrode components, hybrid 1D nanotubes arrays were designed by a simple method using Co_3O_4-MnO_2-NiO ternary hybrid materials. A specific capacitance of 2525 Fg^{-1} was achieved with the capacitance loss of 20% after 5700 cycles. Moreover, the three superior, unique, redox electrode materials as a single unit afforded extraordinary electrochemical behavior with abundant reaction platform [39].

2.2.25   1D NiO Nanowire, Nanoflakes, and Nanosheets
in Supercapacitor Applications

NiO is known as one of the major popular materials among metal oxides used for the preparation of pseudocapacitor transition electrode components. A research group has successfully fabricated the ultrafine hierarchical NiO NWs by the hydrothermal method by developing the same on the mesoporous NiO nanosheets [40]. By using this same effective processing method, another research group produced NiO microspheres and investigated the electrochemical performance in supercapacitor applications [41]. In an experimental study on the development of supercapacitor electrode components, NiO was used in the preparation of nanoflakes. These nanoflakes were developed on nickel foam. It showed a capacitance per unit of area as 870 mF at 1.0 mA [42]. Furthermore, to determine the potential of NiO, a self-assembly method was also applied by a research group to produce hierarchical porous NiO nanosheets. This process was performed at room temperature and pressure, in which substances were not used to reduce surface tension and no support of templates was availed for the assembling of nanosheets [43].

2.2.26   1D Core-Shell Nanocables of ZnO at MoO_3 Nanostructure

Core-shell nanostructures frequently show different chemical and physical characteristics. The electrochemical method, one of the facile methods, is suitable to be performed at room temperature for synthesis. In an experimental study, this method was applied to produce a substantial amount of core-shell nanocables of ZnO at MoO_3. The duration of deposition controlled the thickness of the shell. The characterization by transmission electron microscopy (TEM), X-ray photoelectron spectroscopy, and X-ray diffraction showed that the internal portion of the ZnO nanorod exhibited a single-crystal structure. In addition, both ZnO nanorod and MoO_3 shell showed favorable growth along the [0001] direction. When comparing with the MoO_3 nanoparticles, the fabricated ZnO-MoO_3 core-shell nanocables exhibited much larger specific capacitance of approximately 236 Fg^{-1} at 5 mV/s. Moreover, this combination of electrode components showed highly stable electrochemical properties. It could endure over 1000 cycles without any decrement in specific capacitance [44].

2.2.27   Characteristic and Role of 1D ZnO and MnO_2
Composites in Supercapacitor Applications

Corrosion due to carbon-based components would lead to agglomeration of metal oxides covered over the case of the components. Moreover, it may lead to electrical isolation. The outcome of these effects indicates rapid deceleration of electrochemical behavior of metal oxide electrode components. Hence, to avoid the agglomeration of metal oxides and prevent the corrosion of carbon-based components, looking for sensible and beneficial materials based on non-carbon supports are feasible [45–47]. One-dimensional single-crystal ZnO nanorod is an interesting and desirable functional semiconductor material. Because of its smaller capacity, it provides a better electron-transporting path with effective mechanical flexibility. By nature, it is a chemically stable material with good conductivity; it also increases the electrochemical use of metal oxide. In an experimental study, the modern single-crystal ZnO nanorod and the composite of amorphous-nanoporous metal oxide as shell were used as electrode materials for supercapacitor applications. The combination of these distinct nanomaterials would lead to synergistic integration characteristics and functions. MnO_2 is an optimistic and favorable electrode material for supercapacitor applications with the attributes of eco-friendly and easily available at low cost [48–50]. The core material (single-crystal ZnO nanorods) faces lesser structural effect during the cycling process; owing to its smaller capacity, it provides a better electron-transporting path with effective physical flexibility. The shell material (MnO_2) as amorphous and nanoporous metal oxide increases the uniformity during the electrochemical reaction, decreases the ionic resistance, and expands active regions. This core-shell material combination (ZnO nanorod/MnO_2 shell composites) results in high specific capacitance of 405 Fg^{-1} at 10 mV/s and provides a beneficial way for the preparation of superior performance electrode components for supercapacitors [51].

2.2.28   1D ZnO AND COATED Co_3O_4 NANOROD HETEROJUNCTION COMPOSITES

Because of the inferior electrical conductivity and instability of Co_3O_4 in practical applications, an efficient route has been established for synthesizing Co_3O_4 heterojunction composites with the 1D porous ZnO. This heterostructure supports for increasing the charge transfer and prevents cobalt oxide from corrosion; simultaneously, the structural effect during the cycling process was protected by this 1D porous structure. This structure also enhances ion diffusion. The as-produced ZnO/Co_3O_4 composites displayed superior capacitive behavior and better cycling stability. The behavior of ZnO/Co_3O_4-450 was 4.9 times higher than that of Co_3O_4, and it has 83% retention efficiency after 5000 cycles at 10 Ag^{-1} [52].

2.2.29   1D NEEDLE-LIKE CoO NANOWIRES ON CARBON CLOTH

The hydrothermal method followed by the post-annealing treatment is used to prepare cobalt oxide NWs resembling needle-like structure on the carbon cloth. This combination of nanostructured electrode components shows an excellent retention efficiency of approximately 74% with specific capacitance of 311.8 Fg^{-1} at 1 Ag^{-1} and ensures the longstanding cycle stability. When relating the Co_3O_4 NWs/carbon cloth, CoO microspheres, CoO nanosheets/carbon cloth with the needle-like CoO NWs/carbon cloth, the enhanced electrochemical performance and placed as an interesting material in the supercapacitor applications [53].

2.3   CONCLUSION

The successful implementation of various inorganic 1D nanostructured materials was identified, and their individual characteristics, compositional properties, preparation methodology, and its features were features. Almost, all the forms of 1D structures have been applied in this field for extracting their attractive benefits, especially, high surface area and mechanical flexibility in the forms of activated carbon, activated fiber carbon cloth, carbide-derived carbon, carbon aerogel, graphene-carbon NWs, carbon nanorods, carbon nanosponges, hexagonal carbides and carbonitrides, and perovskites, to enhance and enrich the quality of the system. Fullerenes in the field of energy-storing equipment have to be improved much and creates greatest elicit from researches to applied aspects. Quasi-solid-state supercapacitors have attracted the interest of researchers to meet the requirements for wearable and transferable electronically supported equipment. Nevertheless, the implementation of electrode components in supercapacitor faces various issues and challenges; for instance, volume change upon cycling and slow kinetics intrinsically delay the output and results in unstable electrochemical properties. Multiferroic oxide materials also play a key role in the material design of supercapacitors.

Owing to the distinctive hierarchical nanostructure, positive chemical characteristics and excellent mechanical and electrical properties, novel nanomaterials have been widely scrutinized as effective electrode candidate for the application of supercapacitors. Overall, the redox-active transition metal oxides exhibit tremendous reversible faradic reactions in a rapid way with high specific capacitance, where CoO, Co_3O_4-MnO_2-NiO, $La_{1-x}Sr_xMnO_3$, MnO_2, MoO_2, MoO_3, Mn_3O_4, $MnCo_2O_{4.5}$, NiO, NiV_2O_6, $NiCo_2S_4$, $NiCo_2O_4$, VO_2/TiO_2, V_2O_5, Y_2NiMnO_6, ZnO, ZnO/Co_3O_4, and $ZnCo_2O_4$ are some of the 1D nanomaterials discussed. The advancement of recent electronic devices and energy storing devices extremely relies on the higher productive energy resources, which possesses high power and energy densities with the higher specific surface area. In this regard, 1D inorganic nanomaterials show a greater commitment to the material design of supercapacitors.

REFERENCES

1. Chen, X., Paul, R., & Liming D., Carbon-based supercapacitors for efficient energy storage. *National Science Review*, 2017. **4**(3): p. 37.
2. Yoon, B.J., Jeong, S. H., Lee, K. H., Kim, H. S., Park, C. G., & Han, J. H., Electrical properties of electrical double layer capacitors with integrated carbon nanotube electrodes. *Chemical Physics Letters*, 2004. **388**(1): p. 5.
3. Zukalova, M., Kalbac, M., Kavan, L., Exnar, I., & Graetzel, M., Pseudocapacitive lithium storage in TiO₂ (B). *Chemistry of Materials*, 2005. **17**(5): p. 8.
4. Novoselov, K.S., Geim, A.K., The rise of graphene. *Nature Materials*, 2007. **6**: p. 6.
5. Jie Liu, H.L., Song, J., Sun, Q.-F., & Xie, X.C., Superconductor-graphene-superconductor Josephson junction in the quantum hall regime. *Physical Review*, 2017. **96**(4): 045401.
6. Rycerz, A., J. Worzydło, Beenakker, C. W. J., Valley filter and valley valve in graphene. *Nature Physics*, 2007. **3**: p. 172.
7. Fuentecilla-Carcamoa, I., Palomino-Ovandoa, M., Ramos-Mendietab, F., One dimensional graphene based photonic crystals: Graphene stacks with sequentially-modulated doping for photonic band gap tailoring. *Superlattices and Microstructures*, 2017. **112**: p. 11.
8. Sur, U.K., Graphene: The two-dimensional carbon nanomaterial. *Nano Science and Nano Technology*, 2013. **7**(4): p. 6.
9. Novoselov, K.S., Geim, A.K., Morozov, S. V. et al., Electric field in atomically thin carbon films. *Science*, 2004. **306**(5696): p. 4.

10. Dai, L. & Chen, T., Carbon nanomaterials for highperformance supercapacitors. *Materials Today*, 2013. **16**: p. 9.
11. Bae, J., Song, M.K., Park, Y.J., Kim, J.M., Liu, M., & Wang, Z.L., Fiber supercapacitors made of nanowire-fiber hybrid structures for wearable/flexible energy storage. *Angewandte Chemie International Edition*, 2011. **50**(7): p. 7.
12. Dobry, A., & Costamagna, S., From graphene sheets to graphene nanoribbons: Dimensional crossover signals in structural thermal fluctuations. *Physical Review B*, 2011. **83**(23): p. 233401.
13. Barone, V., Hod, O., Scuseria, G.E., Electronic structure and stability of semiconducting graphene nanoribbons. *Nano Letters*, 2006. **6**(12): p. 7.
14. An, T., Choi, W., Lee, E., Kim, I.-T., Moon, W., Fabrication of functional micro- and nanoneedle electrodes using a carbon nanotube template and electrodeposition. *Nanoscale Research Letters*, 2011. **6**(306): p. 6.
15. Pan, H., Li, J., & Feng, Y., Carbon nanotubes for supercapacitor. *Nanoscale*, 2010: **5**(3): p. 15.
16. Pandey, G.P. & Rastogi, A.C.J., Solid-state supercapacitors based on pulse polymerized poly(3,4-ethylenedioxythiophene) electrodes and ionic liquid gel polymer electrolyte. *Journal of The Electrochemical Society*, 2012. **159**(10): p. 8.
17. Ahuja, P., Sharma, R.K., & Singh, G., Solid-state, high-performance supercapacitor using graphene nanoribbons embedded with zinc manganite. *Materials Chemistry A*, 2015. **3**(9): p. 7.
18. Hao, P., Tian, J., Sang, Y., Tuan, C.-C., Cui, G., Shi, X., Wong, C.P., Tang, B., Liu, H., 1D Ni–Co oxide and sulfide nanoarray/carbon aerogel hybrid nanostructures for asymmetric supercapacitors with high energy density and excellent cycling stability. *Nanoscale*, 2016. **8**(36): p. 10.
19. Lang, X., MO, H., Hu, X., Tian, H., Supercapacitor performance of perovskite La1−xSrxMnO$_3$. *Dalton Transactions*, 2017. **46**(40): p. 11.
20. Alam, M., Mandal, K., Khan, G.G., Double perovskite Y$_2$NiMnO$_6$ nanowires: High temperature ferromagnetic–ferroelectric multiferroic. *RSC Advances*, 2016. **6**(67): p. 6.
21. Alam, M., Karmakar, K., Pal, M., & Mandal, K., Electrochemical supercapacitor based on double perovskite Y$_2$NiMnO$_6$ nanowires. *RSC Advances*, 2016. **6**(115): p. 5.
22. Zhang, H., Liu, Wei, Y., Li, A., Liu, B., Yuan, Y., Zhang, H., Li, G., Zhang, F., Fabrication of a 1D Mn$_3$O$_4$ nano-rod electrode for aqueous asymmetric supercapacitors and capacitive deionization. *Inorganic Chemistry Frontiers*, 2019. **6**(2): p. 11.
23. Bose, V.C. & Biju, V., Mixed valence nanostructured Mn$_3$O$_4$ for supercapacitor applications. *Bulletin of Materials Science*, 2015. **38**(4): p. 9.
24. Hao, P., Zhao, Z., Li, L., Tuan, C.-C., Li, H., Sang, Y., Jiang, H., Wong, C. P. & Liu, H., Hybrid nanostructure of MnCo$_2$O$_{4.5}$ nanoneedle/carbon aerogel for symmetric supercapacitors with high energy density. *Nanoscale*, 2015. **7**(34): p. 7.
25. Prusty, B., Adhikary, M.C, Das, C.K., Supercapacitor electrode material based on nickel vanadium oxide. *American Chemical Science Journal*, 2015. **6**(2): p. 5.
26. Umeshbabu, E., & Ranga Rao, G., Vanadium pentoxide nanochains for high-performance electrochemical supercapacitors. *Journal of Colloid and Interface Science*, 2016. **472**: p. 10.
27. Pan, S., Chen, L., Li, Y., Han, S., Wang, L., Shao, G., Disodium citrate-assisted hydrothermal synthesis of V$_2$O$_5$ nanowires for high performance supercapacitors. *RSC Advances*, 2018. **8**(6): p. 5.
28. Wei, Y., Ryu, C.-W., Kim, K.-B., Cu-doped V$_2$O$_5$ as a high-energy density cathode material for rechargeable lithium batteries. *Journal of Alloys and Compounds*, 2008. **459**(1): p. 5.
29. Hu, C., Xu, H., Liu, X., Zou, F., Qie, L., Huang, Y., Hu, X., VO$_2$/TiO$_2$ Nanosponges as binder-free electrodes for high-performance supercapacitors. *Scientific Reports*, 2015. **5**: p. 8.

30. Boruah, B. D., Maji, A., Misra, A., Synergistic effect in the heterostructure of $ZnCo_2O_4$ and hydrogenated zinc oxide nanorods for high capacitive response. *Nanoscale*, 2017. **9**(27): p. 10.

31. Xue, M., F. Li., J. Zhu, H. Song, M. Zhang, & T. Cao, Structure based enhanced capacitance: In situ growth of highly ordered polyaniline nanorods on reduced graphene oxide patterns. *Advanced Functional Materials*, 2012. **22**(6): p. 7.

32. Zhou, W., Cao, X., Zeng, Z., Shi, W., Zhu, Y., Yan, Q., Liu, H., Wang, J., Zhang, H., One-step synthesis of Ni_3S_2 nanorod@Ni(OH)2nanosheet core–shell nanostructures on a three-dimensional graphene network for high-performance supercapacitors. *Energy and Environmental Science*, 2013. **6**(7): p. 6.

33. Zheng, S., Ju, H., & Lu, X., A high-performance supercapacitor based on KOH activated 1D C 70 microstructures. *Advanced Energy Materials*, 2015. **5**(22): p. 9.

34. Bairi, P., Shrestha, R.G., Hill, J. P., Nishimura, T., Ariga, K., Shrestha, L.K., Mesoporous graphitic carbon microtubes derived from fullerene C70 tubes as a high performance electrode material for advanced supercapacitors. *Journal of Materials Chemistry A*, 2016. **4**(36): p. 8.

35. Liu, S., Yin, Y., Hui, K.S., Hui, K.N., Lee, S.C., Jun, S.C., High-performance flexible quasi-solid-state supercapacitors realized by molybdenum dioxide@nitrogen-doped carbon and copper cobalt sulfide tubular nanostructures. *Advanced Science*, 2018. **5**(10): p. 11.

36. Wang, H.-Y., Xiao, F.X., Yu, L., Liu, B., (David) Lou, X.W., Hierarchical α-MnO_2 Nanowires@Ni1-xMnxOy nanoflakes core–shell nanostructures for supercapacitors. *Small*, 2014. **10**(15): p. 6.

37. Xu, K., Li, S., Yang, J., Hu, J., Hierarchical hollow MnO_2 nanofibers with enhanced supercapacitor. *Journal of Colloid and Interface Science*, 2017. **513**(11): p. 7.

38. Deng, S., Sun, D., Wu, C., Wang, H., Synthesis and electrochemical properties of MnO_2 nanorods/graphene composites for supercapacitor applications. *Electrochimica Acta*, 2013. **111**(08): p. 6.

39. Singh, A.K., Sarkar, D., Karmakar, K., Mandal, K., Khan, G.G., High-performance supercapacitor electrode based on cobalt oxide manganese dioxide-nickel oxide ternary 1D hybrid nanotubes. *ACS Applied Materials and Interfaces*, 2016. **8**(32): p. 7.

40. An, L., Xu, K., Li, W., Liu, Q., Li, B., Zou, R., Chena, Z., Hu, J., Exceptional pseudocapacitive properties of hierarchical NiO ultrafine nanowires grown on mesoporous NiO nanosheets. *Journal of Materials Chemistry A*, 2014. **2**(32): p. 5.

41. Wu, Q., Liu, Y., Hu, Z., Flower-like NiO microspheres prepared by facile method as supercapacitor electrodes. *Journal of Solid State Electrochemistry*, 2013. **17**(6): p. 6.

42. Xiao, H., Qu, F., Wu, X., Ultrathin NiO nanoflakes electrode materials for supercapacitors. *Applied Surface Science*, 2016. **360**: p. 6.

43. Xiao, H., Yao, S., Liu, H., Qu, F., Zhang, X., Wu, X., NiO nanosheet assembles for supercapacitor electrode materials. *Progress in Natural Science: Materials International*, 2016. **26**(3): p. 5.

44. Li, G.-R., Wang, Z.-L., Zheng, F.-L., Ou, Y.-N., & Tong, Y.-X., ZnO@MoO_3 core/shell nanocables: Facile electrochemical synthesis and enhanced supercapacitor performances. *Journal of Materials Chemistry*, 2010. **21**(12): p. 5.

45. Li, G.-R., Wang, Z.-L., Zheng, F.-L., Ou, Y.-N. & Tong, Y.-X., ZnO@MoO_3 core/shell nanocables: Facile electrochemical synthesis and enhanced supercapacitor performances. *Journal of Materials Chemistry*, 2011. **21**(12): p. 5.

46. Hu, C.-C., Hung, C.-Y., Chang, K.-H. & Yang, Y.-L., A hierarchical nanostructure consisting of amorphous MnO_2, Mn_3O_4 nanocrystallites, and single-crystalline MnOOH nanowires for supercapacitors. *Journal of Power Sources*, 2011. **196**(2): p. 4.

47. Hu, C.-C., Guo, H.-Y., Chang, K.-H., Huang, C.-C., Anodic composite deposition of $RuO_2 \cdot xH_2O–TiO_2$ for electrochemical supercapacitors. *Electrochemistry Communications*, 2009. **11**(8): p. 4.

48. Wu, Z.-S., Ren, W., Wang, D.-W., Li, F., Liu, B., Cheng, H.-M., High-energy MnO_2 nanowire/graphene and graphene asymmetric electrochemical capacitors. *ACS Nano*, 2010. **4**(10): p. 8.

49. Roberts, A.J., Slade, R.C., Controlled synthesis of $\varepsilon\text{-}MnO_2$ and its application in hybrid supercapacitor devices. *Journal of Materials Chemistry*, 2010. **20**(16): p. 6.

50. Toupin, M., Brousse, T., & Belanger, D., Charge storage mechanism of MnO_2 electrode used in aqueous electrochemical capacitor. *Chemistry of Materials*, 2004. **16**(16): p. 7.

51. He, Y.-B., Li, G.-R., Wang, Z.-L., Su, C.-Y. & Tong, Y.-X., Single-crystal ZnO nanorod/amorphous and nanoporous metal oxide shell composites: Controllable electrochemical synthesis and enhanced supercapacitor performances. *Energy & Environmental Science*, 2011. **4**(4): p. 5.

52. Gao, M., Wand, W.-K., Rong, Q., Jiang, J., Zhang, Y.-J. & Yu, H.-Q., Porous ZnO-coated Co_3O_4 nanorod as a high-energy-density supercapacitor material. *Applied Materials and Interfaces*, 2018. **10**(27): p. 11.

53. Ji, D.L., Li, J.H., Chen, L.M., Zhang, D., Liu, T., Zhang, N., Ma, R.Z., Qiu, G. Z., Liu, X.H., Needle-like CoO nanowires grown on carbon cloth for enhanced electrochemical properties in supercapacitors. *RSC Advances*, 2012. **5**(52): p. 4.

3 Core-Shell Nanomaterials for Supercapacitors

Nivedhini Iswarya Chandrasekaran
and Manickam Matheswaran

CONTENTS

3.1   INTRODUCTION

Novel approaches allow abundant developments in the synthesis of electrocatalysts with controlled morphology and surface properties. Nanostructures with specific morphology have shown great ability as energy materials with high discrimination have virtual ease of recycling (Min, Kim, and Lee 2010). Researchers have already developed new selective nanocatalysts for application in supercapacitors. Numerous challenges still persist, predominantly in understanding the structural steadiness of the active catalytic part. Core-shell nanomaterials have an enclosed and protective outer cover, which prevents migration and coalescence during the catalytic reactions (Qu et al. 2012). The synthesis and different characteristics of core-shell electrocatalysts are at the epicenter and emphasis of this chapter. Capable core-shell-based electrocatalysts need porous shells, which allow the entry of chemical entities from the external environment to the nanocatalyst surface. Different techniques are available for the synthesis of controllable shell structure with more porosity. Moreover, core-shell nanostructures provide many opportunities for governing the interaction among various components that enhance structural stability or catalytic activity. Core-shell nanomaterials of different shapes have been extensively used in biomedical (Sounderya and Zhang 2008) and pharmaceutical applications (Deutch and Cameron 1992), catalysis (Zhong and Maye 2001), electronics (Peng et al. 1997), photoluminescence (Xiong et al. 2008), photonic crystals, bio-imaging (Chen et al. 2012), controlled drug release, targeted drug delivery (Haag 2004), cell labeling, and tissue engineering applications. The fabricated $CNT/NiCo_2O_4$ core-shell material enhances the capacitance of the electrode by improving the electronic conductivity, surface area, and electron transport efficiently (Liu et al. 2014).

The single-component nanostructure has been studied extensively in various fields due to their observed unique structure and their properties such as secure processing, tunable nature, excellent surface performance, and outstanding catalytic properties (Huang and Jiang 2015). Presently, researchers focus on identifying numerous superior properties using multicomponent nanomaterials. The rapid development in multicomponent nanomaterials is due to diversity in nanostructure and composition, which enable their applications in various fields such as nanoelectronics, catalysis, energy storage, and conversion (Zhao et al. 2016). Hence, core-shell nanostructure has been considered as one of the research highlights in the current era (Chaudhuri and Paria 2012).

The core-shell nanostructures comprise a core (innermost layer) and a shell (exterior layer) material (Luo et al. 2008). The different combinations of core-shell materials are inorganic-inorganic, organic-inorganic, inorganic-organic, and organic-organic materials (Gawande et al. 2015). The choice of materials in core-shell structures depends on their end applications and uses (Chatterjee et al. 2014). Different core-shell nanostructures are shown in Figure 3.1a–m. Among them, spherical-shaped core-shell material is the common structure, which is further classified into single core-shell, multiple core-shell materials (shell layer consists of multiple core particles), concentric core-shell (alternative coating of core and shell), yolk-shell structure (movable core), and hollow shell structure

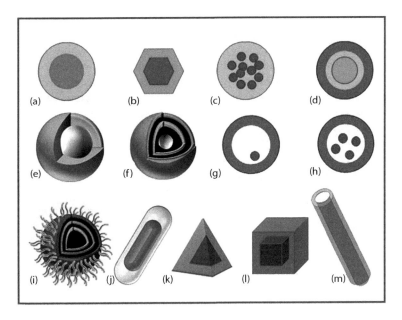

FIGURE 3.1 (a–m) Schematic representations of different shaped core-shell nanostructures: (a) core-shell nanoparticles, (b) polyhedral core-shell nanoparticles, (c) multi-core-single-shell particles, (d) core-multi-shell nanoparticles, (e) 3D core-shell structure, (f) core-double shell nanospheres, (g) yolk-shaped core-shell nanoparticles, (h) movable multi-core-single-shell nanostructures, (i) hollow multi-shell functionalized nanostructure, (j) core-shell nanorods, (k) core-shell nanoprism, (l) core-shell-based nanocubes, and (m) core-shell nanocables.

(absence of core). A different nonspherical core-shell structure is formed due to nonspherical core layer materials (shown in Figure 3.1b and j–m) such as hexagonal core-shell structures.

3.2 APPROACHES FOR THE SYNTHESIS OF CORE-SHELL NANOSTRUCTURE

The synthesis procedure of core-shell nanostructure is categorized into two types. One is a top-down approach such as lithographic technique (e-beam, ion-beam, UV, scanning probe), laser-beam technique, and mechanical processing (milling, grinding, polishing) (Gawande et al. 2015). Another category is the bottom-up approach such as chemical synthesis, self-assembly techniques, colloidal method, electrophoretic deposition, and chemical vapor deposition. Both top-down and bottom-up approaches have their own merits and demerits (Nomoev et al. 2015). However, bottom-up approach is mostly preferred because it is cost-effective and produces smaller sized particles with controllable properties. Here, in the preparation of the core-shell nanostructure, a bottom-up approach is a suitable technique because the constant coating of the shell layer is possible using this approach. Some core-shell structures have been produced using a combination of both top-down and bottom-up approaches.

3.3   APPLICATIONS OF CORE-SHELL NANOSTRUCTURES
IN SUPERCAPACITORS

Depletion of fossil fuels and increasing environmental pollution have compelled researchers to focus on new energy storage devices. Enormous efforts have been devoted to the development of fuel cells and hybrid powers. However, owing to the short life service, high cost, and poor performance at high temperature, it is difficult to resolve all drawbacks simultaneously. Supercapacitor is an emerging energy storage device with rapid and reversible charge-discharge qualities (González et al. 2016). It possesses excellent features such as long cycle life, high energy and power density, good performance at high temperature, and environmental friendly nature, which open a path to overcome all the complications in present energy storage technologies (Liu et al. 2010a). Supercapacitors, otherwise called as ultracapacitors or electrochemical capacitors, have high electrode surface area and thin electrolyte dielectrics to achieve maximum capacitance of several orders of magnitude. In addition, they achieve a high energy density by maintaining their unique power density. The supercapacitor replaces conventional batteries and other energy devices used currently in electric vehicles owing to their greater benefits (Simon, Gogotsi, and Dunn 2014). Additionally, supercapacitors have wide applications when compared with traditional batteries. More research is being conducted to identify suitable electrode materials for supercapacitors to achieve longer life span and lower cost with high energy density and power density. Recent efforts have focused on nanomaterials such as conducting polymers, transition metal oxides, hydroxides, and sulfides for improving specific capacitance (Faraji and Ani 2014). In addition, the pseudocapacitive material should possess large surface area with a proper mesoporous structure, which assists in improving capacitance and lodges the vast quantity of electroactive species to contribute in faradaic redox reactions with a large amount of mass and electron transfer (Huang et al. 2013). The difficulties of presently used electrode materials are short lifetime and low chemical and mechanical stability. In addition, to improve the supercapacitor performance, increased use of a metal substrate can work better. However, this metal loading increases the device weight significantly. Moreover, large volume changes are seen in electrode material during continuous charging and discharging process. There are also many difficulties including low ion transfer efficiency, poor conductivity, volume changes in cycling processes, and loss of electroactive materials because of electrolyte corrosion or breakdown of the structural stability (Zhang et al. 2013). Continuous development in the research area of electrode material designing has motivated the search for a new nanostructure to overcome these drawbacks and improve performance (Bogue 2011). The structural modification of the electrode materials improves mechanical stability, thereby greatly improving their lifetime. Core-shell-based structures provide more electrochemical active sites and enhance rapid ion transmission with low volume expansion (Xia, Tu et al. 2012). The presence of active components in the core-shell structure improves the electronic pathway efficiently in the redox reaction. In addition, this core-shell structure prevents volume expansion on time-continuous charging and discharging cycles (Feng et al. 2018).

The problems of the existing system can be significantly overcome by core-shell nanostructure. The advantages of using the core-shell structure are as follows: protects core material from outside environmental change, reduces volume expansion, prevents aggregation, provides a large surface to volume ratio, provides structural integrity, has high availability of active materials, has fast interfacial transport, and shortens diffusion pathway (Wei et al. 2011). The catalytic activity of the core-shell nanostructure can be improved further by controlling dimension, morphology, and composition of specific materials (Liu et al. 2015; Jia, Wu, and Liu 2018). The electrocatalytic activity depends on its synergistic connections among the core and shell materials. It depends mainly on (a) ion/molecule that disturbs the effect of charge transfer between the substances, (b) composite material characteristics and its effects that manage the adsorption nature, and (c) structural and geometric characteristics (Chen et al. 2011). Because of their different types, core-shell nanomaterials have been used in several areas of electrocatalysis in energy storage applications (Cui et al. 2009; Wu et al. 2011; Hwang et al. 2012). Although, the active substances present in core-shell structures exhibit significant abilities and likely scenarios.

3.4 IMPORTANCE OF CORE-SHELL MATERIALS

Core-shell nanostructures are extremely functional materials with enhanced properties, which have gained attraction in many research fields such as electronics, biomedical, optics, and catalysis. Core-shell materials possess a tunable kind of features, i.e., their properties can be changed by modifying their constituent materials individually (Li, Yuan et al. 2014). Sometimes, the core and shell materials have different features. In addition, the effect of property changes in core material based on shell material coating may increase, stabilize, or decrease (Buchanan et al. 1996). Moreover, the shell coating increases the overall dispersibility and particle stability of core particles. Finally, materials show unique properties of the various materials combined together. To meet the requirements of the different application, the surface needs to be manipulated accordingly (Xie et al. 2013). The main reason for coating is to improve the functionality, dispersibility, stability, and measured discharge of the core materials, decrease the rate of core material consumption, and so on.

Apart from the advantage of enhanced material properties, the economic point is crucial in core-shell nanostructures (Zhang et al. 2009). The first aspect is to minimize the overconsumption of core material in the case of precious metals. The second aspect is the core-shell structure that acts as a prototype for the formation of hollow structure (Serpell et al. 2011). This hollow structure is formed from core-shell structure by eliminating the core materials by dissolution or calcination. This hollow shell structure has been used in various applications such as catalytic supports, adsorbent, micro/nano-vessels, drug delivery, light-weight material, insulator, and so on.

3.5 MECHANISM OF PREPARATION OF CORE-SHELL NANOSTRUCTURE

Preparation procedure of core-shell structure involves a two-step process. The primary step is the formation of core structure, followed by the production of the shell structure. Depending on core availability, the synthesis procedure is classified into two types:

1. The core structure is synthesized individually and then added to the system for shell coating.
2. The core structure is prepared in the same system, followed by a shell coating.

In the first type, the core structure is prepared using the separate reaction procedure, followed by surface modification for shell coating in the different reaction mixture, whereas in the second type, core material is prepared using the appropriate procedure in the presence of a growth inhibitor, followed by shell coating by the addition of more reactants (Pastoriza-Santos et al. 2000). The main advantage of the synthesis of core externally is its purity. However, the synthesis of the core by in situ method has a severe problem of impurity, which is present between core and shell layer formation. A significant step in core-shell structure synthesis is controlling the uniform thickness of the shell coating. Different methods for preparing core-shell structure are precipitation, polymerization, layer-by-layer adsorption techniques, micro-emulsion method, sol–gel condensation, and so on. Proper controlling of uniform coating and thickness of the shell layer over the core is difficult (Malik, O'Brien, and Revaprasadu 2002). The problems that exist in shell coating are as follows:

1. Agglomeration of core materials
2. Shell material forms a separate particle rather than coating them on the core particle
3. The incomplete coating over the core layer
4. Control of the reaction rate.

Generally, these problems can be overcome by performing surface modification using surfactants, polymers, and so on. These surface-active agents modify the core layer surface charges. Therefore, coating of the selectively chosen core material with uniformity is achieved (Ban et al. 2005). Table 3.1 lists different synthetic techniques and reagents involved in the core-shell nanostructure synthesis.

3.6 DIFFERENT SHAPED CORE-SHELL NANOSTRUCTURES

Various shapes of core-shell nanoparticles have equally proven their importance in supercapacitors. Two approaches generally used to synthesize different shaped core-shell nanostructures are

1. Using soft/hard shape core template
2. Using capping agent or suitable reagent to regulate the growth direction.

TABLE 3.1

Different Synthetic Techniques and Reagents Used for Core-Shell Structure Preparation

Core-Shell Structure	Core	Shell		References
	Synthesis Technique	Synthesis Technique	Capacitance	
$NiCo_2O_4@MnO_2$ nanowire	Hydrothermal	Hydrothermal	1.66 F cm^{-2} at 20 mA cm^{-2}	Yu et al. (2013)
$WO_{3-x}@Au@MnO_2$ nanowire	High-temperature thermal evaporation	Sputtering (Au) and anodic electrodeposition (MnO_2)	588 Fg^{-1} at 10 mV s^{-1}	Lu et al. (2012)
$H-TiO_2@MnO_2$ nanowire	Hydrothermal	Anodic electrodeposition	139.6 Fg^{-1} at 1.1 Ag^{-1}	Lu et al. (2013)
Zn_xSnO_4/MnO_2 nanocables	Chemical vapor deposition (CVD)	Spontaneous redox deposition	642.4 Fg^{-1} at 1 Ag^{-1}	Bao, Zang, and Li (2011)
Hydrogenated ZnO-doped MnO_2 nanocables	Wet chemical synthesis	Self-limiting process	1260.9 Fg^{-1} at 1 Ag^{-1}	Yang et al. (2013)
$V_2O_5@$polypyrrole nanoribbon	Hydrothermal	Polymerization	308 Fg^{-1} at 0.1 Ag^{-1}	Qu et al. (2012)
Co_3O_4/NiO nanowire	Hydrothermal	Chemical bath deposition		Xia, Tu et al. (2012)
$ZnO@MoO_3$ nanocables	Electrodeposition	Electrodeposition	236 Fg^{-1} at 5 mV s^{-1}	Li et al. (2011)
$Ni(OH)_2@3DNi$	Ammonium	Electrodeposition	92.8 Fg^{-1} at 1 Ag^{-1}	Su et al. (2014)
$Ni(OH)_2-MnO_2$ nanoflakes	Solvothermal	Hydrothermal	487.4 Fg^{-1} at 1 Ag^{-1}	Jiang et al. (2012)
Homogeneous $NiCo_2S_4$	Hydrothermal	Hydrothermal	252 Fg^{-1} at 2 Ag^{-1}	Kong et al. (2015)
WO_{3-x}/MoO_{3-x}	CVD	Electrochemical deposition	216 mF cm^{-2} at 2 mA cm^{-2}	Xiao et al. (2012)
$Co_3O_4@MnO_2$	Hydrothermal	Hydrothermal	560 Fg^{-1} at 0.2 Ag^{-1}	Huang, Zhang, Li, Zhang et al. (2014)
$NiCo_2O_4@NiCo_2O_4$ nanoflakes	Hydrothermal	CBD	1.55 F cm^{-2} at 2 mA cm^{-2}	Liu et al. (2013)
$CuO@MnO_2$	Solvothermal	Facile and scalable method	276 Fg^{-1} at 0.6 Ag^{-1}	Huang, Zhang, Li, Wang et al. (2014)
Carbon@MnO_2 nanospheres	Hydrothermal	Hydrothermal	252 Fg^{-1} at 2 mV s^{-1}	Zhao, Meng, and Jiang (2014)
$\alpha-MnO_2$ nanowires @ $Ni_{1-x}Mn_xO_y$ nanoflakes	Hydrothermal	Chemical bath deposition	657 Fg^{-1} at 0.25 Ag^{-1}	Wang et al. (2014)

In the first approach, the different shaped core material is used as a template for the formation of a core-shell structure, whereas the shell material is deposited homogeneously onto the core material surface and takes its shape. Usually, by varying the reaction parameter, the synthesis of a core particle template in a different shape is achieved (Liu et al. 2010b). The shell material coating efficiency may vary based on the shape of the core structure. It is also noted that uniformity of coating may decrease with shape alteration (Xia, Tu et al. 2012). Even in the soft template method, the entire particle morphology relies on core and shell thickness.

3.6.1 NANOSPHERES

The core carbon spheres in Carbon@MnO_2 core-shell hybrid nanospheres act as a prototype to the conductive shell material MnO_2, shown in Figure 3.2c–g, which exhibits the highest capacitance of 252 Fg^{-1} at 2 mV s^{-1} (Zhao, Meng, and Jiang 2014). The crystalline Ni_3S_4@MoS_2 nanospheres prepared by one-pot synthesis with tuned shell thickness show a good capacitance of 1440.9 Fg^{-1} at 2 Ag^{-1} and a capacitance retention of 90.7% after 3000 cycles at 10 Ag^{-1} (Zhang et al. 2015). Core–shell ultramicroporous @ microporous carbon nanospheres prepared using polymerization techniques exhibit higher capacitance (411 Fg^{-1} at 1 Ag^{-1}), superior rate capability, outstanding cycle stability (10,000 cycles), and reasonable energy density (5.94 W h kg^{-1} at 50 kW kg^{-1}) (Liu et al. 2015). Hollow core-mesoporous C shell capsule synthesized by polymerization, carbonization and etching techniques could be used as energy storage materials (Yoon et al. 2002). N-doped hollow and yolk-shell-based mesoporous C nanospheres formed by silica-assisted method display outstanding supercapacitor performance with high capacitance of 240 Fg^{-1} and the promising capacitance retention (97.0% capacitive retention after 5000 cycles) (Liu et al. 2016). Double-layer N-doped hollow carbon@MoS_2/MoO_2 nanospheres shown in Figure 3.2a and b display an extraordinary capacitance (569 Fg^{-1} at 1 Ag^{-1}) and exceptional rate performance of 54.8% from 1 (569 Fg^{-1}) to 20 Ag^{-1} (312 Fg^{-1}) (Tian, Zhang, and Li 2018). N-doped hollow mesoporous C capsules prepared by dissolution and capture method to exhibit a superior specific capacitance of 206.0 Fg^{-1} at 1 Ag^{-1} (Chen et al. 2016). Hollow Ni-Al-Mn-layered triple hydroxide nanocomposite prepared by a microwave-assisted technique, shown in Figure 3.2h–j, using urea shows high capacitance of 1756 Fg^{-1} at 4 Ag^{-1} and its capacitance value retains up to 89.5% at the end of 4000 cycles (Chandrasekaran et al. 2017). Fabricated symmetric supercapacitor device using hollow Mn-Cu-Al triple oxide nanomaterials prepared by microwave-assisted methods deliver high energy density of 62.26 Wh kg^{-1} at 461.25 W kg^{-1} (Chandrasekaran et al. 2018). The symmetric supercapacitor constructed using hollow Mn-Cu-Al-layered triple hydroxide nanocomposite provides a high energy density (101.75 Wh kg^{-1}) at 900 W kg^{-1} (Chandrasekaran and Manickam 2017).

3.6.2 NANOTUBES

Figure 3.3a shows the excellent synergistic effect of CNT@polypyrrole@MnO_2 core-double-shell synthesized by electropolymerization and hydrothermal synthesis leads to improved supercapacitor with excellent specific capacitance and cycling stability

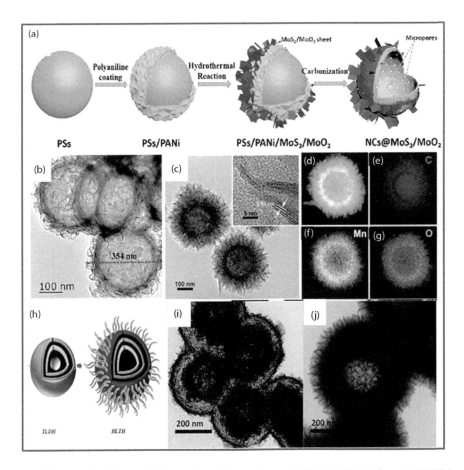

FIGURE 3.2 (a) Schematic illustration of the synthesis of NCs@MoS$_2$/MoO$_2$ and (b) TEM images of NCs@MoS$_2$/MoO$_2$ (Reprinted with permission from Tian, J. et al., *ACS Appl. Mater. Inter.*, 10, 29511–29520, 2018. Copyright 2018 American Chemical Society). (c) High-resolution TEM and (d–g) Energy Dispersive X-Ray Spectroscopy (EDS) mapping results of a single carbon@MnO$_2$ core-shell nanospheres. (Reprinted from *J. Power Sources*, 259, Zhao, Y. et al., Carbon@MnO$_2$ core–shell nanospheres for flexible high-performance supercapacitor electrode materials, 219–226, Copyright (2014), with permission from Elsevier B.V.). (h) Schematic representation of yolk LDH and hollow Layered Triple Hydroxide (LTH), (i) TEM images of Ni-AL yolk LDH, and (j) TEM images of Ni-Al-Mn hollow LTH. (Reprinted from *Mater. Chem. Phys.*, 195, Chandrasekaran, N.I. et al., Hollow nickel-aluminium-manganese layered triple hydroxide nanospheres with tunable architecture for supercapacitor application, 247–258, Copyright 2017, with permission from Elsevier B.V.)

(Li, Yang et al. 2014). NiCo$_2$S$_4$@polypyrrole nanotube array fabricated using the electrodeposition method shows a lower charge-transfer resistance (0.31 Ω) and high areal specific capacitance of 9.781 F cm^{-2} at 5 mA cm^{-2} (Yan et al. 2016). The transmission electron microscopy images of NiCo$_2$S$_4$ (Figure 3.3b–d) and NiCo$_2$S$_4$@PPy nanotubes are shown in Figure 3.3e and f. CNT@ microporous carbon core-shell nanocomposite prepared by resorcinol–formaldehyde resin coating on CNTs deliver

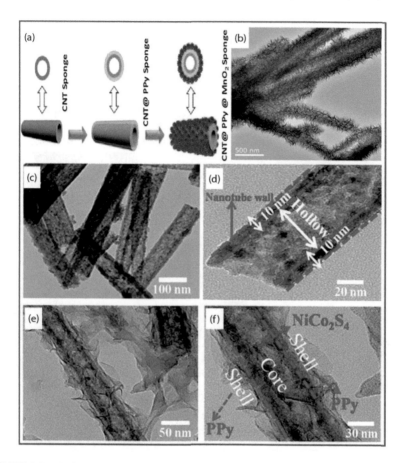

FIGURE 3.3 (a) Schematic representation of the fabrication process from CNT sponge to CNT@PPy@MnO$_2$ core-double-shell sponge. (Reprinted with permission from Li, P. et al., *ACS Appl. Mater. Inter.*, 6, 5228–5234, 2014. Copyright 2014 American Chemical Society). (b) The TEM image of the CoNi$_2$S$_4$ nanotubes. (Reprinted with permission from Li, M, et al., *ACS Appl. Mater. Inter.*, 10, 34254–34264, 2018. Copyright 2018 American Chemical Society). (c, d) TEM images of NiCo$_2$S$_4$ nanotubes and (e, f) TEM images of NiCo$_2$S$_4$@ PPy nanotubes. (Reprinted with permission from Yan, M. et al., *ACS Appl. Mater. Inter.*, 8, 24525–24535, 2016. Copyright 2016 American Chemical Society.)

extraordinary specific capacitance of 237 Fg^{-1}, exceptional rate performance with 75% from 0.1 to 50 Ag^{-1}, and good cyclability (Yao et al. 2015). CNTs@Fe$_2$O$_3$@C were prepared by atomic layer deposition and thermal decomposition, which exhibit excellent electrochemical properties of 678 Fg^{-1} at 1 Ag^{-1} and capacity retention of 82% at 25 Ag^{-1} (Li et al. 2018).

3.6.3 NANOWIRES

A catalyst-free, physical evaporation deposition process has been used to develop WO$_{3-x}$ on carbon fabric to form WO$_{3-x}$@Au@MnO$_2$ nanowires. The specific

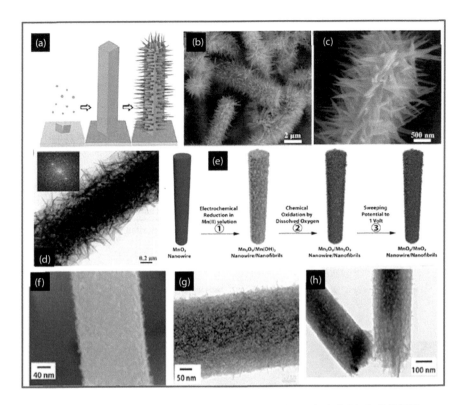

FIGURE 3.4 (a) The synthetic scheme to produce the branched CoMoO$_4$@CoNiO$_2$ core-shell nanowires on Ni foam, (b, c) SEM images, and (d) TEM image of the CoMoO$_4$@CoNiO$_2$ nanowires (*inset* is the EDS of the CoNiO$_2$ nanoneedle). (Reprinted with permission from Ai, Y. et al., *ACS Appl. Mater. Inter.*, 7, 24204–24211, 2015. Copyright 2015 American Chemical Society). (e) Schematic diagram demonstrating the electrochemical synthesis of MnO$_2$ nanofibrils on MnO$_2$ nanowires, (f) high magnification SEM image of nanofibrils on a single nanowire and (g, h) TEM image of MnO$_2$ nanofibrils on MnO$_2$ nanowires. (Reprinted with permission from Duay, J. et al., *ACS Nano*, 7, 1200–1214, 2013. Copyright 2013 American Chemical Society.)

capacitance reaches 588 (10 mV s^{-1}) and 1195 Fg^{-1} (0.75 Ag^{-1}) with a high power density and energy density of 30.6 kW kg^{-1} and 106.4 Wh kg^{-1} respectively (Lu et al. 2012). FeCo$_2$O$_4$@polypyrrole core-shell-based nanowires display a specific capacitance of 2269 Fg^{-1} at a current density of 1 Ag^{-1} and retains 79.2% of initial capacitance at 20 Ag^{-1} (He et al. 2018). MnO$_2$ and M(OH)$_2$/MnO$_2$ nanowires (shown in Figure 3.4e–h) prepared by electrodeposition methods show a high capacitance of 298 and 174 Fg^{-1} at 50 mV s^{-1} and 250 mV s^{-1}, respectively, and also retains 85.2% of initial capacitance after 1000 cycles (Duay et al. 2013). Ultralong V$_2$O$_5$@conducting polypyrrole nanowires synthesized by in situ interfacial methods deliver a high specific capacitance of 334 Fg^{-1} and have larger rate capability and better cycling stability (Wang, Liu et al. 2018). CoMoO$_4$@CoNiO$_2$ core-shell nanowire arrays (shown in Figure 3.4a–d) formed by hydrothermal processing show an excellent areal specific capacitance (5.31 F cm^{-2} at 5 mA cm^{-2}) and higher cycling stability (Ai et al. 2015).

3.6.4 NANOCABLE

Zn$_2$SnO$_4$ nanowires on carbon microfibers (scanning electron microscopy [SEM] images shown in Figure 3.5f and g) are prepared by a simple vapor transport method, in which a coating of MnO$_2$ is performed using simple, spontaneous redox reaction to

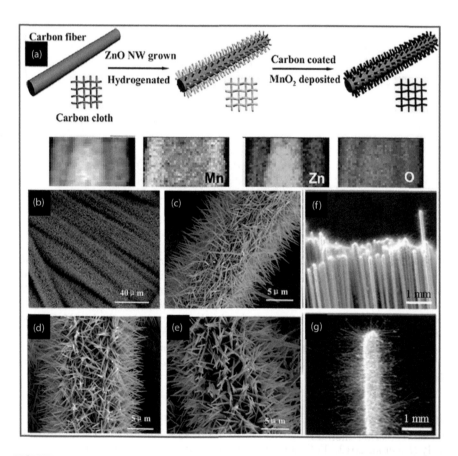

FIGURE 3.5 (a) Schematic representation of HZC deposited with a layer of MnO$_2$ (HZM) core-shell nanocable synthesis. (Reprinted with permission from Yang, P. et al., *ACS Nano*, 7, 2617–2626, 2013. Copyright 2013 American Chemical Society). (b, c) SEM images of HZnO (hydrogenated ZnO nanowires grown on carbon cloth), (d) SEM image of HZnO coated with a layer of carbon (HZC), (e) SEM image of HZM (HZC deposited with a layer of MnO$_2$). (Reprinted with permission from Yang, P. et al., *ACS Nano*, 7, 2617–2626, 2013. Copyright 2013 American Chemical Society). (f, g) SEM images of Zn$_2$SnO$_4$/MnO$_2$ core-shell nanocables grown radially on the woven CMFs. (Reprinted with permission from Bao, L. et al., *Nano Lett.*, 11, 1215–1220, 2011. Copyright 2011 American Chemical Society.)

form Zn_2SnO_4/MnO_2 core-shell nanocables. The nanocables show maximum capacitances of 621.6 Fg^{-1} at 2 mV s^{-1} and 642.4 Fg^{-1} at 1 Ag^{-1} (Bao, Zang, and Li 2011). Single-crystal ZnO@amorphous ZnO-doped MnO_2 nanocables synthesized by the wet chemical method present excellent areal capacitance and specific capacitance of 138.7 mF cm^{-2} and 1260.9 Fg^{-1}, respectively (Yang et al. 2013). The clear schematic representation and SEM images of the layer-by-layer formation of HZC deposited with a layer of MnO_2 (HZM) are shown in Figure 3.5a–e.

3.6.5 NANORODS

Chemical bath deposition, electrodeposition, and hydrothermal techniques have been used consecutively to prepare a bespoke metal @ carbon sphere nanoarray (Xia et al. 2015). The core-shell structure begins with ZnO micro-rod as a template, followed by self-assembly of CNSs, which exhibits a good capacitance of 227 Fg^{-1} (at 2.5 Ag^{-1}) and exceptional cycling stability and retains 97% of capacitance after 40,000 cycles. The SEM images and graphical representation of Ni microtubes-carbon nanospheres hollow structure are shown clearly in Figure 3.6a–f. Core-shell $NiMoO_4$@Ni-Co-S nanorods provided a good specific capacitance of 2.27 F cm^{-2} (1892 Fg^{-1}) at 5 mA cm^{-2}. Also, it retained 91.7% of the initial capacitance at the end of 6000 cycles (Chen et al. 2018). Likewise, Fe_3O_4@Fe_2O_3 core-shell nanorod arrays formed by hydrothermal and electrodeposition methods exhibit superior supercapacitive performance representing large volumetric capacitance of 1206 F cm^{-3} with 1.25 mg cm^{-2} of mass loading (Tang et al. 2015). Unique core-shell $NiCo_2O_4$@PANI nanorod (shown in Figure 3.6g–i) achieved by hydrothermal and electrodeposition methods displays 901 Fg^{-1} of specific capacitance at 1 Ag^{-1} in the presence of H_2SO_4 electrolyte and retains 91% of capacitance after 3000 cycles at 10 Ag^{-1} (Jabeen et al. 2016).

3.6.6 NANOFLOWERS

α-FeOOH/Fe_2O_3 core-shell particles produced by controlled synthesis in a semi-aqueous-organic medium using a simple ligand mediation method exhibit excellent specific capacitance (200 Fg^{-1} at 5 mAg^{-1}) and long-standing cycling stability (99.9% at 5 mAg^{-1} at the end of 500 cycles) (Barik et al. 2014). High-performance $CoSe_2$@PPy core-shell structured electrodes prepared using the electrodeposition method show an output of 1.6 V and attain a high energy density of 2.63 mWh cm^{-3} at 14 mW cm^{-3} (Wang, Ma et al. 2018). $MnCo_2O_4$@Ni(OH)$_2$ core-shell nanoflowers synthesized by a simple and economic hydrothermal method shows a specific capacitance of 2154 Fg^{-1} at 5 Ag^{-1} (Zhao et al. 2016). 3D-based Co_3O_4@$NiMoO_4$ nanoflowers on Ni foam prepared via hydrothermal and post-annealing treatment deliver 636.8 Cg^{-1} at 5 mA cm^{-2}, which retains 84.1% at the end of 2000 cycles at 20 mA cm^{-2} (Zhang et al. 2016). NiSe@NiOOH core-shell-based hyacinth-like nanomaterials on Ni foam prepared by solvothermal and electrochemical oxidation techniques display high electrochemical performance (Li et al. 2016).

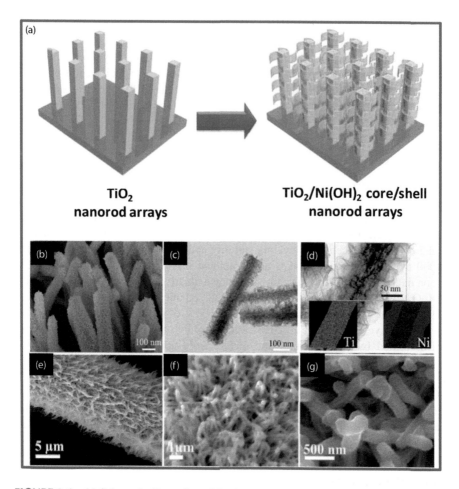

FIGURE 3.6 (a) Schematic illustration of the formation of $TiO_2/Ni(OH)_2$ core-shell nanorod, (b and c) SEM and TEM images of $TiO_2/Ni(OH)_2$ core-shell nanorod, and (d) EDS elemental mappings of Ti and Ni in $TiO_2/Ni(OH)_2$ core-shell nanorod. (With kind permission from Springer Science+Business Media: *Sci. Rep.*, Xia, X. et al. Integrated photoelectrochemical energy storage: Solar hydrogen generation and supercapacitor, 2, 2012, 981) (e) Field-emission SEM image of carbon fiber after the coating of PANI on $NiCo_2O_4$ nanorods and (f, g) high magnification images of $NiCo_2O_4$@PANI nanorods. (Reprinted with permission from Jabeen, N. et al., *ACS Appl. Mater. Inter.*, 8, 6093–6100, 2016. Copyright 2016 American Chemical Society.)

3.6.7 NANOCONES

$ZnCo_2O_4/MnO_2$ core-shell nanocones with a mesoporous structure grown hydrothermally on a 3D nickel foam, as shown in Figure 3.7a–c, exhibit exceptional specific capacitance of 2339 Fg^{-1} at 1 Ag^{-1} with long-term capacitance retention of 95.9% at the end of 3000 cycles (Qiu et al. 2015). 3D Co_3O_4@$CoMoO_4$ core-shell nanopine structures grown on Ni foam using hydrothermal procedure exhibit 1902 Fg^{-1} at the current density of 1 Ag^{-1} with good rate capability and retain 99% of capacitance after 5000 cycles (Wang et al. 2016).

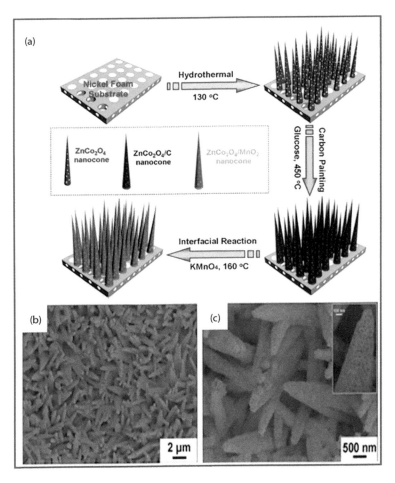

FIGURE 3.7 (a) Schematic representation of the $ZnCo_2O_4/MnO_2$ core-shell nanocone formation and (b, c) SEM images of $ZnCo_2O_4/MnO_2$ obtained at different magnifications. (Reprinted from *Nano Energy*, 11, Qiu, K. et al., Mesoporous, hierarchical core/shell structured $ZnCo_2O_4/MnO_2$ nanocone forests for high-performance supercapacitors, 687–696, Copyright 2014, with permission from Elsevier Ltd.)

3.6.8 NANOPRISM

Mesoporous Ni_xS_y@CoS polyhedra synthesized through a metal-organic-framework route (Figure 3.8a–f) exhibit an extraordinarily good capacitance of 2291 Fg^{-1} at 1.0 Ag^{-1} (Gao et al. 2017). Hollow Mn-Co layered double hydroxide (LDH) fabricated by the solvothermal method exhibits excellent capacitor behavior, with specific capacitance of 511 Fg^{-1} at 2 Ag^{-1} and good stability over 2000 cycles at 5 Ag^{-1} (Wu et al. 2017). Yolk-shelled Ni-Co oxide-based nanoprisms synthesized by the thermal annealing process deliver a high capacitance of 1000 Fg^{-1} at 10 Ag^{-1} and retains 98% of capacitance after 15,000 cycles (Yu et al. 2015).

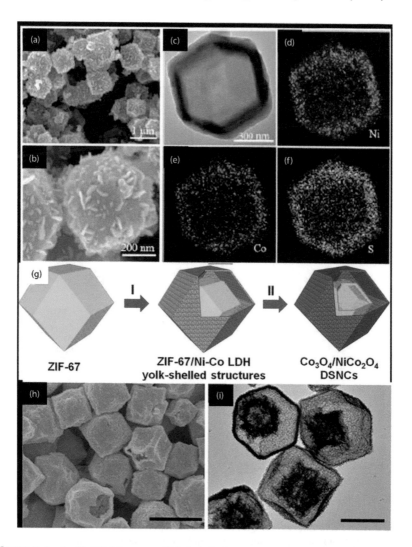

FIGURE 3.8　(a, b) SEM images of Ni_xS_y@CoS double-shelled nanocages, (c) Scanning Transmission Electron Microscope (STEM) image, and (d–f) the elemental mappings of Ni, Co, and S of Ni_xS_y@CoS double-shelled nanocages, respectively. (Reprinted from *Electrochim. Acta*, 237, Gao, R. et al. Novel amorphous nickel sulfide@CoS double-shelled polyhedral nanocages for supercapacitor electrode materials with superior electrochemical properties, 94–101, Copyright 2017, with permission from Elsevier Ltd.). (g) Schematic illustration of the formation process of Co_3O_4/$NiCo_2O_4$ double-shelled nanocages, (h) and (i) Field-emission SEM and TEM images of Co_3O_4/$NiCo_2O_4$ double-shell nanocages. (Reprinted with permission from Hu, H. et al., *J. Am. Chem. Soc.*, 137, 5590–5595, 2015. Copyright 2015 American Chemical Society.)

3.7 FACTOR DISTURBING THE DIMENSION AND DISTRIBUTION OF CORE-SHELL NANOSTRUCTURE

The magnitude and distribution are equally significant parameters in applications of materials. It is a well-known fact that the surface area and bandgap energy are inversely proportional to particle size, which results in property changes. Smaller particle size with narrow size distribution and preventon of agglomeration are other important factors in particle synthesis. The abovementioned factors can be controlled by the proper choice of reaction media, reaction parameters (temperature, concentration, and solution pH), reaction nature, and the applied external force.

3.7.1 REACTION MEDIA

Synthesis media determine the particle size and distribution of particles. Generally, bulk phase or micro-emulsion technique in "bottom-up" approaches is used in the synthesis of particles.

3.7.1.1 Bulk Phase Media

The advantages of bulk phase media are easy particle separation, simplicity in up-scaling, gentle control of reactant concentration, and effortless post-treatment of a particle that make this phase an easy process compared to other methods. The major disadvantage is controlling the external parameters, i.e., difficult to maintain the particle size and its distribution. It is also difficult to achieve uniform coating of shell material over core particles (Barik et al. 2014). The core substances are coated with a suitable opposite-charged surfactant to achieve a uniform surface. Another difficulty with this reaction media is the manufacturing of large particles with a broad distribution.

3.7.1.2 Microemulsion Method

The benefits of the microemulsion method are ease of controlling the size, distribution, and uniformity of particles. The disadvantages are difficulties in separation or purification of nanoparticles because the microemulsion process uses vast surfactant and oil phases (Lim et al. 2016). There are also problems in generating huge amounts of nanoparticles. In a microemulsion system, the nanoparticle size, morphology, and dispersity depend on the droplet size, surfactant nature, concentration and nature of reactants, volume fraction, and mixing ratio. The effects of micro-emulsion system are as follows:

- Mole ratio of water and surfactant
- Reagent effect
- Polar volume fraction
- Mixing effect
- Effect of the oil phase.

3.7.2 TEMPERATURE EFFECTS

The range of working temperature relies mostly on the reaction mechanism in the formation of core or shell materials. Generally, a low temperature provides a favorable environment for core-shell formation. This supports for coating of shell layer on the core material apart from the development of distinct nuclei, by maintaining the formation of unsaturated shell material inside the reaction mixture (Chen et al. 2007). Moreover, a low temperature prime to a poor blending of core and shell atomic structure at the boundary due to surface modification by adsorption of a suitable surfactant on the core surface for the deposition of a charged shell material. This surface modification is impossible at high temperatures owing to reduced adsorption of surface molecules, which results in uneven coatings.

3.7.3   REACTANT CONCENTRATION EFFECTS

The final product size (including core dimension or shell thickness) depends on two stages: (i) reaction among the reactants to form nuclei and (ii) collision among nuclei or molecule diffused on the surface of nuclei and deposited there to create a final product. The primary stage is a reaction step followed by a growth process, which depends on the reaction rate. The relationship between the nanoparticle size and the reactant concentration has not yet been reported. Differently, the particle size depends on the types and synthesis media. Generally, particle size increases with an increase in product amount. At a lower reactant concentration, the slow reaction rate results in the formation of less number of nuclei. However, at high reactant concentration, a more significant number of nuclei are developed owing to the high reaction rate. As a result, ultimately, particle size decreases (Guo et al. 2015).

Shell formation over the core particles usually occurs by heterogeneous nucleation. The shell substance is first placed on the core structure, which is followed by nuclei construction and development on the surface. Therefore, the slower reaction rate results in even coating. Hence, the lower concentration of reactant is feasible for core-shell structure formation. Generally, the size and thickness of core-shell nanoparticle depend on reactant concentration.

3.7.4   EFFECT OF SURFACE MODIFIER

The dimension and morphology of core-shell particles are controlled by surface modifier concentration. The surface modifier usually adsorbs on the core surface and is responsible for important aspects such as controlling the particle size; it also acts as a powerful agent for shell construction by surface modification in core-shell structure preparation. The concentration of the surface modifier defines the adsorption amount of the surface modifier on core structure (Wang, Shi, and Jiang 2008). The adsorption amount is directly proportional to concentration until a particular limit is reached and then becomes constant. Hence, the size distribution of nanoparticles mostly depends on the surface charge.

3.7.5   EFFECT OF pH

The reaction mechanism involved in a reaction determines the effect of pH on the nanoparticle dimension. In a redox method, the reduction of a material with the highest reduction potential is achieved by performing oxidation of a material possessing lesser reduction potential. Generally, the half-cell reduction potential of a redox pair depends on the pH of the reaction solution (Cai et al. 2014). Therefore, pH is an extremely desired parameter to control the reaction process and to know the extent of surface modification. Hence, the solution pH was controlled to increase the formation of a uniform shell coating.

3.7.6 EFFECT OF EXTERNAL FORCES

External forces are used in the reaction method to control the nanoparticle size. Among various external forces, the electrical and sonication forces are the most commonly used for the formation of nanoparticles.

3.7.6.1 Sonication

The three main effects involved in the sonication process are as follows:

- The transient heating of liquid results from bubble collapse
- Agglomeration is prevented by the formation of shock waves through bubble implosion
- Acoustic cavitation results in uniform mixing.

All these effects help in achieving reduced particle size with increased homogeneity by controlling the nuclei population (De Castro and Priego-Capote 2007).

3.7.6.2 Electric Field

Shell formation using charged or inorganic substances on the core materials is performed by applying external electric potential. A precise shell thickness is achieved using this method. Generally, the electric field is a mixed wave with positive and negative cycles. The inorganic material deposits on the core structure in a negative cycle, whereas the charged substance deposits in a positive cycle. The shell thickness is controlled by controlling the cycle period (Yuan et al. 2015).

3.8 CONCLUSION

Core-shell nanostructure is a significant material and is a major pivotal topic in energy storage applications. Recently, core-shell structures have been extensively applied in various fields such as batteries, supercapacitors, hydrogen generation/evolution, biomedicine, therapeutics, oxygen reduction and evolution reactions, and solar cells. Core-shell nanoparticles show excellent specific capacitance, improved conductivity, and favorable redox behaviour, which plays a significant role as electrode materials in supercapacitor research. The better conductivity with high redox capacity of core-shell composite materials has enormous potential in energy storage applications. Taking inspiration from core-shell structures as electrode materials, many attempts have been made to use the benefits of these structures to a maximum extent by blending different properties in a single structure based on applications. The structure supports for long cycling stability by resisting morphological deformation on successive charging-discharging cycles. Different synthetic approaches to change morphologies in order to fine tune the electrochemical activity and performance of core-shell structures have also been studied. Parameters such as porosity, crystallinity, stoichiometry, shape, size, and surface area strongly influence the catalytic nature of core-shell structures. In addition, the subtleties of the formation of a uniform shell structure have also been studied. The synthetic procedures involved in controlling the shell thickness, lower the processing temperature, and uniform deposition of shell on to the core structure result in better electron transport, short diffusion length, and reduced morphological deformation. This contributes to the overall performance and value parameters to be considered in large-scale applications. With compatible electrolytes, the core-shell nanoforms can greatly improvise the performance of a supercapacitor. Discussion on various forms of core-shell structure that supports the improvement in capacitance properties is provided elaborately with some typical literature. The study of the effects of various parameters involved in the formation of core-shell structures is also described clearly. Thus, it will be more significant to start a large-scale synthesis of core-shell nanostructures with the proper controlling system to identify and overcome any inadequacies.

REFERENCES

Ai, Yuanfei, Xuewen Geng, Zheng Lou, Zhiming M Wang, and Guozhen Shen. 2015. "Rational Synthesis of Branched $CoMoO_4$@$CoNiO_2$ Core/Shell Nanowire Arrays for All-Solid-State Supercapacitors with Improved Performance." *ACS Applied Materials & Interfaces* 7 (43): 24204–24211. doi:10.1021/acsami.5b07599.

Ban, Zhihui, Yuri A. Barnakov, Feng Li, Vladimir O. Golub, and Charles J. O'Connor. 2005. "The Synthesis of Core–shell Iron@gold Nanoparticles and Their Characterization." *Journal of Materials Chemistry* 15 (43): 4660–4662. doi:10.1039/B504304B.

Bao, Lihong, Jianfeng Zang, and Xiaodong Li. 2011. "Flexible Zn_2SnO_4/MnO_2 Core/Shell Nanocable–Carbon Microfiber Hybrid Composites for High-Performance Supercapacitor Electrodes." *Nano Letters* 11 (3): 1215–1220. doi:10.1021/nl104205s.

Barik, Rasmita, Bikash Kumar Jena, Ajit Dash, and Mamata Mohapatra. 2014. "In Situ Synthesis of Flowery-Shaped α-FeOOH/Fe_2O_3 Nanoparticles and Their Phase Dependent Supercapacitive Behaviour." *RSC Advances* 4 (36): 18827–18834. doi:10.1039/C4RA01258E.

Bogue, Robert. 2011. "Nanoc omposites: A Review of Technology and Applications." *Assembly Automation* 31 (2): 106–112. doi:10.1108/01445151111117683.

Buchanan, R. C., R. D. Roseman, and K. R. Eufinger. 1996. "Electrical and Mechanical Properties of Core-Shell Type Structures in Doped BaTiO/Sub 3/." In *ISAF'96. Proceedings of the Tenth IEEE International Symposium on Applications of Ferroelectrics* 2: 887–890. doi:10.1109/ISAF.1996.598168.

Cai, Daoping, Hui Huang, Dandan Wang, Bin Liu, Lingling Wang, Yuan Liu, Qiuhong Li, and Taihong Wang. 2014. "High-Performance Supercapacitor Electrode Based on the Unique ZnO@Co_3O_4 Core/Shell Heterostructures on Nickel Foam." *ACS Applied Materials & Interfaces* 6 (18): 15905–15912. doi:10.1021/am5035494.

Chandrasekaran, Nivedhini Iswarya, Meena Kumari, Harshiny Muthukumar, and Manickam Matheswaran. 2018. "Strategy for Multifunctional Hollow Shelled Triple Oxide Mn–Cu–Al Nanocomposite Synthesis via Microwave-Assisted Technique." *ACS Sustainable Chemistry & Engineering* 6 (1): 1009–1021. doi:10.1021/acssuschemeng.7b03339.

Chandrasekaran, Nivedhini Iswarya, and Matheswaran Manickam. 2017. "Mesoporous Hollow MnCuAl Layered Triple Hydroxides Nanocomposite Synthesized via Microwave Assisted Technique for Symmetrical Supercapacitor." *International Journal of Hydrogen Energy* 42 (42): 26475–26487. doi:10.1016/j.ijhydene.2017.08.007.

Chandrasekaran, Nivedhini Iswarya, Harshiny Muthukumar, Aiswarya Devi Sekar, and Matheswaran Manickam. 2017. "Hollow Nickel-Aluminium-Manganese Layered Triple Hydroxide Nanospheres with Tunable Architecture for Supercapacitor Application." *Materials Chemistry and Physics* 195: 247–258. doi:10.1016/j.matchemphys.2017.04.027.

Chatterjee, Krishnendu, Sreerupa Sarkar, K. Jagajjanani Rao, and Santanu Paria. 2014. "Core/Shell Nanoparticles in Biomedical Applications." *Advances in Colloid and Interface Science* 209: 8–39. doi:10.1016/j.cis.2013.12.008.

Chaudhuri, Rajib Ghosh, and Santanu Paria. 2012. "Core/Shell Nanoparticles: Classes, Properties, Synthesis Mechanisms, Characterization, and Applications." *Chemical Reviews* 112: 2373–2433. doi:10.1021/cr100449n.

Chen, Aibing, Kechan Xia, Linsong Zhang, Yifeng Yu, Yuetong Li, Hexu Sun, Yuying Wang, Yunqian Li, and Shuhui Li. 2016. "Fabrication of Nitrogen-Doped Hollow Mesoporous Spherical Carbon Capsules for Supercapacitors." *Langmuir* 32 (35): 8934–8941. doi:10.1021/acs.langmuir.6b02250.

Chen, Chao, Dan Yan, Xin Luo, Wenjia Gao, Guanjie Huang, Ziwu Han, Yan Zeng, and Zhihong Zhu. 2018. "Construction of Core–Shell $NiMoO_4$@Ni-Co-S Nanorods as Advanced Electrodes for High-Performance Asymmetric Supercapacitors." *ACS Applied Materials & Interfaces* 10 (5): 4662–4671. doi:10.1021/acsami.7b16271.

Chen, Guanying, Jie Shen, Tymish Y. Ohulchanskyy, Nayan J. Patel, Artem Kutikov, Zhipeng Li, Jie Song et al. 2012. "(α-NaYbF4:Tm^{3+})/CaF_2 Core/Shell Nanoparticles with Efficient Near-Infrared to Near-Infrared Upconversion for High-Contrast Deep Tissue Bioimaging." *ACS Nano* 6 (9): 8280–8287. doi:10.1021/nn302972r.

Chen, Shouhu Xuan and Lingyun Hao and Wanquan Jiang and Xinglong Gong and Yuan Hu and Zuyao. 2007. "A Facile Method to Fabricate Carbon-Encapsulated Fe_3O_4 Core/Shell Composites." *Nanotechnology* 18 (3): 35602. http://stacks.iop.org/0957-4484/18/i=3/a=035602.

Chen, Zhebo, Dustin Cummins, Benjamin N. Reinecke, Ezra Clark, Mahendra K. Sunkara, and Thomas F. Jaramillo. 2011. "Core–shell MoO_3–MoS_2 Nanowires for Hydrogen Evolution: A Functional Design for Electrocatalytic Materials." *Nano Letters* 11 (10): 4168–4175. doi:10.1021/nl2020476.

Cui, Li-Feng, Riccardo Ruffo, Candace K. Chan, Hailin Peng, and Yi Cui. 2009. "Crystalline-Amorphous Core−Shell Silicon Nanowires for High Capacity and High Current Battery Electrodes." *Nano Letters* 9 (1): 491–495. doi:10.1021/nl8036323.

De Castro, M. D. Luque, and F. Priego-Capote. 2007. "Ultrasound-Assisted Crystallization (Sonocrystallization)." *Ultrasonics Sonochemistry* 14 (6): 717–724.

Deutch, A. Y., and D. S. Cameron. 1992. "Pharmacological Characterization of Dopamine Systems in the Nucleus Accumbens Core and Shell." *Neuroscience* 46 (1): 49–56. doi:10.1016/0306-4522(92)90007-O.

Duay, Jonathon, Stefanie A. Sherrill, Zhe Gui, Eleanor Gillette, and Sang Bok Lee. 2013. "Self-Limiting Electrodeposition of Hierarchical MnO_2 and M(OH)$_2$/MnO_2 Nanofibril/Nanowires: Mechanism and Supercapacitor Properties." *ACS Nano* 7 (2): 1200–1214. doi:10.1021/nn3056077.

Faraji, Soheila, and Farid Nasir Ani. 2014. "Microwave-Assisted Synthesis of Metal Oxide/Hydroxide Composite Electrodes for High Power Supercapacitors-a Review." *Journal of Power Sources* 263: 338–360.

Feng, Hao-Peng, Lin Tang, Guang-Ming Zeng, Jing Tang, Yao-Cheng Deng, Ming Yan, Ya-Ni Liu, Yao-Yu Zhou, Xiao-Ya Ren, and Song Chen. 2018. "Carbon-Based Core–shell Nanostructured Materials for Electrochemical Energy Storage." *Journal of Materials Chemistry A* 6 (17): 7310–7337. doi:10.1039/C8TA01257A.

Gao, Runsheng, Qiugen Zhang, Faizal Soyekwo, Chenxiao Lin, Ruixue Lv, Yan Qu, Mengmeng Chen, Aimei Zhu, and Qinglin Liu. 2017. "Novel Amorphous Nickel Sulfide@CoS Double-Shelled Polyhedral Nanocages for Supercapacitor Electrode Materials with Superior Electrochemical Properties." *Electrochimica Acta* 237: 94–101. doi:10.1016/j.electacta.2017.03.214.

Gawande, Manoj B., Anandarup Goswami, Tewodros Asefa, Huizhang Guo, Ankush V. Biradar, Dong-Liang Peng, Radek Zboril, and Rajender S. Varma. 2015. "Core–Shell Nanoparticles: Synthesis and Applications in Catalysis and Electrocatalysis." *Chemical Society Reviews* 44 (21): 7540–7590. doi:10.1039/C5CS00343A.

González, Ander, Eider Goikolea, Jon Andoni Barrena, and Roman Mysyk. 2016. "Review on Supercapacitors: Technologies and Materials." *Renewable and Sustainable Energy Reviews* 58: 1189–1206.

Guo, Xiao Long, Gang Li, Min Kuang, Liang Yu, and Yu Xin Zhang. 2015. "Tailoring Kirkendall Effect of the KCu_7S_4 Microwires towards $CuO@MnO_2$ Core-Shell Nanostructures for Supercapacitors." *Electrochimica Acta* 174: 87–92. doi:10.1016/j.electacta.2015.05.157.

Haag, Rainer. 2004. "Supramolecular Drug-Delivery Systems Based on Polymeric Core–Shell Architectures." *Angewandte Chemie International Edition* 43 (3): 278–282. doi:10.1002/anie.200301694.

He, Xinyi, Yunhe Zhao, Rongrong Chen, Hongsen Zhang, Jingyuan Liu, Qi Liu, Dalei Song, Rumin Li, and Jun Wang. 2018. "Hierarchical $FeCo_2O_4$@polypyrrole Core/Shell Nanowires on Carbon Cloth for High-Performance Flexible All-Solid-State Asymmetric Supercapacitors." *ACS Sustainable Chemistry & Engineering* 6 (11): 14945–14954. doi:10.1021/acssuschemeng.8b03440.

Hu, Han, Buyuan Guan, Baoyu Xia, and Xiong Wen (David) Lou. 2015. "Designed Formation of $Co_3O_4/NiCo_2O_4$ Double-Shelled Nanocages with Enhanced Pseudocapacitive and Electrocatalytic Properties." *Journal of the American Chemical Society* 137 (16): 5590–5595. doi:10.1021/jacs.5b02465.

Huang, Liang, Dongchang Chen, Yong Ding, Shi Feng, Zhong Lin Wang, and Meilin Liu. 2013. "Nickel–Cobalt Hydroxide Nanosheets Coated on $NiCo_2O_4$ Nanowires Grown on Carbon Fiber Paper for High-Performance Pseudocapacitors." *Nano Letters* 13 (7): 3135–3139.

Huang, Ming, Yuxin Zhang, Fei Li, Zhongchang Wang, Alamusi, Ning Hu, Zhiyu Wen, and Qing Liu. 2014. "Merging of Kirkendall Growth and Ostwald Ripening: $CuO@MnO_2$ Core-Shell Architectures for Asymmetric Supercapacitors." *Scientific Reports* 4: 4518. doi:10.1038/srep04518.

Huang, Ming, Yuxin Zhang, Fei Li, Lili Zhang, Zhiyu Wen, and Qing Liu. 2014. "Facile Synthesis of Hierarchical $Co_3O_4@MnO_2$ Core–shell Arrays on Ni Foam for Asymmetric Supercapacitors." *Journal of Power Sources* 252: 98–106. doi:10.1016/j.jpowsour.2013.12.030.

Huang, Xingyi, and Pingkai Jiang. 2015. "Core–Shell Structured High-k Polymer Nanocomposites for Energy Storage and Dielectric Applications." *Advanced Materials* 27 (3): 546–54. https://doi.org/10.1002/adma.201401310.

Hwang, Tae Hoon, Yong Min Lee, Byung-Seon Kong, Jin-Seok Seo, and Jang Wook Choi. 2012. "Electrospun Core–Shell Fibers for Robust Silicon Nanoparticle-Based Lithium Ion Battery Anodes." *Nano Letters* 12 (2): 802–807. doi:10.1021/nl203817r.

Jabeen, Nawishta, Qiuying Xia, Mei Yang, and Hui Xia. 2016. "Unique Core–Shell Nanorod Arrays with Polyaniline Deposited into Mesoporous $NiCo_2O_4$ Support for High-Performance Supercapacitor Electrodes." *ACS Applied Materials & Interfaces* 8 (9): 6093–6100. doi:10.1021/acsami.6b00207.

Jia, Xinxu, Xiang Wu, and Baodan Liu. 2018. "Formation of $ZnCo_2O_4@MnO_2$ Core–Shell Electrode Materials for Hybrid Supercapacitor." *Dalton Transactions* 47 (43): 15506–15511. doi:10.1039/C8DT03298J.

Jiang, Hao, Chunzhong Li, Ting Sun, and Jan Ma. 2012. "High-Performance Supercapacitor Material Based on $Ni(OH)_2$ Nanowire-MnO_2 Nanoflakes Core–Shell Nanostructures." *Chemical Communications* 48 (20): 2606–2608. doi:10.1039/C2CC18079K.

Kong, Wei, Chenchen Lu, Wu Zhang, Jun Pu, and Zhenghua Wang. 2015. "Homogeneous Core–Shell $NiCo_2S_4$ Nanostructures Supported on Nickel Foam for Supercapacitors." *Journal of Materials Chemistry A* 3 (23): 12452–12460. doi:10.1039/C5TA02432C.

Li, Gao-Ren, Zi-Long Wang, Fu-Lin Zheng, Yan-Nan Ou, and Ye-Xiang Tong. 2011. "ZnO@ MoO_3 Core/Shell Nanocables: Facile Electrochemical Synthesis and Enhanced Supercapacitor Performances." *Journal of Materials Chemistry* 21 (12): 4217–4221. doi:10.1039/C0JM03500A.

Li, Meng, Xinyue Li, Zhizhang Li, and Yihui Wu. 2018. "Hierarchical Nanosheet-Built $CoNi_2S_4$ Nanotubes Coupled with Carbon-Encapsulated Carbon Nanotubes@ Fe_2O_3 Composites toward High-Performance Aqueous Hybrid Supercapacitor Devices." *ACS Applied Materials & Interfaces* 10 (40): 34254–34264. doi:10.1021/acsami.8b11416.

Li, Peixu, Yanbing Yang, Enzheng Shi, Qicang Shen, Yuanyuan Shang, Shiting Wu, Jinquan Wei et al. 2014. "Core-Double-Shell, Carbon Nanotube@ Polypyrrole@ MnO_2 Sponge as Freestanding, Compressible Supercapacitor Electrode." *ACS Applied Materials & Interfaces* 6 (7): 5228–5234.

Li, Qinghua, Yongbiao Yuan, Zihan Chen, Xiao Jin, Tai-huei Wei, Yue Li, Yuancheng Qin, and Weifu Sun. 2014. "Core–Shell Nanophosphor Architecture: Toward Efficient Energy Transport in Inorganic/Organic Hybrid Solar Cells." *ACS Applied Materials & Interfaces* 6 (15): 12798–12807. doi:10.1021/am5027709.

Li, Xiao, Guan-Qun Han, Yan-Ru Liu, Bin Dong, Wen-Hui Hu, Xiao Shang, Yong-Ming Chai, and Chen-Guang Liu. 2016. "NiSe@ NiOOH Core–Shell Hyacinth-like Nanostructures on Nickel Foam Synthesized by in Situ Electrochemical Oxidation as an Efficient Electrocatalyst for the Oxygen Evolution Reaction." *ACS Applied Materials & Interfaces* 8 (31): 20057–20066.

Lim, Eunho, Changshin Jo, Min Su Kim, Mok-Hwa Kim, Jinyoung Chun, Haegyeom Kim, Jongnam Park et al. 2016. "High-Performance Sodium-Ion Hybrid Supercapacitor Based on $Nb_2O_5@Carbon$ Core–Shell Nanoparticles and Reduced Graphene Oxide Nanocomposites." *Advanced Functional Materials* 26 (21): 3711–3719. doi:10.1002/adfm.201505548.

Liu, Chao, Jing Wang, Jiansheng Li, Mengli Zeng, Rui Luo, Jinyou Shen, Xiuyun Sun, Weiqing Han, and Lianjun Wang. 2016. "Synthesis of N-Doped Hollow-Structured Mesoporous Carbon Nanospheres for High-Performance Supercapacitors." *ACS Applied Materials & Interfaces* 8 (11): 7194–7204. doi:10.1021/acsami.6b02404.

Liu, Chenguang, Zhenning Yu, David Neff, Aruna Zhamu, and Bor Z Jang. 2010a. "Graphene-Based Supercapacitor with an Ultrahigh Energy Density." *Nano Letters* 10 (12): 4863–4868.

Liu, Jian, Shi Zhang Qiao, Sandy Budi Hartono, and Gao Qing (Max) Lu. 2010b. "Monodisperse Yolk–Shell Nanoparticles with a Hierarchical Porous Structure for Delivery Vehicles and Nanoreactors." *Angewandte Chemie International Edition* 49 (29): 4981–4985. doi:10.1002/anie.201001252.

Liu, Mingxian, Jiasheng Qian, Yunhui Zhao, Dazhang Zhu, Lihua Gan, and Longwu Chen. 2015. "Core–Shell Ultramicroporous@microporous Carbon Nanospheres as Advanced Supercapacitor Electrodes." *Journal of Materials Chemistry A* 3 (21): 11517–11526. doi:10.1039/C5TA02224J.

Liu, Wen-Wen, Congxiang Lu, Kun Liang, and Beng Kang Tay. 2014. "A Three Dimensional Vertically Aligned Multiwall Carbon Nanotube/NiCo$_2$O$_4$ Core/Shell Structure for Novel High-Performance Supercapacitors." *Journal of Materials Chemistry A* 2 (14): 5100–5107. doi:10.1039/C4TA00107A.

Liu, Xiayuan, Shaojun Shi, Qinqin Xiong, Lu Li, Yijun Zhang, Hong Tang, Changdong Gu, Xiuli Wang, and Jiangping Tu. 2013. "Hierarchical NiCo$_2$O$_4$@NiCo$_2$O$_4$ Core/Shell Nanoflake Arrays as High-Performance Supercapacitor Materials." *ACS Applied Materials & Interfaces* 5 (17): 8790–8795. doi:10.1021/am402681m.

Lu, Xihong, Minghao Yu, Gongming Wang, Teng Zhai, Shilei Xie, Yichuan Ling, Yexiang Tong, and Yat Li. 2013. "H-TiO$_2$@MnO$_2$//H-TiO$_2$@C Core–Shell Nanowires for High Performance and Flexible Asymmetric Supercapacitors." *Advanced Materials* 25 (2): 267–272. doi:10.1002/adma.201203410.

Lu, Xihong, Teng Zhai, Xianghui Zhang, Yongqi Shen, Longyan Yuan, Bin Hu, Li Gong et al. 2012. "WO$_{3-x}$@Au@MnO$_2$ Core–Shell Nanowires on Carbon Fabric for High-Performance Flexible Supercapacitors." *Advanced Materials* 24 (7): 938–944. doi:10.1002/adma.201104113.

Luo, Jin, Lingyan Wang, Derrick Mott, Peter N. Njoki, Yan Lin, Ting He, Zhichuan Xu, Bridgid N Wanjana, I.-Im S. Lim, and Chuan-Jian Zhong. 2008. "Core/Shell Nanoparticles as Electrocatalysts for Fuel Cell Reactions." *Advanced Materials* 20 (22): 4342–4347. doi:10.1002/adma.200703009.

Malik, M. Azad, Paul O'Brien, and Neerish Revaprasadu. 2002. "A Simple Route to the Synthesis of Core/Shell Nanoparticles of Chalcogenides." *Chemistry of Materials* 14 (5): 2004–2010. doi:10.1021/cm011154w.

Min, Minkyu, Cheonghee Kim, and Hyunjoo Lee. 2010. "Electrocatalytic Properties of Platinum Overgrown on Various Shapes of Gold Nanocrystals." *Journal of Molecular Catalysis A: Chemical* 333 (1): 6–10. doi:10.1016/j.molcata.2010.09.020.

Nomoev, Andrey V., Sergey P. Bardakhanov, Makoto Schreiber, Dashima G. Bazarova, Nikolai A. Romanov, Boris B. Baldanov, Bair R. Radnaev, and Viacheslav V. Syzrantsev. 2015. "Structure and Mechanism of the Formation of Core–Shell Nanoparticles Obtained through a One-Step Gas-Phase Synthesis by Electron Beam Evaporation." *Beilstein Journal of Nanotechnology* 6: 874.

Pastoriza-Santos, Isabel, Dmitry S. Koktysh, Arif A. Mamedov, Michael Giersig, Nicholas A. Kotov, and Luis M. Liz-Marzán. 2000. "One-Pot Synthesis of Ag@TiO$_2$ Core–Shell Nanoparticles and Their Layer-by-Layer Assembly." *Langmuir* 16 (6): 2731–2735. doi:10.1021/la991212g.

Peng, Xiaogang, Michael C. Schlamp, Andreas V. Kadavanich, and A. P. Alivisatos. 1997. "Epitaxial Growth of Highly Luminescent CdSe/CdS Core/Shell Nanocrystals with Photostability and Electronic Accessibility." *Journal of the American Chemical Society* 119 (30): 7019–7029. doi:10.1021/ja970754m.

Qiu, Kangwen, Yang Lu, Deyang Zhang, Jinbing Cheng, Hailong Yan, Jinyou Xu, Xianming Liu, Jang-Kyo Kim, and Yongsong Luo. 2015. "Mesoporous, Hierarchical Core/Shell Structured ZnCo$_2$O$_4$/MnO$_2$ Nanocone Forests for High-Performance Supercapacitors." *Nano Energy* 11: 687–696. doi:10.1016/j.nanoen.2014.11.063.

Qu, Qunting, Yusong Zhu, Xiangwen Gao, and Yuping Wu. 2012. "Core–Shell Structure of Polypyrrole Grown on V$_2$O$_5$ Nanoribbon as High Performance Anode Material for Supercapacitors." *Advanced Energy Materials* 2 (8): 950–955. doi:10.1002/aenm.201200088.

Serpell, Christopher J., James Cookson, Dogan Ozkaya, and Paul D. Beer. 2011. "Core@shell Bimetallic Nanoparticle Synthesis via Anion Coordination." *Nature Chemistry* 3: 478. doi:10.1038/nchem.1030.

Simon, Patrice, Yury Gogotsi, and Bruce Dunn. 2014. "Where Do Batteries End and Supercapacitors Begin?" *Science* 343 (6176): 1210LP–1211LP. doi:10.1126/science.1249625.

Sounderya, Nagarajan, and Yong Zhang. 2008. "Use of Core/Shell Structured Nanoparticles for Biomedical Applications." *Recent Patents on Biomedical Engineering* 1 (1): 34–42.

Su, Yu-Zhi, Kang Xiao, Nan Li, Zhao-Qing Liu, and Shi-Zhang Qiao. 2014. "Amorphous Ni $(OH)_2$@ Three-Dimensional Ni Core–Shell Nanostructures for High Capacitance Pseudocapacitors and Asymmetric Supercapacitors." *Journal of Materials Chemistry A* 2 (34): 13845–13853.

Tang, Xiao, Ruyue Jia, Teng Zhai, and Hui Xia. 2015. "Hierarchical Fe_3O_4@Fe_2O_3 Core–Shell Nanorod Arrays as High-Performance Anodes for Asymmetric Supercapacitors." *ACS Applied Materials & Interfaces* 7 (49): 27518–27525. doi:10.1021/acsami.5b09766.

Tian, Jingyang, Haiyan Zhang, and Zhenghui Li. 2018. "Synthesis of Double-Layer Nitrogen-Doped Microporous Hollow Carbon@MoS_2/MoO_2 Nanospheres for Supercapacitors." *ACS Applied Materials & Interfaces* 10 (35): 29511–29520. doi:10.1021/acsami.8b08534.

Wang, Hsin-Yi, Fang-Xing Xiao, Le Yu, Bin Liu, and Xiong Wen (David) Lou. 2014. "Hierarchical α-MnO_2 Nanowires@Ni1-XMnxOy Nanoflakes Core–Shell Nanostructures for Supercapacitors." *Small* 10 (15): 3181–3186. doi:10.1002/smll.201303836.

Wang, J, Tj Shi, and Xc Jiang. 2008. "Synthesis and Characterization of Core-Shell ZrO_2/PAAEM/PS Nanoparticles." *Nanoscale Research Letters* 4 (3): 240–246. doi:10.1007/s11671-008-9232-3.

Wang, Jian-Gan, Huanyan Liu, Hongzhen Liu, Wei Hua, and Minhua Shao. 2018. "Interfacial Constructing Flexible V_2O_5@Polypyrrole Core–Shell Nanowire Membrane with Superior Supercapacitive Performance." *ACS Applied Materials & Interfaces* 10 (22): 18816–18823. doi:10.1021/acsami.8b05660.

Wang, Jing, Xiang Zhang, Qiulong Wei, Haiming Lv, Yanlong Tian, Zhongqiu Tong, Xusong Liu et al. 2016. "3D Self-Supported Nanopine Forest-like Co_3O_4@$CoMoO_4$ Core–shell Architectures for High-Energy Solid State Supercapacitors." *Nano Energy* 19: 222–233. doi:10.1016/j.nanoen.2015.10.036.

Wang, Qiufan, Yun Ma, Xiao Liang, Daohong Zhang, and Menghe Miao. 2018. "Novel Core/Shell $CoSe_2$@PPy Nanoflowers for High-Performance Fiber Asymmetric Supercapacitors." *Journal of Materials Chemistry A* 6 (22): 10361–10369. doi:10.1039/C8TA02056F.

Wei, Suying, Qiang Wang, Jiahua Zhu, Luyi Sun, Hongfei Lin, and Zhanhu Guo. 2011. "Multifunctional Composite Core–shell Nanoparticles." *Nanoscale* 3 (11): 4474–4502. doi:10.1039/C1NR11000D.

Wu, Feng, Junzheng Chen, Renjie Chen, Shengxian Wu, Li Li, Shi Chen, and Teng Zhao. 2011. "Sulfur/Polythiophene with a Core/Shell Structure: Synthesis and Electrochemical Properties of the Cathode for Rechargeable Lithium Batteries." *The Journal of Physical Chemistry C* 115 (13): 6057–6063. doi:10.1021/jp1114724.

Wu, Nanshi, Jingxiang Low, Tao Liu, Jiaguo Yu, and Shaowen Cao. 2017. "Hierarchical Hollow Cages of Mn-Co Layered Double Hydroxide as Supercapacitor Electrode Materials." *Applied Surface Science* 413: 35–40. doi:10.1016/j.apsusc.2017.03.297.

Xia, Xinhui, Jingshan Luo, Zhiyuan Zeng, Cao Guan, Yongqi Zhang, Jiangping Tu, Hua Zhang, and Hong Jin Fan. 2012. "Integrated Photoelectrochemical Energy Storage: Solar Hydrogen Generation and Supercapacitor." *Scientific Reports* 2: 981. doi:10.1038/srep00981.

Xia, Xinhui, Jiangping Tu, Yongqi Zhang, Xiuli Wang, Changdong Gu, Xin-bing Zhao, and Hong Jin Fan. 2012. "High-Quality Metal Oxide Core/Shell Nanowire Arrays on Conductive Substrates for Electrochemical Energy Storage." *ACS Nano* 6 (6): 5531–5538.

Xia, Xinhui, Yongqi Zhang, Zhanxi Fan, Dongliang Chao, Qinqin Xiong, Jiangping Tu, Hua Zhang, and Hong Jin Fan. 2015. "Novel Metal@Carbon Spheres Core–Shell Arrays by Controlled Self-Assembly of Carbon Nanospheres: A Stable and Flexible Supercapacitor Electrode." *Advanced Energy Materials* 5 (6): 1401709. doi:10.1002/aenm.201401709.

Xiao, Xu, Tianpeng Ding, Longyan Yuan, Yongqi Shen, Qize Zhong, Xianghui Zhang, Yuanzhi Cao, et al. 2012. "WO_{3-x}/MoO_{3-x} Core/Shell Nanowires on Carbon Fabric as an Anode for All-Solid-State Asymmetric Supercapacitors." *Advanced Energy Materials* 2 (11): 1328–1332. doi:10.1002/aenm.201200380.

Xie, Dini, Hongshang Peng, Shihua Huang, and Fangtian You. 2013. "Core-Shell Structure in Doped Inorganic Nanoparticles: Approaches for Optimizing Luminescence Properties." *Journal of Nanomaterials* 2013: 4.

Xiong, Huan-Ming, Yang Xu, Qing-Guang Ren, and Yong-Yao Xia. 2008. "Stable Aqueous ZnO@Polymer Core–Shell Nanoparticles with Tunable Photoluminescence and Their Application in Cell Imaging." *Journal of the American Chemical Society* 130 (24): 7522–7523. doi:10.1021/ja800999u.

Yan, Minglei, Yadong Yao, Jiqiu Wen, Lu Long, Menglai Kong, Guanggao Zhang, Xiaoming Liao, Guangfu Yin, and Zhongbing Huang. 2016. "Construction of a Hierarchical $NiCo_2S_4$@PPy Core–Shell Heterostructure Nanotube Array on Ni Foam for a High-Performance Asymmetric Supercapacitor." *ACS Applied Materials & Interfaces* 8 (37): 24525–24535. doi:10.1021/acsami.6b05618.

Yang, Peihua, Xu Xiao, Yuzhi Li, Yong Ding, Pengfei Qiang, Xinghua Tan, Wenjie Mai et al. 2013. "Hydrogenated ZnO Core–Shell Nanocables for Flexible Supercapacitors and Self-Powered Systems." *ACS Nano* 7 (3): 2617–2626. doi:10.1021/nn306044d.

Yao, Yuanyuan, Cheng Ma, Jitong Wang, Wenming Qiao, Licheng Ling, and Donghui Long. 2015. "Rational Design of High-Surface-Area Carbon Nanotube/Microporous Carbon Core–Shell Nanocomposites for Supercapacitor Electrodes." *ACS Applied Materials & Interfaces* 7 (8): 4817–4825. doi:10.1021/am5087374.

Yoon, S. B., K. Sohn, J. Y. Kim, C.-H. Shin, J.-S. Yu, and T. Hyeon. 2002. "Fabrication of Carbon Capsules with Hollow Macroporous Core/Mesoporous Shell Structures." *Advanced Materials* 14 (1): 19–21. doi:10.1002/1521-4095(20020104)14:1<19::AID-ADMA19>3.0.CO;2-X.

Yu, Le, Buyuan Guan, Wei Xiao, and Xiong Wen (David) Lou. 2015. "Formation of Yolk-Shelled Ni–Co Mixed Oxide Nanoprisms with Enhanced Electrochemical Performance for Hybrid Supercapacitors and Lithium Ion Batteries." *Advanced Energy Materials* 5 (21): 1500981. doi:10.1002/aenm.201500981.

Yu, Le, Genqiang Zhang, Changzhou Yuan, and Xiong Wen (David) Lou. 2013. "Hierarchical $NiCo_2O_4$@MnO_2 Core–shell Heterostructured Nanowire Arrays on Ni Foam as High-Performance Supercapacitor Electrodes." *Chemical Communications* 49 (2): 137–139. doi:10.1039/C2CC37117K.

Yuan, Chuanjun, Haibo Lin, Haiyan Lu, Endong Xing, Yusi Zhang, and Bingyao Xie. 2015. "Electrodeposition of Three-Dimensional ZnO@MnO_2 Core–shell Nanocables as High-Performance Electrode Material for Supercapacitors." *Energy* 93: 1259–1266. doi:10.1016/j.energy.2015.09.103.

Zhang, Fan, Tengfei Zhang, Xi Yang, Long Zhang, Kai Leng, Yi Huang, and Yongsheng Chen. 2013. "A High-Performance Supercapacitor-Battery Hybrid Energy Storage Device Based on Graphene-Enhanced Electrode Materials with Ultrahigh Energy Density." *Energy & Environmental Science* 6 (5): 1623–1632. doi:10.1039/C3EE40509E.

Zhang, Xin-Bo, Jun-Min Yan, Song Han, Hiroshi Shioyama, and Qiang Xu. 2009. "Magnetically Recyclable Fe@Pt Core–Shell Nanoparticles and Their Use as Electrocatalysts for Ammonia Borane Oxidation: The Role of Crystallinity of the Core." *Journal of the American Chemical Society* 131 (8): 2778–2779. doi:10.1021/ja808830a.

Zhang, Yiran, Yefeng Yang, Lingxiao Mao, Ding Cheng, Ziyue Zhan, and Jie Xiong. 2016. "Growth of Three-Dimensional Hierarchical Co_3O_4@$NiMoO_4$ Core-Shell Nanoflowers on Ni Foam as Electrode Materials for Hybrid Supercapacitors." *Materials Letters* 182: 298–301. doi:10.1016/j.matlet.2016.07.011.

Zhang, Yu, Wenping Sun, Xianhong Rui, Bing Li, Hui Teng Tan, Guilue Guo, Srinivasan Madhavi, Yun Zong, and Qingyu Yan. 2015. "One-Pot Synthesis of Tunable Crystalline Ni_3S_4@Amorphous MoS_2 Core/Shell Nanospheres for High-Performance Supercapacitors." *Small* 11 (30): 3694–3702. doi:10.1002/smll.201403772.

Zhao, Yan, Linfeng Hu, Shuyan Zhao, and Limin Wu. 2016. "Preparation of $MnCo_2O_4$@$Ni(OH)_2$ Core–Shell Flowers for Asymmetric Supercapacitor Materials with Ultrahigh Specific Capacitance." *Advanced Functional Materials* 26 (23): 4085–4093. doi:10.1002/adfm.201600494.

Zhao, Yong, Yuena Meng, and Peng Jiang. 2014. "Carbon@MnO_2 Core–Shell Nanospheres for Flexible High-Performance Supercapacitor Electrode Materials." *Journal of Power Sources* 259: 219–226. doi:10.1016/j.jpowsour.2014.02.086.

Zhong, C. J., and M. M. Maye. 2001. "Core–Shell Assembled Nanoparticles as Catalysts." *Advanced Materials* 13 (19): 1507–1511. doi:10.1002/1521-4095(200110)13:19<1507:: AID-ADMA1507>3.0.CO;2-#.

4 Hierarchical Nanostructures for Supercapacitors

Xiangchao Meng, Mengqing Wang,
Liang Wang, and Zisheng Zhang

CONTENTS

4.1 INTRODUCTION

With the depletion of fossil fuels and the rapid development of the global economy, it is urgent to find alternative technologies for energy storage and conversion. Electrochemistry has rapidly expanded into application areas of energy storage and conversion. Electrochemical technologies primarily include three categories, namely batteries, fuel cells, and electrochemical supercapacitors (ES). A significant attention has drawn on the development of ES in recent years, primarily due to their high power density, long lifecycle, and bridging function for the power/energy gap between traditional dielectric capacitors (high power output) and batteries/fuel cells (high energy storage) [1]. The major challenges of ES in practical use are low energy density and high production cost. One of the approaches to overcome these challenges is to develop new electrode materials.

The capacitance, C, can be defined as follows (Equation 4.1) [2]:

$$C = \frac{\varepsilon_r \varepsilon_0 A}{d} \tag{4.1}$$

where ε_r, ε_0, A, and d represent the dielectric constant of the electrolyte, the dielectric constant of the vacuum, the accessible surface area, and the effective thickness of the double layer, respectively. The maximum storage energy, E, within an ES can be described as follows (Equation 4.2):

$$E = \frac{1}{2}CV^2$$

(4.2)

where V is the cell voltage. It can be deduced that with an increase in the accessible surface area of the electrodes in an ES, both the capacitance and energy stored in the ES increase. One of the approaches to improve the surface area of the electrodes is to prepare materials with novel morphologies. Hierarchical nanostructures have the potential to be applied as electrode materials. As shown in Figure 4.1a, in the last two decades, the number of yearly publications on the subject of supercapacitors (SCs) has significantly increased, indicating the significance of SCs, and, as shown in Figure 4.1b, hierarchical materials used in the fabrication of ES have been massively reported in the last decade. However, compared with ES, the number of yearly publications on "hierarchical" materials-based SCs is still limited.

4.2 HIERARCHICAL MATERIALS

According to the energy storage mechanism, there are two categories of SCs, electrical double-layer capacitors (EDLCs) and pseudocapacitors [3]. In fact, most of the supercapacitor materials can be used as both EDLCs and pseudocapacitors. However, the capacity offered by the two kinds of capacitors varies greatly for different materials. The different hierarchical materials will be categorized (Table 4.1) and discussed in the next subsections.

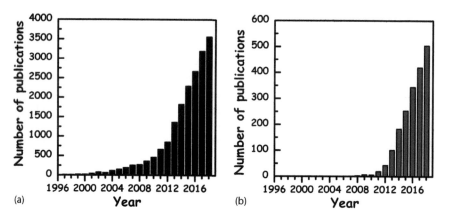

FIGURE 4.1   Number of yearly publications with topic keywords of (a) "supercapacitor" and (b) "supercapacitor" + "hierarchical." (Searched on February 14, 2019, using the SCOPUS database).

TABLE 4.1

Hierarchical Materials as Electrodes of SCs

Materials	Preparation Method	SSA (m^2g^{-1})	Electrolyte	Capacitance @ Current Density or Scan Rate	Retention (%)	Stability Cycles	Current Density or Scan Rate	References
HCSs @ CT-60[a]	Encapsulating heteroatom-doped hollow carbon spheres into one carbonaceous nanotube	318	6 M KOH	235 Fg^{-1} @ 0.2 Ag^{-1}	103	1000	5 Ag^{-1}	[6]
HPC[b]	Using shrimp shell intrinsic mineral scaffold ($CaCO_3$) as the self-template combined with KOH activation	1113	6 M KOH	348 Fg^{-1} @ 0.05 Ag^{-1}	93.6	1000	1 Ag^{-1}	[7]
HPNC-NS[c]	The metal salt activation-graphitization process	2494	1 M $LiPF_6$	242 Fg^{-1} @ 0.1 Ag^{-1}	92	10,000	2 Ag^{-1}	[8]
3D-HPCFs[d]	-	1286	6 M KOH	379 Fg^{-1} @ 0.2 Ag^{-1}	94	1000	5 Ag^{-1}	[9]
Mesoporous δ-MnO_2	Single-step, template-free chemical route	238	1 M Na_2SO_4	364 Fg^{-1} @ 1 Ag^{-1}	98	3000	20 Ag^{-1}	[13]
Ultrathin birnessite-type MnO_2 nanosheets	Rapid hydrothermal method	160	1 M Na_2SO_4	269 Fg^{-1} @ 0.3 Ag^{-1}	94	2000	1 Ag^{-1}	[14]
MnO_2 nanofibril/nanowire array	-	-	1 M $LiClO_4$	298 Fg^{-1} @ 50 mV s^{-1}; 329 Fg^{-1} @ 250 mV s^{-1}	85.2	1000	50–250 mV	[15]
Hydrated α-MnO_2 microspheres	-	-	1 M Na_2SO_4	365 Fg^{-1} @ 2 Ag^{-1}	100	2000	2 Ag^{-1}	[16]
HFC[e] Co_3O_4	Calcination of $Co(OH)_2$ hollow cages at 250°C	245.5	2 M KOH	948.9 Fg^{-1} @ 1 Ag^{-1}	High	10,000	10 Ag^{-1}	[20]

(*Continued*)

TABLE 4.1 (Continued)
Hierarchical Materials as Electrodes of SCs

Materials	Preparation Method	SSA (m^{2}g^{-1})	Electrolyte	Capacitance @ Current Density or Scan Rate	Stability			References
					Retention (%)	Cycles	Current Density or Scan Rate	
Enoki mushroom-like Co$_3$O$_4$ hierarchitectures	Simple reflux	-	6 M KOH	787 Fg^{-1} @ 1 Ag^{-1}	94.5	1000	10 Ag^{-1}	[21]
Hierarchical porous NiO nanotube arrays	Facile successive electrodeposition (ED) method	~165	2 M KOH	675 Fg^{-1} @ 2 Ag^{-1}	93.2	10,000	2 Ag^{-1}	[22]
NiO hollow nanospheres	Layer-by-layer self-assembly	92.99	2 M KOH	612.5 Fg^{-1} @ 0.5 Ag^{-1}	90	1000	0.5 Ag^{-1}	[23]
Flower-shape CuO	Alkaline solution oxidation method	119.6	1 M KOH	520 Fg^{-1} @ 1 Ag^{-1}	95.2	5000	1 Ag^{-1}	[24]
Rod-like Bi$_2$O$_3$	A facile one-step precipitation method	-	6 M KOH	1352 Fg^{-1} @ 0.1 Ag^{-1}	97.6	1000	20 mV s^{-1}	[26]
Flower-shaped NiCo$_2$O$_4$ microspheres	Rapid and template-free microwave-assisted heating reflux approach	148.5	6 M KOH	1006 Fg^{-1} @ 1 Ag^{-1}	93.2	1000	8 Ag^{-1}	[27]
Hierarchical NiCoO$_2$ nanotubes comprised ultrathin nanosheets	Template of polymeric nanotubes followed by a thermal annealing treatment	98.9	2 M KOH	1468 Fg^{-1} @ 2 Ag^{-1}	99.2	3000	10 Ag^{-1}	[29]
NiCo$_2$O$_4$@MnO$_2$ core–shell nanosheet	Two-step hydrothermal method	75.06	1 M NaOH	1595.1 Fg^{-1} @ 3 mA cm^{-2}	92.6	2000	40 m Ag^{-1}	[30]
Ni@FeCo$_2$O$_4$@MnO$_2$ core–shell nanosheet arrays	Two-step hydrothermal synthesis	-	PVA: poly(vinyl alcohol)-LiCl gel	2491.8 Fg^{-1} @ 4 mA cm^{-2} 2112.9 Fg^{-1} @ 40 mA cm^{-2}	87.2	5000	8 mA cm^{-2}	[31]

(Continued)

TABLE 4.1 (Continued)
Hierarchical Materials as Electrodes of SCs

Materials	Preparation Method	SSA (m^2g^{-1})	Electrolyte	Capacitance @ Current Density or Scan Rate	Stability			References
					Retention (%)	Cycles	Current Density or Scan Rate	
RuO_2-CNTs	Simple thermal treatment from the hydrated RuO_2 precursors at 250°C	-	1 M H_2SO_4	80.6 F cm^{-2} @ 2 mA cm^{-2}	-	-	-	[32]
H-PANI^f	-	-	1 M H_2SO_4	520 Fg^{-1} @ 0.5 Ag^{-1}	57	1000	2 Ag^{-1}	[33]
Molybdenum sulfide hierarchical nanospheres	Hydrothermal method	-	-	368 Fg^{-1} @ 5 mV s^{-1}	96.5	5000	0.8 m Ag^{-1}	[34]
Hierarchical porous carbon aerogel	A five-step synthesis method	1892.4	6 M KOH	142.1 Fg^{-1} @ 0.5 Ag^{-1}	93.9	5000	1 Ag^{-1}	[35]
Graphene /PANI^g nanostructure free-standing composite films	1. Microemulsion polymerization 2. Hierarchical organization 3. Targeted self-assembly 4. In situ chemical reduction	-	-	448 Fg^{-1} @ 0.5 Ag^{-1}	81	5000	2 Ag^{-1}	[36]
Hierarchical $NiCo_2O_4$@ $NiCo_2O_4$ core–shell nanoflake arrays	Two-step solution-based method^h	-	2 M KOH	2.20 F cm^{-2} @ 5 m Ag^{-1} 1.55 F cm^{-2} @ 40 m Ag^{-1}	98.6	4000	5 m Ag^{-1}	[37]
$NiCo_2O_4$@MnO_2 core–shell heterostructured nanowire arrays on nickel foam	Two-step solution route coupled with a post-calcination treatment	-	1 M LiOH	3.31 F cm^{-2} @ 2 mA cm^{-2}	88	1000	10 mA cm^{-2}	[38]
HPCF based on banana peel Hierarchical porous carbon foams	Self-template strategy	1650	6 M KOH	206 Fg^{-1} @ 1 Ag^{-1}	98.3	1000	1 Ag^{-1}	[39]

(Continued)

TABLE 4.1 (Continued)
Hierarchical Materials as Electrodes of SCs

Materials	Preparation Method	SSA (m²g⁻¹)	Electrolyte	Capacitance @ Current Density or Scan Rate	Stability Retention (%)	Stability Cycles	Stability Current Density or Scan Rate	References
Amorphous MnO_2, Mn_3O_4 nanocrystallites hierarchical nanostructure	Two-step electrochemical deposition	-	0.5 M $CaCl_2$	470 Fg^{-1} @ 0.5 V	95.7	10,000	25 mV s⁻¹	[40]
$MnMoO_4$/$CoMoO_4$	Simple refluxing method	54.06	2 M NaOH	187.1 Fg^{-1} @ 1 Ag^{-1}	98	1000	1 Ag^{-1}	[41]
Functionalized carbon spheres/graphene nanosheets	Self-assembly approach	-	6 M KOH	198 Fg^{-1} @ 175 m Ag^{-1}	95	1000	175 m Ag^{-1}	[42]
Bi_2O_3 nanobelts	ED	-	1 M Na_2SO_4	250 Fg^{-1} @ 100 mV s⁻¹	~100	1000	100 mV s⁻¹	[43]
Nanocomposites of interpenetrating CNT @ V_2O_5 nanowire networks	Simple in situ hydrothermal process	125	1 M Na_2SO_4	440 Fg^{-1} @ 0.25 Ag^{-1}	-	-	-	[44]

a Hollow carbon spheres carbonaceous nanotubes.
b Nitrogen-doped graded porous carbon.
c Hierarchical porous nitrogen-doped carbon nanosheets.
d 3D hierarchically porous carbon-CNT-graphene ternary all-carbon foam.
e Hollow fluffy cages.
f 3D hierarchical PANI micro/nanostructure.
g Polyaniline.
h Hydrothermal process and chemical bath deposition.

4.2.1 HIERARCHICALLY NANOSTRUCTURED CARBON MATERIALS FOR SCs

Carbon materials have a long history of use as a material for SCs [4]. According to the characteristics of EDLC, the capacity of a SC is related to the electrode potential and specific surface area (SSA). Therefore, carbon materials with high SSA can significantly increase the capacity of the SC [5]. In addition, carbon materials are cheaper than transition metal, are environmentally friendly, and have excellent cycle stability and a wide electrochemical window. Thus, over these years, researchers have continuously studied of carbon materials in SCs. Because of the discovery of graphene in 2004, there seems to be a new upsurge in the research in this direction in recent years. At present, well-developed carbon materials are activated carbon, carbon fiber, carbon aerogel, carbon nanotubes, and graphene. However, due to the limitations of EDLC characteristics, the defect of the carbon-based EDLC is also remarkable [5]; the specific capacitance is lower than that of the pseudocapacitors, which will be discussed later.

In the modification of carbon materials, two main methods are used; the first is changing their morphology to increase the SSA and second is activating or doping the surface of the carbon material to introduce a functional group.

Chen et al. [6] designed a spheres-in-tube carbon nanostructure with hierarchical porosity through confined assembly of hollow carbon spheres in carbonaceous nanotube. They used different dual templates and synthesized three different products: HCSs@CT-60, HCSs@CT-90, and HCSs@CT-110, where the number refers to the particle size of SiO_2. HCSs@CT-60 has the highest S_{BET}, which is 318 m^2g^{-1}, followed by HCSs@CT-90 (241 m^2g^{-1}) and HCSs@CT-110 (208 m^2g^{-1}). Moreover, these three products show the same trend in electrical cyclic voltammetry (CV) measurements. At both low current density (0.2 Ag^{-1}) and high current density (20 Ag^{-1}), HCSs@CT-60 exhibits excellent specific capacitance of 235 Fg^{-1} and 156 Fg^{-1}, respectively (6 M KOH electrolyte), which is much higher than the other two products. In terms of electrochemical stability, at the 1000th cycle, the specific capacity of HCSs@CT-60 increases to 103% of the first cycle, which indicates that the continuous impregnation of electrolyte ions in the micro/mesopore of the electrode increases during charge and discharge; after 10,000 cycles, it showed very good stability and no significant drop in specific capacity was observed.

Gao et al. [7] using shrimp shell as a material and their intrinsic mineral scaffold as a template prepared a nitrogen-doped hierarchical porous carbon (HPC). They prepared three samples at 600°C, 700°C, and 800°C, respectively. Among them, the sample prepared at 800°C has the largest SSA of 1343 m^2g^{-1}, while C/KOH-700 shows the highest specific capacities, which are 348, 328, 320, 300, 295, and 290 Fg^{-1} at 0.05, 0.1, 0.2, 0.5, 1.0, and 2.0 Ag^{-1}, respectively. Similarly, Hou et al. [8] prepared hierarchical porous nitrogen-doped carbon (HPNC) nanosheets (NS) through simultaneous activation and graphitization of biomass-derived natural silk. This material exhibits an extremely high SSA of 2494 m^2g^{-1} and a high volume of hierarchical pores of 2.28 cm^3g^{-1}. This HPNC-NS has a specific capacitance of 242 Fg^{-1} (113 F/cm^3) at 0.1 Ag^{-1} in 1M $LiPF_6$, which is much larger than that of commercial ACs. Regarding the stability, HPNC-NS retains 92% capacity of the first cycle after 10,000 cycles at 2 Ag^{-1}.

You et al. [9] prepared a ternary full carbon foam with three-dimensional (3D) microporous and mesoporous structures. They combined the hierarchically porous carbon, CNT, and graphene in a unique process to produce this new material. It has an SSA of 1286 m^2g^{-1} and a specific capacitance of 379 Fg^{-1} at a current density of 0.2 Ag^{-1} in a 6 M KOH aqueous solution.

4.2.2 HIERARCHICALLY NANOSTRUCTURED TMOs FOR SCs

Different from the mechanism of EDLC, pseudocapacitors mainly originate from the rapid redox reaction on the electrode surface. The specific capacitance of pseudocapacitors is much higher than that of EDLC, and the transition metal oxides (TMOs) are the main material for pseudocapacitors [10]. However, the TMO electrode has poor conductivity and high internal resistance; thus, it cannot be applied in alternating current. In addition, as the morphology changes after charge and discharge, the cycle performance is not good. In recent years, due to the unique structure of hierarchically nanostructured materials that can help overcome the shortcomings of traditional TMO SCs, it has gradually become a hotspot in the research on SCs [11]. Next, we will systematically introduce the synthetic method and electrochemical properties of hierarchically nanostructured materials: single-phase TMO materials and multiphase TMO materials.

4.2.2.1 Single-Phase TMOs Materials

Although RuO_2 has been considered as the best electrode material of TMOs, its application is limited due to the high cost. Therefore, manganese oxides, cobalt oxides, copper oxide, nickel oxide and so on, which are cheap and rich resources but have similar electrochemical properties with RuO_2, have gradually received researchers' attention.

4.2.2.1.1 Manganese Oxides

Manganese dioxide is abundant in natural resources, low in price, and has high theoretical specific capacitance [12]. It is one of the interesting materials in the research on TMOs for SCs. In addition, it is rich in polymorphisms, which has also drawn researchers's attention to study it as an ideal SC material.

Bag et al. [13] used a single-step, template-free chemical route to synthesize hierarchical mesoporous δ-MnO_2. This kind of MnO_2 has a 3D mesoporous structure, and its growth kinetics and surface morphology can be controlled by adjusting the concentration of Br^-. It has a large SSA of 238 m^2g^{-1}, and the average pore size and volume are 36.14 Å and 0.567 ccg^{-1}, respectively. In electrochemical testing, this kind of material exhibits a specific capacitance of 364 Fg^{-1} at a current density of 1 Ag^{-1} in 1 M Na_2SO_4 aqueous solution and retains 100% initial specific capacitance after 3000 cycles.

Zhang et al. [14] developed a rapid hydrothermal method to synthesize hierarchical porous nanostructures assembled from ultrathin birnessite-type MnO_2 nanosheets. This method does not require the use of template and surfactant. They heated the mixture of $MnSO_4 \cdot H_2O$ (2 mmol) and $K_2S_2O_8$ (2 mmol) in a Teflon-lined stainless steel autoclave at 160°C for 30 min, and the obtained product shows good properties. The as-prepared MnO_2 has a large $SSA_{(BET)}$ of 160 m^2g^{-1}, and uniform pore size distribution of 5.4 nm. Moreover, the specific capacitance can reach 269 Fg^{-1} at a current density of 0.3 Ag^{-1} in 1 M Na_2SO_4 solution. The stability is also excellent, which remain at 94% after 2000 cycles at 1 Ag^{-1}.

Duay's group [15] synthesized a special hierarchical MnO_2 nanofibril/nanowire array. This kind of material shows different charge storage mechanisms in aqueous electrolyte and acetonitrile electrolyte. In aqueous electrolyte, it exhibits a specific capacitance of 298 Fg^{-1} at 50 mV s^{-1} and 170 Fg^{-1} at 250 mV s^{-1} and retains 85.2% capacitance of the first cycle after 1000 charge and discharge cycles.

Sumboja's group [16] prepared hierarchical nanostructures of hydrated α-manganese dioxide and characterized their properties. At 2 Ag^{-1}, the results showed that the specific capacitance of this material is 356 Fg^{-1}, and there is no capacitance degradation observed in the first 2000 cycles.

4.2.2.1.2 Cobalt Oxides

Similar to manganese oxide, cobalt oxides are also an ideal SC material [17]. They are environment-friendly, have low cost, have high catalytic activity, and so on, and cobalt oxides have an ultra-high theoretical specific capacitance of 3560 Fg^{-1} [18]. In recent years, scientists have synthesized a series form of cobalt oxides, such as nanoparticles, nanorods, nanowires, nanosheets, and porous nanostructures [19]. However, the specific capacitance of cobalt oxide SCs is difficult to reach the theoretical value in practice.

Zhou et al. [20] synthesized hollow fluffy cages (HFC) constructed of ultrathin nanosheets through the formation of $Co(OH)_2$ hollow cages, and then calcine at 250°C. The product annealed at 250°C has a large SSA of 245.2 m^2g^{-1}. In the electrochemical evaluation, this material shows a good specific capacitance of 948.9 and 536.8 Fg^{-1} at 1 and 40 Ag^{-1}, respectively, and an excellent stability of 10,000 cycles at 10 Ag^{-1} in 2 KOH electrolyte.

Luo's group [21] used a simple reflux method to prepare a 3D enoki mushroom-like Co_3O_3 hierarchitectures, which is constructed by one-dimensional (1D) nanowires with a diameter of 3.2 μm. In 6 M KOH solution, this material shows a high specific capacitance of 787 Fg^{-1} at 1 Ag^{-1}. Even at 10 Ag^{-1}, the capacitance remains at 94.5% and 76.6% of the first cycle after 1000 and 4000 deep cycles, respectively. The 3D enoki mushroom-like structure significantly increases the ion diffusion rate and electrode stability.

4.2.2.1.3 Nickel Oxide

Similar to the TMOs mentioned above, nickel oxides are also inexpensive, environmentally friendly, and have high theoretical capacity; they are also one of the most widely studied materials in the SCs field in recent years.

By using ZnO nanorod template, Gao et al. [22] fabricated a 3D hierarchical porous NiO nanotube arrays on a nickel foam by a facile successive electrodeposition method. The as-prepared nanotube arrays, which consist of interconnected branch nanoflakes of approximately 10 nm, have a diameter of approximately 170 nm. The electrochemical evaluation was performed in a 2 M KOH solution, and the result showed that it has a specific capacitance of 675 and 569 Fg^{-1} at 2 and 40 Ag^{-1}, respectively, and excellent capacitance retention of 93.2% after 10,000 cycles at 2 Fg^{-1}.

Yang's group [23] prepared a kind of multishelled NiO hollow nanospheres by a layer-by-layer self-assembly method. The results showed that when 19 mg sample is coated on the foamed nickel, the prepared product has the largest SSA (92.99 m^2g^{-1}) and exhibits the best electrochemical performance. The highest specific capacitance of the material is 612.5 Fg^{-1} at 0.5 Ag^{-1}, and the stability test shows that capacitance remains at 90.1% after 1000 cycles.

4.2.2.1.4 Copper Oxide

Copper oxide is another candidate material for SCs. It is nontoxic, abundant in nature, and easy to prepare. As a result, the cost of copper oxides is low.

Lu et al. [24] used an alkaline solution oxidation method to synthesize a high-performance copper oxide electrode that can be used in SCs. The as-prepared material has a unique flower-shape nanostructure, which contributes to a large SSA. The highest SSA of the material is 119.6 m^2g^{-1}. In the electrochemical evaluation, this material exhibits a specific capacitance of 520 and 405 Fg^{-1} at 1 and 60 Ag^{-1}, respectively. It can also retain 95.2% capacitance of the first cycle after 5000 charge and discharge cycles.

4.2.2.1.5 Bismuth Oxide

Similar to manganese oxide, bismuth oxide also has a very rich crystal morphology and unique properties [25]. In recent years, researchers have been interested in applying bismuth oxide to SCs, but, currently, there are only a few reports in the literature.

By using a surfactant, Su et al. [26] synthesized a rod-like Bi_2O_3 by a facile one-step precipitation method. The test results showed that the material prepared by using P123 as a surfactant has good electrochemical properties. The specific capacitance of this rod-like Bi_2O_3 can reach 1352 Fg^{-1} at 0.1 Ag^{-1}. Moreover, in the stability test, the capacitance reduced only by 2.4% of the first cycle after 1000 charge and discharge cycles.

4.2.2.1.6 Multi-Metal Oxide Materials

The interaction of different metal ions is very complex; thus, they sometimes provide better electrochemical properties than single metal oxides. Because of the combination of improved electrical conductivity and an effective porous structure, binary metal oxides with a 3D superstructure are considered as the ideal electrode material for SCs [27]. Nickel-cobalt binary metal oxides are one of the most common bimetallic oxides with excellent electrochemical properties [28].

Lei's group [27] synthesized a 3D hierarchical flower-shaped nickel cobaltite microspheres through a rapid and template-free microwave-assisted heating reflux approach. The as-prepared microspheres are constructed by ultrathin nanopetals with a thickness of approximately 15 nm. They also have a large SSA of 148.5 m^2g^{-1} and narrow pore size distribution (5–10 nm). This material shows a specific capacitance of 1006 Fg^{-1} at 1 Ag^{-1} and relatively good stability, which retains 93.2% capacitance after 1000 cycles at 8 Ag^{-1}.

Xu et al. [29] used polymeric nanotubes as a template synthesizing hierarchical $NiCoO_2$ nanosheets nanotubes. These unique nanosheets nanotubes show an excellent specific capacitance of 1468, 1352, 1233, 1178, 1020, and 672 at current densities of 2, 4, 8, 10, 20, and 40 Ag^{-1}, respectively. Regarding stability, after 3000 charge and discharge cycles at a high current density of 10 Ag^{-1}, the capacitance remains at 99.2%. This enhanced performance is attributed to the porous and layered structure of $NiCoO_2$ nanosheet nanotubes, which provides more contact area between the active material and the electrolyte ions while buffering large volume changes during rapid charge/discharge.

4.2.2.2 Multi-Phase TMO Materials

Composites prepared by combining two or more TMOs may exhibit better electrochemical performance than a single material due to the synergistic effect of different TMOs. Therefore, some researchers have attempted to compound a variety of TMOs and perform electrochemical evaluation. This section will present some of the current results.

Bao's group [30] designed a facile two-step hydrothermal method to synthesize novel $NiCo_2O_4@MnO_2$ core–shell nanosheet arrays on hybrid composite/nickel foam integrated electrode. Characterization test results showed that the synthesized material has a complex core–shell nanostructure. Because of this special structure, this composite exhibits a high specific capacitance, which is 2.39 F cm^{-2} (1595.1 Fg^{-1}) at a current density of 3 mA cm^{-2}. The capacitance loss is only 7.4% after 2000 cycles. Similarly, Yuan et al. [28] synthesized $NiCo_2O_4@MnO_2$ core–shell heterostructured nanowire arrays on a nickel foam using a two-step solution route coupled with a post-calcination treatment. The specific capacitance of this material is 3.31 F cm^{-2} at 2 mA cm^{-2}, and it can remain 88% after 1000 cycles at 10mA cm^{-2}.

Lin et al. [31] synthesized a composite of three compounds. They developed nickel foam-hierarchical 3D iron cobaltate nanosheet arrays framework, covering the surface with thin nanosheets of manganese dioxide ($Ni@FeCo_2O_4@MnO_2$). This structure increases the energy density of SCs made of this material. The specific capacitance can reach 2.20 and 1.55 cm^{-2} at 5 and 40 m Ag^{-1}, respectively. There is only 1.4% capacitance loss observed after 4000 cycles at 5 m Ag^{-1}.

4.2.3 CARBON/TMO COMPOSITE MATERIALS

As carbon-based materials and TMOs have their own advantages and disadvantages as SC materials, researchers have combined these two materials to prepare a new type of carbon/TMO SC. This type of material improves the conductivity and cycle stability of materials by adding carbon-based materials to TMOs, while TMOs can provide considerable pseudocapacitors for high energy density SCs. Currently, carbon/TMOs composites materials are a direction of SCs research.

Lee' group [32] used a hydrated cerium oxide precursor as a raw material and performed a simple heat treatment at 250°C. Successfully synthesized multistage-driven single-crystal Ruthenium oxide (RuO_2) nanorods on a small number of wall carbon nanotubes (CNTs). This material shows a specific capacitance of 80.6 F cm^{-2}

at 2 mA cm^{-2}. It can be noted that such a well-structured 1D heterostructure nanostructure will be a promising direction for preparing highly active electrochemical electrode materials.

4.3   CONCLUSIONS

Hierarchical nanostructures have been widely applied in the field of SCs. Recent publications on SCs were summarized and discussed in this chapter. Various materials, especially TMOs, are fabricated with the hierarchical morphology, which were comprehensively introduced in this chapter. It can be concluded that hierarchical nanostructures prepared for SCs exhibit promising prospects. However, more efforts are required to further improve their performance. To fabricate a SC with high energy density, long cycling life, low cost, and high safety will be of interest for future research.

REFERENCES

1. G. Wang, L. Zhang, J. Zhang, A review of electrode materials for electrochemical supercapacitors, *Chem. Soc. Rev.* 41 (2012) 797–828. doi:10.1039/C1CS15060J.
2. A. González, E. Goikolea, J.A. Barrena, R. Mysyk, Review on supercapacitors: Technologies and materials, *Renew. Sustain. Energy Rev.* 58 (2016) 1189–1206. doi:10.1016/j.rser.2015.12.249.
3. Y. Wang, Y. Song, Y. Xia, Electrochemical capacitors: Mechanism, materials, systems, characterization and applications, *Chem. Soc. Rev.* 45 (2016) 5925–5950. doi:10.1039/C5CS00580A.
4. E. Frackowiak, Carbon materials for supercapacitor application, *Phys. Chem. Chem. Phys.* 9 (2007) 1774–1785. doi:10.1039/b618139m.
5. L.L. Zhang, X.S. Zhao, Carbon-based materials as supercapacitor electrodes, *Chem. Soc. Rev.* 38 (2009) 2520–2531. doi:10.1039/b813846j.
6. X. Yang, Z. Chen, Y. Ge, Z. Zhu, S. Ye, S.D. Evans, Y. Tu, Energy storage: confined assembly of hollow carbon spheres in carbonaceous nanotube: A spheres-in-tube carbon nanostructure with hierarchical porosity for high-performance supercapacitor (Small 19/2018), *Small.* 14 (2018) 1870089. doi:10.1002/smll.201870089.
7. F. Gao, J. Qu, C. Geng, G. Shao, M. Wu, Self-templating synthesis of nitrogen-decorated hierarchical porous carbon from shrimp shell for supercapacitors, *J. Mater. Chem. A* 4 (2016) 7445–7452. doi:10.1039/c6ta01314g.
8. J. Hou, C. Cao, F. Idrees, X. Ma, Hierarchical porous nitrogen-doped carbon nanosheets derived from silk for ultrahigh-capacity battery anodes and supercapacitors, *ACS Nano.* 9 (2015) 2556–2564. doi:10.1021/nn506394r.
9. B. You, J. Jiang, S. Fan, Three-dimensional hierarchically porous all-carbon foams for supercapacitor, *ACS Appl. Mater. Interfaces* 6 (2014) 15302–15308. doi:10.1021/am503783t.
10. X.F. Lu, G.R. Li, Y.X. Tong, A review of negative electrode materials for electrochemical supercapacitors, *Sci. China Technol. Sci.* 58 (2015) 1799–1808. doi:10.1007/s11431-015-5931-z.
11. Z. Yu, L. Tetard, L. Zhai, J. Thomas, Supercapacitor electrode materials: Nanostructures from 0 to 3 dimensions, *Energy Environ. Sci.* 8 (2015) 702–730. doi:10.1039/c4ee03229b.
12. W. Wei, X. Cui, W. Chen, D.G. Ivey, Manganese oxide-based materials as electrochemical supercapacitor electrodes, *Chem. Soc. Rev.* 40 (2011) 1697–1721. doi:10.1039/c0cs00127a.

13. S. Bag, C.R. Raj, Facile shape-controlled growth of hierarchical mesoporous δ-MnO$_2$ for the development of asymmetric supercapacitors, *J. Mater. Chem. A*. 4 (2016) 8384–8394. doi:10.1039/c6ta01879c.

14. X. Zhang, P. Yu, H. Zhang, D. Zhang, X. Sun, Y. Ma, Rapid hydrothermal synthesis of hierarchical nanostructures assembled from ultrathin birnessite-type MnO$_2$ nanosheets for supercapacitor applications, *Electrochim. Acta* 89 (2013) 523–529. doi:10.1016/j.electacta.2012.11.089.

15. J. Duay, S.A. Sherrill, Z. Gui, E. Gillette, S.B. Lee, Self-limiting electrodeposition of hierarchical MnO$_2$ and M(OH)$_2$/MnO$_2$ nanofibril/nanowires: Mechanism and supercapacitor properties, *ACS Nano* 7 (2013) 1200–1214. doi:10.1021/nn3056077.

16. A. Sumboja, U.M. Tefashe, G. Wittstock, P.S. Lee, Monitoring electroactive ions at manganese dioxide pseudocapacitive electrodes with scanning electrochemical microscope for supercapacitor electrodes, *J. Power Sources* 207 (2012) 205–211. doi:10.1016/j.jpowsour.2012.01.153.

17. X.C. Dong, H. Xu, X.W. Wang, Y.X. Huang, M.B. Chan-Park, H. Zhang, L.H. Wang, W. Huang, P. Chen, 3D graphene-cobalt oxide electrode for high-performance supercapacitor and enzymeless glucose detection, *ACS Nano* 6 (2012) 3206–3213. doi:10.1021/nn300097q.

18. Y. Gao, G. Wang, J. Yin, D. Cao, S. Chen, Electrochemical capacitance of Co$_3$O$_4$ nanowire arrays supported on nickel foam, *J. Power Sources* 195 (2009) 1757–1760. doi:10.1016/j.jpowsour.2009.09.048.

19. M. Zheng, X. Xiao, L. Li, P. Gu, X. Dai, H. Tang, Q. Hu, H. Xue, H. Pang, Hierarchically nanostructured transition metal oxides for supercapacitors, *Sci. China Mater*. 61 (2018) 185–209. doi:10.1007/s40843-017-9095-4.

20. X. Zhou, X. Shen, Z. Xia, Z. Zhang, J. Li, Y. Ma, Y. Qu, Hollow fluffy Co$_3$O$_4$ cages as efficient electroactive materials for supercapacitors and oxygen evolution reaction, *ACS Appl. Mater. Interfaces* 7 (2015) 20322–20331. doi:10.1021/acsami.5b05989.

21. H. Yuan, D. Xiao, W. Yang, Y. Lei, J. Li, F. Luo, Three-dimensional enoki mushroom-like Co$_3$O$_4$ hierarchitectures constructed by one-dimension nanowires for high-performance supercapacitors, *Electrochim. Acta* 135 (2014) 495–502. doi:10.1016/j.electacta.2014.04.075.

22. F. Cao, G.X. Pan, X.H. Xia, P.S. Tang, H.F. Chen, Synthesis of hierarchical porous NiO nanotube arrays for supercapacitor application, *J. Power Sources* 264 (2014) 161–167. doi:10.1016/j.jpowsour.2014.04.103.

23. W. Zhang, F. Xu, B. Pei, Z. Yang, X. Zhu, Z. Mei, Controllable preparation of multishelled NiO hollow nanospheres via layer-by-layer self-assembly for supercapacitor application, *J. Power Sources* 246 (2013) 24–31. doi:10.1016/j.jpowsour.2013.07.057.

24. Y. Lu, H. Yan, K. Qiu, J. Cheng, W. Wang, X. Liu, C. Tang, J.K. Kim, Y. Luo, Hierarchical porous CuO nanostructures with tunable properties for high performance supercapacitors, *RSC Adv*. 5 (2015) 10773–10781. doi:10.1039/c4ra16924g.

25. L. Li, X. Zhang, Z. Zhang, M. Zhang, L. Cong, Y. Pan, S. Lin, A bismuth oxide nanosheet-coated electrospun carbon nanofiber film: A free-standing negative electrode for flexible asymmetric supercapacitors, *J. Mater. Chem. A*. 4 (2016) 16635–16644. doi:10.1039/c6ta06755g.

26. D. Yuan, N. Xia, Q. Liang, S. Cao, X. Huang, H. Su, J. Yan, Controllable growth of Bi$_2$O$_3$ with rod-like structures via the surfactants and its electrochemical properties, *J. Appl. Electrochem*. 44 (2014) 735–740. doi:10.1007/s10800-014-0681-3.

27. Y. Lei, J. Li, Y. Wang, L. Gu, Y. Chang, H. Yuan, D. Xiao, Rapid microwave-assisted green synthesis of 3D hierarchical flower-shaped NiCo$_2$O$_4$ microsphere for high-performance supercapacitor, *ACS Appl. Mater. Interfaces* 6 (2014) 1773–1780. doi:10.1021/am404765y.

28. C. Yuan, J. Li, L. Hou, X. Zhang, L. Shen, X.W. Lou, Ultrathin mesoporous $NiCo_2O_4$ nanosheets supported on Ni foam as advanced electrodes for supercapacitors, *Adv. Funct. Mater.* 22 (2012) 4592–4597. doi:10.1002/adfm.201200994.

29. X. Xu, H. Zhou, S. Ding, J. Li, B. Li, D. Yu, The facile synthesis of hierarchical $NiCoO_2$ nanotubes comprised ultrathin nanosheets for supercapacitors, *J. Power Sources* 267 (2014) 641–647. doi:10.1016/j.jpowsour.2014.05.077.

30. F. Bao, Z. Zhang, W. Guo, X. Liu, Facile synthesis of three dimensional $NiCo_2O_4$@ MnO_2 core-shell nanosheet arrays and its supercapacitive performance, *Electrochim. Acta* 157 (2015) 31–40. doi:10.1016/j.electacta.2015.01.060.

31. L. Lin, S. Tang, S. Zhao, X. Peng, N. Hu, Hierarchical three-dimensional $FeCo_2O_4$@ MnO_2 core-shell nanosheet arrays on nickel foam for high-performance supercapacitor, *Electrochim. Acta* 228 (2017) 175–182. doi:10.1016/j.electacta.2017.01.022.

32. Y. Lee, B. Kim, H.J. Jung, J.H. Shim, Y. Lee, C. Lee, J.M. Baik, W. Kim, M.H. Kim, Hierarchically grown single crystalline RuO_2 nanorods on vertically aligned few-walled carbon nanotubes, *Mater. Lett.* 89 (2012) 115–117. doi:10.1016/j.matlet.2012.08.097.

33. X. Wang, M. Xu, Y. Fu, S. Wang, T. Yang, K. Jiao, A highly conductive and hierarchical PANI micro/nanostructure and its supercapacitor application, *Electrochim. Acta* 222 (2016) 701–708. doi:10.1016/j.electacta.2016.11.026.

34. M.S. Javed, S. Dai, M. Wang, D. Guo, L. Chen, X. Wang, C. Hu, Y. Xi, High performance solid state flexible supercapacitor based on molybdenum sulfide hierarchical nanospheres, *J. Power Sources* 285 (2015) 63–69. doi:10.1016/j.jpowsour.2015.03.079.

35. P. Hao, Z. Zhao, J. Tian, H. Li, Y. Sang, G. Yu, H. Cai, H. Liu, C.P. Wong, A. Umar, Hierarchical porous carbon aerogel derived from bagasse for high performance supercapacitor electrode, *Nanoscale* 6 (2014) 12120–12129. doi:10.1039/c4nr03574g.

36. M. Hassan, K.R. Reddy, E. Haque, S.N. Faisal, S. Ghasemi, A.I. Minett, V.G. Gomes, Hierarchical assembly of graphene/polyaniline nanostructures to synthesize free-standing supercapacitor electrode, *Compos. Sci. Technol.* 98 (2014) 1–8. doi:10.1016/j.compscitech.2014.04.007.

37. X. Liu, S. Shi, Q. Xiong, L. Li, Y. Zhang, H. Tang, C. Gu, X. Wang, J. Tu, Hierarchical $NiCo_2O_4$@$NiCo_2O_4$ core/shell nanoflake arrays as high-performance supercapacitor materials, *ACS Appl. Mater. Interfaces* 5 (2013) 8790–8795. doi:10.1021/am402681m.

38. L. Yu, G. Zhang, C. Yuan, X.W. (David) Lou, Hierarchical $NiCo_2O_4$@MnO_2 core–shell heterostructured nanowire arrays on Ni foam as high-performance supercapacitor electrodes, *Chem. Commun.* 49 (2012) 137–139. doi:10.1039/c2cc37117k.

39. Y. Lv, L. Gan, M. Liu, W. Xiong, Z. Xu, D. Zhu, D.S. Wright, A self-template synthesis of hierarchical porous carbon foams based on banana peel for supercapacitor electrodes, *J. Power Sources* 209 (2012) 152–157. doi:10.1016/j.jpowsour.2012.02.089.

40. C.C. Hu, C.Y. Hung, K.H. Chang, Y.L. Yang, A hierarchical nanostructure consisting of amorphous MnO_2, Mn_3O_4 nanocrystallites, and single-crystalline MnOOH nanowires for supercapacitors, *J. Power Sources* 196 (2011) 847–850. doi:10.1016/j.jpowsour.2010.08.001.

41. L.Q. Mai, F. Yang, Y.L. Zhao, X. Xu, L. Xu, Y.Z. Luo, Hierarchical $MnMoO_4$/$CoMoO_4$ heterostructured nanowires with enhanced supercapacitor performance, *Nat. Commun.* 2 (2011). doi:10.1038/ncomms1387.

42. C.X. Guo, C.M. Li, A self-assembled hierarchical nanostructure comprising carbon spheres and graphene nanosheets for enhanced supercapacitor performance, *Energy Environ. Sci.* 4 (2011) 4504–4507. doi:10.1039/c1ee01676h.

43. F.L. Zheng, G.R. Li, Y.N. Ou, Z.L. Wang, C.Y. Su, Y.X. Tong, Synthesis of hierarchical rippled Bi_2O_3 nanobelts for supercapacitor applications, *Chem. Commun.* 46 (2010) 5021–5023. doi:10.1039/c002126a.

44. Y. Peng, D. Weng, Q. Xiao, Y. Qin, F. Wei, H. Li, Y. Lu, Z. Chen, X. Wang, Design and synthesis of hierarchical nanowire composites for electrochemical energy storage, *Adv. Funct. Mater.* 19 (2009) 3420–3426. doi:10.1002/adfm.200900971.

5 Vertically Aligned 1D and 2D Nanomaterials for High-Frequency Supercapacitors

Wenyue Li, Xuan Pan, and Zhaoyang Fan

CONTENTS

5.1 INTRODUCTION

Electrochemical supercapacitors (ECs) generally have much higher rate capability than batteries, making them as ideal high-power sources either operating independently or cooperating with batteries as a pulse power buffer. However, in comparison to electrolytic capacitors that have high frequency responses (from kHz to MHz), these conventional ECs work at very low frequencies, normally below 1 Hz, preventing them from being used in many applications such as current ripple filtering and very short pulse generation, which have been fulfilled by the traditional aluminum electrolytic capacitors (AECs). However, with a low capacitance density and thus a bulky volume, AECs have become an issue for downscaling the circuit board profile to developing compacted electronics. Thus, developing ultrafast ECs that maintain their relatively high capacitance density while efficiently working at tens or even kilohertz (kHz) frequencies, herein, called as high-frequency ECs (HF-ECs), has been considered as an effective strategy to address the aforementioned issues and have attracted more attention of researchers (Fan, Islam, and Bayne 2017, Li, Islam et al. 2019). This chapter introduces studies in this exciting niche area of EC development, with a focus on novel electrodes based on 1D and 2D vertically aligned nanostructures.

Depending on the attainable capacitance density and frequency response that usually contradicts with each other, it is envisioned that HF-ECs might find different applications. For those HF-ECs that can efficiently work at above hundreds of hertz and show reasonable capacitance densities, they will be a promising candidate in the substitution of the bulky AECs for line-frequency alternating current (AC) filtering; while for those relatively slower HF-ECs that have large capacitance densities, they will be competitive in high-current pulse applications such as flashing light, communication, and many others.

In conventional ECs, activated carbon and carbon black are commonly used as electrode materials because of their large specific surface area (SSA), low cost, and chemical inertness. Although the tortuous micropores in these carbon materials provide them a large SSA, these same micropores also limit the electronic conductivity of the electrodes and the electrolytic ion migration rate in the pores, leading to frequency responses of the conventional ECs below 1 Hz. Early attempts on fast ECs were carried out using membrane electrodes fabricated from well-dispersed carbon nanotubes (CNTs), which provided a characteristic frequency, defined as the frequency when the phase angle reaches −45°, of approximately 500 Hz (Niu et al. 1997, Du and Pan 2006). In 2010, there was a breakthrough in the development of kHz-frequency ECs when vertically oriented graphene grown on Ni foil was used as the electrode, demonstrating a characteristic frequency as high as 15 kHz (Miller, Outlaw, and Holloway 2010). Since then, steady progresses have been made in this HF-EC niche area (Islam et al. 2017).

A variety of carbon-based nanostructures have been investigated for HF-ECs by exploiting their large SSA, high electronic conductivity, and regular pore geometry that facilitate fast migration of electrolytic ions to achieve high-frequency response with a reasonably large capacitance density. Surprisingly, nanostructures based on metallic-like oxides, nitrides, carbides, and others, because of their extremely large electronic conductivity, might also be electrode candidates for power HF-ECs. Though, the pseudocapacitive charge storing in these materials, based on Faraday redox reaction mechanism, is intrinsically slower than the electrical double-based charge storing in carbon-based materials. One possibility is that at such high frequencies, charges are only stored in the electrical double layer without charge transfer.

This chapter mainly discusses carbon nanostructures, and conductive compounds are also highlighted. Section 5.2 covers general procedure for HF-EC design. Application of vertically aligned carbon nanotubes (VACNTs) is covered in Section 5.3. Vertical graphene (VG), including edge-oriented graphene (EOG) in 3D scaffolds, is discussed in Section 5.4. Section 5.5 highlights other oriented 2D materials for developing HF-ECs, followed by a summary and perspectives.

5.2 GENERAL CONSIDERATION

5.2.1 CHARACTERISTICS OF CAPACITOR FREQUENCY RESPONSE

The difference in frequency response between the conventional EC and the AEC and how the HF-EC could bridge this gap are schematically shown in Figure 5.1a. The complex-plane impedance plots of ECs are illustrated in Figure 5.1b–e.

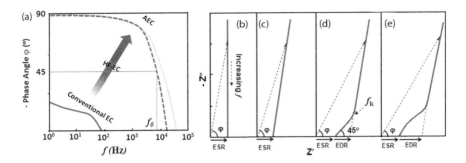

FIGURE 5.1 (a) Schematic of phase angle spectra of conventional ECs, HF-ECs, and aluminum electrolytic capacitor (AECs). Schematics of complex-plane impedance of (b) ideal capacitor in series with ESR and (c) CPE in series with ESR, (d) with the 45° slope feature, and (e) with the semicircle feature. (Reprinted from *Nano Energy*, 39, Fan, Z. et al., Towards kilohertz electrochemical capacitors for filtering and pulse energy harvesting, 306–320, for (a, d, e), Copyright 2017, with permission from Elsevier.)

Due to their inherently low-frequency characteristic, parasitic inductance is not considered here. Figure 5.1b represents an ideal capacitor (constant C) with a series resistance, called as an equivalent series resistance (ESR). It is the impedance when the impedance spectrum intersects with the Z'-axis. As frequency (f) increases, the absolute value of the phase angle (φ) will decrease accordingly. For a more realistic EC, its impedance spectrum is composed of a tilted line, and the capacitance is modeled as a constant phase element (CPE) in series with its ESR (Figure 5.1c). As a simplified model when the tilt is small (closer to vertical), the CPE is treated as an ideal capacitor. Then, the cell capacitance can be calculated according to

$$C(\omega) = -1 / \left[\omega \operatorname{Im}(Z) \right] \qquad (5.1)$$

where $C(\omega)$ is frequency-dependent capacitance, ω is the angular frequency, and Im(Z) is the imaginary part of impedance Z.

As HF-ECs discussed here are based on nanostructured electrodes for achieving a large SSA, porous effect caused by slow ion diffusion is often unavoidable. If a regular or simple pore geometry is assumed, as modeled by De Levie (1963), the 45° slope feature region will appear on account of distributed capacitance and resistance in a porous electrode (Figure 5.1d). The so-called knee frequency (f_k) can be subsequently defined. Obviously, Eq. (5.1) can only be applied when frequency is below f_k where the impedance curve is approximated by a vertical line, while above f_k, the phase angle dramatically decreases. In other more complex situations, such as irregular and dead pores presenting in the nanostructure, an interfacial resistance existing between the electrode and the current collector, or pseudocapacitive effect appearing from the electrode materials (Gamby et al. 2001), the semicircle feature, followed by or merged with the 45° slope region, will emerge (Figure 5.1e). Clearly, this feature will reduce f_k and hence decrease the effective response frequency of

HF-ECs significantly. In such cases, ESR cannot be used to evaluate the frequency response in the simplified RC model. It can have a very small value caused by the 45° slope region and the semicircle region in the impedance spectrum, but the device still shows sluggish response. In this circumstance, the resistance (R) should be derived by extrapolating the near-vertical linear section of the impedance curve to intersect with the Z'-axis. The difference between this R and the ESR is therefore defined as an equivalent distributed resistance (EDR) (Kötz and Carlen 2000), which accounts for the slowed frequency response caused by the 45° slope region and the semicircle region in the impedance spectrum.

When HF-ECs are designed for line-frequency (60 Hz) AC filtering applications, several key parameters are used to evaluate their performance, including volumetric and areal capacitance densities and phase angle at 120 Hz, and the characteristic frequency when the phase angle reaches −45° that delineates the frontier between the capacitive impedance dominance and the resistive impedance dominance. Another way to analyze the frequency response of HF-ECs is to introduce the complex capacitance (Taberna, Simon, and Fauvarque 2003), which is defined as

$$C = C' - jC''$$

(5.2)

$$Z = Z' + jZ'' = 1/(j2\pi C)$$

(5.3)

where Z is the complex impedance, C' is the useable capacitance, and C'' accounts for energy dissipation, i.e., resistive loss at the given frequency. The dependence of C'' vs. frequency reaches a maximum at a particular f_0, corresponding to the relaxation time constant $\tau_0 = 1/f_0$. This characteristic frequency is the same as the one obtained in the Bode plot when phase angle equals to −45°.

5.2.2 ELECTRODE NANOSTRUCTURES

Extremely small ESR and EDR are essential for HF-ECs to have high-frequency responses. The following recommendations should be generally followed when designing electrodes for HF-ECs (Fan, Islam, and Bayne 2017):

- Highly conductive materials with aligned structures, such as intrinsically connected 1D and 2D materials with well-defined electron and ion pathways, are good candidates.
- The contact resistance between electrode and current collector should be minimized by directly growing or closely coating/pressing the active materials on the current collector.
- Electrodes with thin-layer structure are preferred to reduce the porous effect, which becomes severe as the electrode thickness increases.
- Pores should be large and well interconnected for facile electrolyte transportation, while micropores, dead pores, and tortuous pores must be avoided. (De Levie 1963, Ghosh et al. 2013).

Based on these considerations, vertically aligned carbon nanotubes (VACNTs), VG or EOG, as well as some other oriented 2D materials, with merits of intrinsically high conductivity and simple pore structures, are very suitable for developing HF-ECs.

It is noteworthy to point out that aqueous electrolytes, due to their much smaller ion sizes and much higher ionic conductivities than organic electrolytes, are commonly used in HF-ECs. Their much smaller voltage window (< ~0.8 V), however, severely restrains their practical applications. Therefore, organic electrolyte-based cell developments are more interesting and will be particularly emphasized.

5.3 VERTICALLY ALIGNED CNTs (VACNTs)

VACNTs, when compared to randomly dispersed CNT films, are more ideal for HF-ECs. The oriented pores between aligned CNTs are simple, facilitating ion transportation between the two electrodes, while the large conductance of CNT itself ensures a small electrode resistance. However, early studies were not very promising (Schindall, Kassakian, and Signorelli 2007, Honda et al. 2007). In the typically thick VACNT forests, the deep and narrow pores between densely packed CNTs obviously introduce distributed capacitance and resistance, leading to a characteristic frequency much less than 1 kHz (Dörfler et al. 2012, 2013, Wang, Ozkan, and Ozkan 2016). In one study, the effect of VACNT film thickness on the frequency response was elucidated by using 5, 10, and 15 μm thick VACNT films grown on graphene films, which exhibited a characteristic frequency of 1343, 754, and 460 Hz, respectively (Lin et al. 2013). At 120 Hz, the 5 μm thick VACNT film gave a phase angle of $-81.5°$, while it was increased to $-73.4°$ for the 10 μm thick one. In another very informative study that combined experimental data with numerical modeling, pore size distribution was considered in the transmission line model to simulate the frequency response in an interdigitated planar cell configuration (Ghosh et al. 2013). The effect of CNT density, electrode thickness, pore length and diameter, and electrolyte conductivity on the cell frequency response were explicitly analyzed.

Besides issues caused by the dense packing of VACNTs, another challenging issue is the relatively large interfacial resistance rendered by the nonmetallic substrates used for VACNTs growth. Commonly, they are grown by chemical vapor deposition (CVD) method with dispersed nanoparticles catalyzing individual CNT growth. The substrate choice is therefore critical in preventing the catalyst agglomeration, and Al_2O_3, SiO_2, or other insulating materials are typically used. Although VACNTs can be transferred after growth onto a metal foil, a severe interfacial resistance will be present. Exceptionally, VACNTs were successfully grown on conductive Al substrates (Dörfler et al. 2013), but corrosive electrolytes can cause degradation of the Al current collector and hence the cell performance. The very few successes in using VACNTs as electrode for HF-ECs were those with conductive graphite as a substrate for direct VACNTs growth to achieve a minimum interfacial resistance. Due to these limiting factors, most reports use dispersed CNT thin films (sub-μm thick) on a current collector. For such a structure, again the interfacial resistance from CNT-CNT and CNT-current collector become critical issues to high-frequency response.

A recent prominent study on VACNTs-based HF-EC (Li, Sun et al. 2019) may spark interest in this structure again. In this study, a graphite thin layer was used as a flexible and conductive substrate, on which 2 nm thick Fe was deposited as a catalyst, and further covered by 3 nm Al_2O_3 to facilitate tip-grown VACNTs in a CVD process. With 6 M KOH electrolyte, the fabricated EC exhibited very promising electrochemical performances. At 120 Hz, its phase angle, areal capacitance, and volumetric capacitance were $-84.8°$, 1.38 mF cm^{-2}, and 345 mF cm^{-3}, respectively. A characteristic frequency of 1.98 kHz was also measured. The large areal capacitance at the cell level is exceptional.

5.4   VERTICAL GRAPHENE

Unlike VACNTs whose growth depends on catalysts, VG, including EOG with 3D orientations, can be grown directly on conductive substrates by plasma-enhanced chemical vapor deposition (PECVD). In such a process, hydrocarbons (e.g., CH_4 and C_2H_2) or fluorocarbons are used as carbon sources and reductive gases such as H_2 or NH_3 are used as etching reagents (Hiramatsu and Hori 2010, Bo, Mao et al. 2015). With the assistance of plasma activation, the carbon-containing fragments are decomposed and deposited on a substrate, leading to the interconnected graphene sheets growing perpendicular to the substrate surface (Figure 5.2a, b).

The closely interconnected network, the high intrinsic conductivity of graphene, and the solid connection between VG network and the underlying conducting

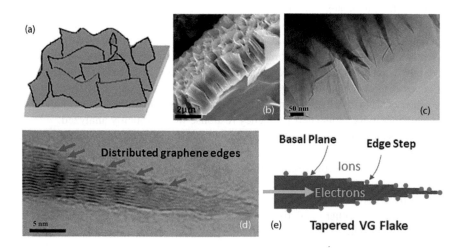

FIGURE 5.2   (a) Schematic of VG network on a conductive substrate. (b) SEM cross-sectional view of VG structure. (c) TEM image of vertical graphene sheets. (d) High-resolution TEM image of an individual sheet showing its tapered geometry with fully exposed graphene edges. (e) Schematic showing the tapered shape of VG flakes with exposed graphene edges and basal planes. (Reprinted from *Carbon*, 71, Ren, G. et al., Kilohertz ultrafast electrochemical supercapacitors based on perpendicularly-oriented graphene grown inside of nickel foam, 94–101, for (b), Copyright 2014, with permission from Elsevier; Reprinted from *J. Power Sources*, 325, Ren, G. et al., Ultrahigh-rate supercapacitors with large capacitance based on edge-oriented graphene-coated carbonized cellulous paper as flexible freestanding electrodes, 152–160, for (c, d), Copyright 2016, with permission from Elsevier.)

substrate ensure a very low resistance in the VG-based electrodes. Another striking feature of the VG is that electrolytic ions can easily access graphene edges and basal planes through simple channels between VG sheets, resulting in a rapid mass transport process during HF-EC operation. Therefore, compared to the aforementioned VACNTs based electrode, a much smaller ESR can be achieved by VG-based electrode as long as the conductivity of the electrolyte is reasonable. In addition, as revealed by the transmission electron microscopy (TEM) images (Figure 5.2c, d), VG sheets have tapered shapes with thick bases and sharp tips. Along their walls, abundant graphene steps or edges are well distributed, which can offer sufficient absorption sites for electrolyte ions. Thus, this unique structure is also expected to provide a large capacitance density. Interestingly, if scaffolds with 3D porous structures such as metal foams are used during the PECVD process, EOG networks can be grown on the scaffold surface, which further increase the available surface area of the electrodes. All these merits collectively offer the possibility of creating VG-based HF-ECs with a much reduced ESR and a high level of charge storage capability.

In the pioneering work reported by Miller et al. (2010, 2011), a 0.6 µm VG layer was deposited on a Ni foil, which was directly applied as the electrode in a KOH electrolyte-based cell. A characteristic frequency of 15 kHz was obtained, which is close to the frequency response of a commercial AEC. Since then, interests in VG materials have led to improved electrochemical performances of HF-ECs (Cai et al. 2012, Bo et al. 2012). For example, VG-based cells (Cai et al. 2014) exhibited a 120 Hz phase angle of $-85°$ and an areal capacitance of 0.265 mF cm^{-2}, as well as 20 kHz characteristic frequency. Other studies referring to some specific parameters of VG such as the wettability and edge density were also carried out to further boost the electrochemical performance of HF-ECs. The strategy of reducing the channel spacing between VG sheets to improve their wettability also increased the accessible surface area by as large as three times (Shuai et al. 2017). Compared to its basal plane, graphene edges show much better charge storage capability. By controlling the edge density in a VG layer, the HF-EC performance parameters such as frequency response and capacitance density are expected to be enhanced significantly (Yang et al. 2016, Bo, Yang et al. 2015). These findings provide very useful guidelines in optimizing VG structures for better HF-EC performances.

In general, VG film thickness is limited to ~1 µm, above which the depth of those easily accessible channels will not further increase because of the continuous growth from the channel bottom. In such a thin VG film that also has a low density of graphene, the available area and therefore the footprint-based areal capacitance are not very large. The composite of VG and carbon black was therefore tested to increase capacitance density by filling carbon black nanoparticles into the void spaces between VG flakes, although frequency response was sacrificed (Miller and Outlaw 2015).

Another way to solve the above issue is to deposit EOG on 3D scaffolds to achieve a larger surface area than their counterparts on 2D foils. For instance, in contrast to lateral growth of graphene (Cao et al. 2011), EOG deposited inside of Ni foam, with each Ni strut encircled by perpendicular graphene nanoflakes, was successfully prepared (Ren et al. 2014). At 120 Hz, the measured electrode areal capacitance was 0.72 mF cm^{-2}, corresponding to a cell capacitance of 0.36 mF cm^{-2}. A 4 kHz characteristic frequency was obtained (Ren et al. 2014). By etching off the Ni foam, freestanding EOG foam could be obtained, which would increase the specific capacitance significantly (Ren et al. 2016). In another example, carbon nanofiber (CNF) films were used as the scaffold for EOG growth, due to their high conductive, light weight, chemical inertness, and low cost (Islam, Warzywoda, and Fan 2018). A porous CNF film was first prepared by spin-coating well-dispersed CNF solution onto a Ni foil, and then EOG was grown into this porous layer with graphene nanosheets wrapped around each CNF, thus forming a highly conductive 3D electrode. In an aqueous electrolyte cell, the characteristic frequency was found to be as high as 22 kHz and the areal capacitance for a single electrode was approximately 0.37 mF cm^{-2} at 120 Hz.

More interesting studies focused on EOG growth on carbonized cellulose microfibers (CMFs) films and their applications in HF-EC (Ren et al. 2016, Islam et al. 2019). Cellulose-derived carbon materials have been investigated in different electrochemical applications (Li, Ren et al. 2017, Li and Fan 2019, Li, Mou et al. 2017), and their carbonization processes were commonly conducted in a high-temperature furnace for a long time. In contrast, the freestanding EOG/CMF electrodes used in HF-ECs were synthesized by PECVD process that only lasted for 5–10 min. In this process, cellulose fibers, such as tissue sheets and filter papers, were pyrolyzed into CMFs rapidly by the high-temperature plasma, on which EOG was synthesized simultaneously. As schematically shown in Figure 5.3a, by using the cellulose wiper (Kimwipes™) as the raw material, a thin (~10 μm) EOG/CMF electrode was obtained

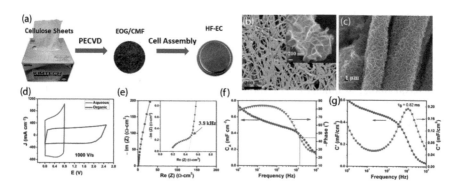

FIGURE 5.3 (a) Schematic illustrating the process using cellulose sheets to fabricate EOG/CMF electrode and HF-ECs in a coin cell configuration. (b, c) SEM images of EOG/CMF at different magnifications. Electrochemical performances of EOG/CMF-based HF-ECs: (d) CV curves at the rate of 1000 V s^{-1} in aqueous and organic electrolyte cells, (e) Nyquist plot, (f) the curve of the electrode areal capacitance vs. frequency and phase diagram, and (g) real and imaginary components of the complex electrode capacitances in an organic cell. (Reprinted from *Carbon*, 141, Islam, N. et al., Vertically edge-oriented graphene on plasma pyrolyzed cellulose fibers and demonstration of kilohertz high-frequency filtering electrical double layer capacitors, 523–530, for (a, d–g), Copyright 2019, with permission from Elsevier; Reprinted from *J. Power Sources*, 325, Ren, G. et al., Ultrahigh-rate supercapacitors with large capacitance based on edge-oriented graphene-coated carbonized cellulous paper as flexible freestanding electrodes, 152–160, for (b), Copyright 2016, with permission from Elsevier.)

after the PECVD process. Scanning electron microscopy images in Figure 5.3b, c show the macroporous structure of the CMF sheet and the microscale morphology of EOG around each CMF fiber. After assembling EOG/CMF electrodes into coin cells with KOH electrolyte, the electrode areal capacitance at 120 Hz, the phase angle at 120 Hz, and the characteristic frequency were measured to be 0.59 mF cm^{-2}, $-83°$, and 12 kHz, respectively. Due to the macro-size pores, the thin EOG/CMF films were further stacked together to obtain thick electrodes with enhanced areal capacitance without compromising their frequency response and volumetric capacitance. Electrodes composed of three EOG/CMF layers with a thickness of 30 μm exhibited 1.5 mF cm^{-2} areal capacitance and 0.5 F cm^{-3} volumetric capacitance at 120 Hz. Its characteristic frequency was still maintained at approximately 5.6 kHz. After 200,000 cycles of fully charge-discharge, the electrode capacitance showed no degradation, indicating its structural stability.

EOG/CMF electrodes with 10 μm thickness were also investigated using 1 M tetraethylammonium tetrafluoroborate (TEABF$_4$) in anhydrous acetonitrile solution as the electrolyte. The cyclic voltammogram (CV) curve recorded at 1000 V s^{-1} (Figure 5.3d) indicates that the potential window of the organic cell is widened by a factor of ~3 in comparison with that of the aqueous cell. The desired quasi-rectangular shape at such a high scan rate suggests the high-speed capability of these cells. Figure 5.3e presents the complex-plane impedance curve of the organic cell. A small semicircle and a near 45° linear region appear above a knee frequency of 3.8 kHz, below which the impedance spectrum is a nearly vertical line. Figure 5.3f shows the electrode areal capacitance and phase angle dependence on frequency. At 120 Hz, an electrode capacitance of 0.49 mF cm^{-2} and a phase angle of $-80.4°$ are found, as well as a characteristic frequency of 1.3 kHz. The complex capacitance was also derived from the impedance spectrum as shown in Figure 5.3g. The modeled capacitance (C′) agrees with the results extracted from Figure 5.3f. The value of C″ reaches a maximum at the characteristic frequency of 1.3 kHz, from which a relaxation time constant of 0.8 ms was derived.

As proof of concepts, EOG/CMF-based organic-electrolyte HF-ECs were further tested as a filtering capacitor in a 60 Hz AC/DC converter and as a pulse storage device in an energy harvester. For AC/DC conversion (Figure 5.4a–c), the input is a 60 Hz sinusoidal wave voltage with a 4 V amplitude, which is converted to DC by a full-bridge rectifier and then filtered by an HF-EC filter capacitor before being delivered to a load. As shown in Figure 5.4c, a near ripple-free 2.8 V DC voltage is obtained. This clearly demonstrates that HF-ECs, with their larger capacitance densities, are promising to replace bulky AECs for AC line-filtering in electronics and power applications after further development. HF-ECs may also be used to store and smooth electric pulses scavenged from ambient vibrations for powering small autonomous devices. Here, a microwatt load was used to simulate a microsensor. As shown in Figure 5.4d, a piezoelectric element converts environmental vibration into electric pulses (Figure 5.4e), which, after rectification, are stored in the HF-EC. A 2 V smooth voltage is outputted to drive the load (Figure 5.4f). This result demonstrates that HF-ECs, if their leakage could be reduced, might be used in energy harvesters to store and smooth pulses for powering low-power autonomous devices.

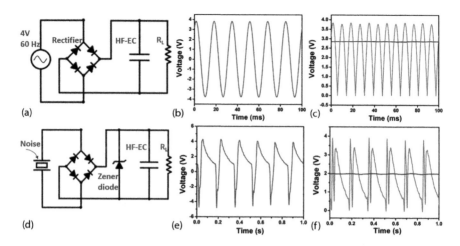

FIGURE 5.4 Demonstration of HF-EC as the filtering capacitor in AC/DC conversion: (a) circuit diagram, (b) input 60 Hz sine wave, and (c) the output from the rectifier without filtering (green curve) and the smooth DC output after using HF-EC filter (blue line). Demonstration of pulse power storing and smoothing using an HF-EC: (d) circuit diagram, (e) the pulse from the piezo-element, and (f) the output from the rectifier without smoothing (green curve) and the constant DC output after HF-EC smoothing (blue line). (Reprinted from *Carbon*, 141, Islam, N. et al., Vertically edge-oriented graphene on plasma pyrolyzed cellulose fibers and demonstration of kilohertz high-frequency filtering electrical double layer capacitors, 523–530, for (a–f), Copyright 2019, with permission from Elsevier.)

5.5 OTHER VERTICALLY ALIGNED NANOMATERIALS

In addition to carbon-based nanostructures, many other materials, particularly those with metallic conductivities or 2D sheet-like structure, have the potential to improve the frequency response of typical slow pseudocapacitors. When integrated with a suitable conductive framework (Ren, Zhang, and Fan 2018), they may also be candidates to develop compact high-frequency supercapacitors. These materials include transition metal oxides (Pan et al. 2014), nitrides (Lu et al. 2012), carbides, and dichalcogenides. In particular, metal dichalcogenides such as VS_2 (Feng et al. 2011), MoS_2 (Acerce, Voiry, and Chhowalla 2015), and WS_2 (Khalil et al. 2016), combining their 2D sheet-like morphology with exceptionally low resistance (Peng et al. 2016), have already been studied as electrodes for high-rate supercapacitors (Bissett et al. 2016). However, the random dispersion of the nanoflakes in the electrode leads to a large ESR and hence a very low frequency response. Aligned 2D nanostructures based on these materials are required to construct fast ion transportation pathways (Ren et al. 2015).

As an example, ultrathin MoS_2 nanosheets grown on highly conductive CMFs were prepared to form binder-free freestanding films and tested as electrodes for fast-rate supercapacitors (Islam et al. 2018). CMF sheets used here were prepared by a rapid plasma pyrolysis process to avoid the formation of undesired micropores. The growth process of vertically oriented MoS_2 nanosheets was conducted

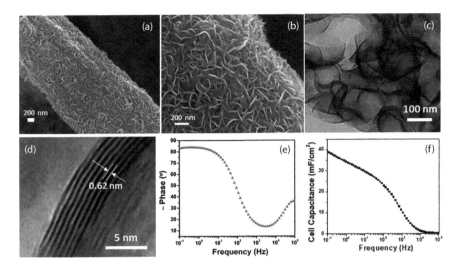

FIGURE 5.5 (a, b) SEM images of edge-oriented MoS$_2$ network grown circling around carbon fibers. (c) TEM image of MoS$_2$ nanoflakes. (d) High-resolution TEM image shows the few layered flakes. (e) Bode diagram of the cell. (f) The curve of cell capacitance vs. frequency. (Reprinted from *J. Power Sources*, 400, Islam, N. et al., Fast supercapacitors based on vertically oriented MoS$_2$ nanosheets on plasma pyrolyzed cellulose filter paper, 277–283, for (a–f), Copyright 2018, with permission from Elsevier.)

by the hydrothermal method using sodium molybdate and thiourea as precursors (Miao et al. 2016, Wang et al. 2017). As shown in Figure 5.5a and b, CMFs are fully wrapped by a few layered vertical MoS$_2$ nanosheets that can be clearly observed from high-resolution TEM images shown in Figure 5.5c. The vertical structure of MoS$_2$ yields a large surface area for electrolytic ion absorption and provides a rapid ionic diffusion route, leading to a fast capacitive charge-discharge response. As shown by the impedance Bode diagram of the assembled cells in Figure 5.5d, a characteristic frequency of 103 Hz is achieved. The cell also exhibited an absolute phase angle above 80° for frequencies below 10 Hz. At 10 Hz, the cell capacitance was ~30 mF cm^{-2}, which was maintained at ~25 mF cm^{-2} at 100 Hz, indicating their great potential for high-power pulse applications.

5.6 CONCLUSION

The development of HF-ECs requires electrodes with extremely low resistance and a nanostructure that facilitates electrolytic ion transportation while maintaining a large capacitance density. Due to their simple porous structure and intrinsically high conductivity, vertically aligned 1D and 2D nanomaterials have been proved to have great potential for developing such HF-ECs. Preliminarily functionable HF-ECs based on VACNT and VG have been successfully demonstrated, but their capacitance density needs to be further improved to satisfy practical requirements. How to further increase their easily accessible surface area without compromising the frequency response becomes critical. Studies that integrated carbon black into VG network or VG into 3D microfiber network have been reported with improved performance. It is expected that integration of 1D, 2D, and 3D carbon components into ordered or aligned structures might further boost the capacitance density while maintaining the necessary frequency response. In addition, combining these carbon materials with other 2D materials into an organized structure might provide another route to significantly boost the capacitance density. More efforts in structural design and compositional optimization are still needed.

ACKNOWLEDGMENTS

Z.F was supported by the National Science Foundation (1611060).

REFERENCES

Acerce, M., D. Voiry, and M. Chhowalla. 2015. "Metallic 1T phase MoS_2 nanosheets as supercapacitor electrode materials." *Nature Nanotechnology* 10 (4):313–318.
Bissett, M., S. Worrall, I. Kinloch, and R. Dryfe. 2016. "Comparison of two-dimensional transition metal dichalcogenides for electrochemical supercapacitors." *Electrochimica Acta* 201:30–37.
Bo, Z., H. Yang, S. Zhang, J. Yang, J. Yan, and K. Cen. 2015. "Molecular insights into aqueous NaCl electrolytes confined within vertically-oriented graphenes." *Scientific Reports* 5:14652.
Bo, Z., S. Mao, Z. Han, K. Cen, J. Chen, and K. Ostrikov. 2015. "Emerging energy and environmental applications of vertically-oriented graphenes." *Chemical Society Reviews* 44 (8):2108–2121.
Bo, Z., Z. Wen, H. Kim, G. Lu, K. Yu, and J. Chen. 2012. "One-step fabrication and capacitive behavior of electrochemical double layer capacitor electrodes using vertically-oriented graphene directly grown on metal." *Carbon* 50 (12):4379–4387.
Cai, M., R. A. Outlaw, R. A. Quinlan, D. Premathilake, S. M. Butler, and J. R. Miller. 2014. "Fast response, vertically oriented graphene nanosheet electric double layer capacitors synthesized from C(2)H(2)." *ACS Nano* 8 (6):5873–5882. doi: 10.1021/nn5009319.
Cai, M., R. A. Outlaw, S. M. Butler, and J. R. Miller. 2012. "A high density of vertically-oriented graphenes for use in electric double layer capacitors." *Carbon* 50 (15):5481–5488.
Cao, X., Y. Shi, W. Shi, G. Lu, X. Huang, Q. Yan, Q. Zhang, and H. Zhang. 2011. "Preparation of novel 3D graphene networks for supercapacitor applications." *Small* 7 (22):3163–3168. doi:10.1002/smll.201100990.
Dörfler, S., I. Felhösi, I. Kek, T. Marek, H. Althues, S. Kaskel, and L. Nyikos. 2012. "Tailoring structural and electrochemical properties of vertical aligned carbon nanotubes on metal foil using scalable wet-chemical catalyst deposition." *Journal of Power Sources* 208:426–433.
Dörfler, S., I. Felhösi, T. Marek, S. Thieme, H. Althues, L. Nyikos, and S. Kaskel. 2013. "High power supercap electrodes based on vertical aligned carbon nanotubes on aluminum." *Journal of Power Sources* 227:218–228.
Du, C., and N. Pan. 2006. "Supercapacitors using carbon nanotubes films by electrophoretic deposition." *Journal of Power Sources* 160 (2):1487–1494.
Fan, Z., N. Islam, and S. Bayne. 2017. "Towards kilohertz electrochemical capacitors for filtering and pulse energy harvesting." *Nano Energy* 39:306–320.
Feng, J., X. Sun, C. Wu, L. Peng, C. Lin, S. Hu, J. Yang, and Y. Xie. 2011. "Metallic few-layered VS2 ultrathin nanosheets: High two-dimensional conductivity for in-plane supercapacitors." *Journal of the American Chemical Society* 133 (44):17832–17838.

Gamby, J., P. L. Taberna, P. Simon, J. F. Fauvarque, and M. Chesneau. 2001. "Studies and characterisations of various activated carbons used for carbon/carbon supercapacitors." *Journal of Power Sources* 101 (1):109–116.

Ghosh, A., V. Le, J. Bae, and Y. Lee. 2013. "TLM-PSD model for optimization of energy and power density of vertically aligned carbon nanotube supercapacitor." *Scientific Reports* 3:2939.

Hiramatsu, M., and M. Hori. 2010. *Carbon Nanowalls: Synthesis and Emerging Applications.* Springer Science & Business Media, New York.

Honda, Y., T. Haramoto, M. Takeshige, H. Shiozaki, T. Kitamura, and M. Ishikawa. 2007. "Aligned MWCNT sheet electrodes prepared by transfer methodology providing high-power capacitor performance." *Electrochemical and Solid-State Letters* 10 (4):A106–A110.

Islam, N., J. Warzywoda, and Z. Fan. 2018. "Edge-oriented graphene on carbon nanofiber for high-frequency supercapacitors." *Nano-Micro Letters* 10 (1):9.

Islam, N., M. Hoque, W. Li, S. Wang, J. Warzywoda, and Z. Fan. 2019. "Vertically edge-oriented graphene on plasma pyrolyzed cellulose fibers and demonstration of kilohertz high-frequency filtering electrical double layer capacitors." *Carbon* 141:523–530.

Islam, N., S. Li, G. Ren, Y. Zu, J. Warzywoda, S. Wang, and Z. Fan. 2017. "High-frequency electrochemical capacitors based on plasma pyrolyzed bacterial cellulose aerogel for current ripple filtering and pulse energy storage." *Nano Energy* 40:107–114.

Islam, N., S. Wang, J. Warzywoda, and Z. Fan. 2018. "Fast supercapacitors based on vertically oriented MoS_2 nanosheets on plasma pyrolyzed cellulose filter paper." *Journal of Power Sources* 400:277–283.

Khalil, A., Q. Liu, Q. He, T. Xiang, D. Liu, C. Wang, Q. Fang, and L. Song. 2016. "Metallic 1T-WS_2 nanoribbons as highly conductive electrodes for supercapacitors." *RSC Advances* 6 (54):48788–48791.

Kötz, R., and M. Carlen. 2000. "Principles and applications of electrochemical capacitors." *Electrochimica Acta* 45 (15):2483–2498.

De Levie, R. 1963. "On porous electrodes in electrolyte solutions: I. Capacitance effects." *Electrochimica Acta* 8 (10):751–780.

Li, Q., S. Sun, A. Smith, P. Lundgren, Y. Fu, P. Su, T. Xu, L. Ye, L. Sun, and J. Liu. 2019. "Compact and low loss electrochemical capacitors using a graphite/carbon nanotube hybrid material for miniaturized systems." *Journal of Power Sources* 412:374–383.

Li, S., and Z. Fan. 2019. "Nitrogen-doped carbon mesh from pyrolysis of cotton in ammonia as binder-free electrodes of supercapacitors." *Microporous and Mesoporous Materials* 274:313–317.

Li, S., G. Ren, M. Hoque, Z. Dong, J. Warzywoda, and Z. Fan. 2017. "Carbonized cellulose paper as an effective interlayer in lithium-sulfur batteries." *Applied Surface Science* 396:637–643.

Li, S., T. Mou, G. Ren, J. Warzywoda, Z. Wei, B. Wang, and Z. Fan. 2017. "Gel based sulfur cathodes with a high sulfur content and large mass loading for high-performance lithium–sulfur batteries." *Journal of Materials Chemistry A* 5 (4):1650–1657.

Li, W., N. Islam, G. Ren, S. Li, and Z. Fan. 2019. "AC-filtering supercapacitors based on edge oriented vertical graphene and cross-linked carbon nanofiber." *Materials* 12:604.

Lin, J., C. Zhang, Z. Yan, Y. Zhu, Z. Peng, R. H. Hauge, D. Natelson, and J. M. Tour. 2013. "3-Dimensional graphene carbon nanotube carpet-based microsupercapacitors with high electrochemical performance." *Nano Letters* 13 (1):72–78. doi: 10.1021/nl3034976.

Lu, X., G. Wang, T. Zhai, M. Yu, S. Xie, Y. Ling, C. Liang, Y. Tong, and Y. Li. 2012. "Stabilized TiN nanowire arrays for high-performance and flexible supercapacitors." *Nano Letters* 12 (10):5376–5381.

Miao, H., X. Hu, Q. Sun, Y. Hao, H. Wu, D. Zhang, J. Bai, E. Liu, J. Fan, and X. Hou. 2016. "Hydrothermal synthesis of MoS_2 nanosheets films: Microstructure and formation mechanism research." *Materials Letters* 166:121–124.

Miller, J. R., and R. A. Outlaw. 2015. "Vertically-oriented graphene electric double layer capacitor designs." *Journal of the Electrochemical Society* 162 (5):A5077–A5082.

Miller, J. R., R. A. Outlaw, and B. C. Holloway. 2010. "Graphene double-layer capacitor with ac line-filtering performance." *Science* 329 (5999):1637–1639. doi: 10.1126/science.1194372.

Miller, J. R., R. A. Outlaw, and B. C. Holloway. 2011. "Graphene electric double layer capacitor with ultra-high-power performance." *Electrochimica Acta* 56 (28):10443–10449.

Niu, C., E. Sichel, R. Hoch, D. Moy, and H. Tennent. 1997. "High power electrochemical capacitors based on carbon nanotube electrodes." *Applied Physics Letters* 70 (11):1480–1482.

Pan, X., G. Ren, M. Hoque, S. Bayne, K. Zhu, and Z. Fan. 2014. "Fast supercapacitors based on graphene-bridged V_2O_3/VOx core–shell nanostructure electrodes with a power density of 1 MW kg– 1." *Advanced Materials Interfaces* 1 (9):1400398.

Peng, L., Y. Zhu, D. Chen, R. S. Ruoff, and G. Yu. 2016. "Two-dimensional materials for beyond-lithium-ion batteries." *Advanced Energy Materials* 6 (11):1600025-n/a.

Ren, G., M. Hoque, X. Pan, J. Warzywoda, and Z. Fan. 2015. "Vertically aligned VO 2 (B) nanobelt forest and its three-dimensional structure on oriented graphene for energy storage." *Journal of Materials Chemistry A* 3 (20):10787–10794.

Ren, G., R. Zhang, and Z. Fan. 2018. "VO2 nanoparticles on edge oriented graphene foam for high rate lithium ion batteries and supercapacitors." *Applied Surface Science* 441:466–473.

Ren, G., S. Li, Z. Fan, M. Hoque, and Z. Fan. 2016. "Ultrahigh-rate supercapacitors with large capacitance based on edge oriented graphene coated carbonized cellulous paper as flexible freestanding electrodes." *Journal of Power Sources* 325:152–160.

Ren, G., X. Pan, S. Bayne, and Z. Fan. 2014. "Kilohertz ultrafast electrochemical supercapacitors based on perpendicularly-oriented graphene grown inside of nickel foam." *Carbon* 71:94–101.

Schindall, J., J. Kassakian, and R. Signorelli. 2007. *"Paper Presented at 2007 Advanced Capacitor World Summit."* 2007 Advanced Capacitor World Summit, San Diego, CA.

Shuai, X., Z. Bo, J. Kong, J. Yan, and K. Cen. 2017. "Wettability of vertically-oriented graphenes with different intersheet distances." *RSC Advances* 7 (5):2667–2675.

Taberna, P. L., P. Simon, and J. F. Fauvarque. 2003. "Electrochemical characteristics and impedance spectroscopy studies of carbon-carbon supercapacitors." *Journal of the Electrochemical Society* 150 (3):A292. doi: 10.1149/1.1543948.

Wang, L., Y. Ma, M. Yang, and Y. Qi. 2017. "Titanium plate supported MoS_2 nanosheet arrays for supercapacitor application." *Applied Surface Science* 396:1466–1471.

Wang, W., M. Ozkan, and C. S. Ozkan. 2016. "Ultrafast high energy supercapacitors based on pillared graphene nanostructures." *Journal of Materials Chemistry A* 4 (9):3356–3361.

Yang, H., J. Yang, Z. Bo, S. Zhang, J. Yan, and K. Cen. 2016. "Edge effects in vertically-oriented graphene based electric double-layer capacitors." *Journal of Power Sources* 324:309–316.

6 Mesoporous Electrodes for Supercapacitors

Godlisten N. Shao and Talam E. Kibona

CONTENTS

6.1 INTRODUCTION

6.1.1 BACKGROUND

Presently, materials scientists and engineers are compelled to design highly efficient electrical energy storage devices because of enormous demand of such devices in portable devices, hybrid vehicles, power distributions, telecommunication, and railways infrastructure. Supercapacitors have been widely used in advanced hybrid energy storage systems due to their high power capabilities and long cycle-life, making them the best candidates suitable for efficient electrical storage devices (Simon and Gogotsi 2013). Thus, the development of supercapacitors with high-performance parameters such as low cost, high energy density, high power density, environment-friendly, and a commendable cycle-life is in great demand (Divya and Østergaard 2009, González et al. 2016). The reports by the Ryoo and Hyeon groups in 1999 revealed a new insight regarding ordered mesoporous carbons (OMCs) and a new gateway to improve the electrochemical performances of supercapacitors (Ryoo, Joo, and Jun 1999, Oh and Kim 1999). Therefore, extensive investigations involving the utilization of OMCs have been carried out to improve the structure

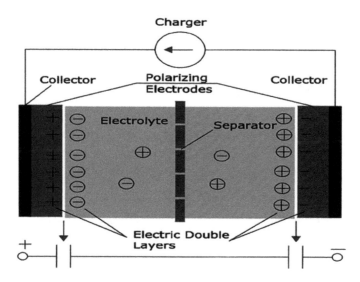

FIGURE 6.1 Typical configuration of an EDLC. (Adapted from Jayalakshmi, M. and Balasubramanian, K., *Int. J. Electrochem. Sci.*, 3, 1196–1217, 2008.)

and applications of supercapacitors for large scale production and commercialization of energy storage materials. The current chapter intends to survey recent advances that have been achieved in the development of mesoporous electrodes for supercapacitor applications.

6.1.2 STRUCTURE AND CATEGORIES OF SUPERCAPACITORS

Supercapacitor cells are electric double-layer capacitors (EDLCs) with no dielectric. They possess two high surface area porous carbon electrodes and a porous separator between them. The electrical interaction between the electrodes is inhibited by the separator; thus, the charge transport is mainly due to ions dissolved in an electrolyte. Figure 6.1 presents a general configuration of EDLC. Supercapacitors can be categorized based on the storage mechanism and cell configuration (González et al. 2016, Xia et al. 2008). Most studies have pointed out two main categories of supercapacitors, namely EDLCs and pseudocapacitors. However, the combination of these two classes can generate novel materials (hybrid supercapacitors) with outstanding physicochemical properties and applications.

6.1.3 OVERVIEW OF MESOPOROUS ELECTRODES FOR SUPERCAPACITORS

Carbon is one of the materials that can be used as electrodes in supercapacitors. Customarily, carbonaceous materials are very abundant and exhibit large specific surface area, high conductivity, and outstanding chemical stability; hence, they can be used as excellent energy storage materials (Yang and Zhou 2017, Wang and Xia 2013). This signifies that the structural, electronic, and textural properties of materials are very essential attributes that govern the electrochemical performances of electrodes.

The relationship between the electrical energy (E), of a capacitor; the capacitance (C), and voltage (V) can be seen in the equation: $E = 1/2CV^2$. Thus, maximizing C and V can unquestionably improve the energy of a given supercapacitor. It has been pointed out that the specific capacitance of a capacitor can be enhanced by (i) forming electrodes with porous structure and large surface area and (ii) forming hybrid supercapacitors with a carbon electrode and pseudocapacitors (Wang and Xia 2013). Therefore, materials with a large surface area generate high capacitance. Thus, designing of suitable preparation methods to improve the surface area and the pore structure of electrode materials is very essential. The existence of a favorable porous structure improves ion transport process. This advantage can facilitate the fabrication of supercapacitors with improved power density and energy density.

As it has been pointed out early, since the introduction of OMCs by the Ryoo and Hyeon group in 1999, extensive research has been performed to improve the composition, morphologies, performance, and applications of OMCs (Ryoo, Joo, and Jun 1999, Oh and Kim 1999). OMCs exhibit better electrochemical performance than conventional activated carbons. OMCs possess well-connected ordered mesopore channels, which provide a more favorable path for the transportation of organic ions (Wang and Xia 2013).

It should be noted that before the introduction of the Ryoo and Hyeon reports, most of the investigations focused on increasing the surface area of carbon materials. At that time, carbonization was used as the major technique to improve the surface area of carbonaceous electrode materials. Hence, there are considerable findings that have reported on carbonization as shown in Section 6.1.3.1.

6.1.3.1 Carbonization and Electrochemical Properties

Carbonization is the thermal conversion of organic materials to carbon (Lewis 1982). The general understanding of the chemistry of carbonization and the way mesospheres form in carbonaceous materials are very fundamental in fabricating mesoporous electrodes with desirable properties for supercapacitors applications (Imamura, Nakamizo, and Honda 1978, Lewis 1982, Xu et al. 2010, Du et al. 2010). It is noteworthy that various investigations have been carried out to study the influence of carbonization on the electrochemical properties of supercapacitors. For instance, Xu et al. (2010) reported the fabrication of an activated carbon with high capacitance using NaOH activation for supercapacitors. The authors noted that the specific capacitance and textural properties (surface area, pore volume, and pore size) can be decreased by increasing the carbonization temperature of apricot shells. The findings showed that the surface area and specific capacitance were decreased upon increasing carbonization temperature. The surface area was decreased from 2335 to 1342 m^{2}g^{-1} upon carbonization at a temperature of 400°C and 800°C, respectively. A decrease in the specific capacitance from 348 (400°C) to 235 Fg^{-1} (800°C) was also noted. Similarly, further reports regarding the influence of carbonization temperature on the specific surface area and specific capacitance using phenolic resin-based sphere also demonstrated that these parameters decreased upon increasing carbonization temperature (Du et al. 2010). It is thought that high carbonization temperature yields a more ordered carbon structure that is more likely to hamper etching of the activating agent.

Xu et al. (2007) also investigated the optimization of carbonization temperatures from 400°C to 900°C on polyacrylonitrile (PAN) fabrics. It was revealed that high surface area, pore size, and specific capacitance were exhibited by the samples carbonized at 600°C. Lee et al. (2014) reported the effects of carbonization temperature on pore development in PAN-based activated carbon nanofibers. It was revealed that the capacitance was harnessed by increasing the carbonization temperature. Similarly, Lanlan and co-workers (2013) also revealed that the textural properties and specific capacitance of metal organic framework (MOF-5) increase upon increasing the carbonization temperature from 700°C to 900°C. Nevertheless, the specific capacitance and surface area were decreased upon increasing the carbonization temperature to 950°C.

Therefore, these considerable efforts signify that the electrochemical properties and applications of electrode materials can be influenced by the carbonization temperatures. However, it is noteworthy that the electrochemical properties also depend on the nature of the carbonized materials. The abovementioned reports indicate that the structural properties of the carbonized materials varied due to their differences in the compositions of their biomasses (Daud, Ali, and Sulaiman 2000, Shafizadeh 1982). Hence, the pyrolytic behavior differs from one carbonaceous material to another. Moreover, the nature of the carbonaceous material governs the preparation method and applications of the final products.

6.1.3.2   Principles and Fabrication of Mesoporous Electrodes for Supercapacitors

Regardless of the different categories of supercapacitors, primarily supercapacitors store energy based on two principles, namely electrostatic interaction (electrical double layer, EDL) and faradic reaction. In pseudocapacitance, the redox reactions occur between the electrode materials and electrolyte (pseudocapacitance). The redox reactions generate electrons, which are then transferred via electrode/electrolyte boundaries (Conway 2013). The benefit of EDL includes easy electrochemical kinetics, implying that polarization resistance has no effect and active materials experience no swelling. For planar capacitors model, the number of charges available in the electrode is presumed to be directly proportional to the surface area (A) of the electrodes, permittivity in free space (ε_0) and dielectric constant (ε_r) and inversely proportional to the effective thickness (d) of the EDL according to Equation 6.1 (Huang, Sumpter, and Meunier 2008a). There is proportionality between energy stored and the amount of charge stored. It is therefore stretched that the higher the surface area of an electrode, the better supercapacitor behavior. However, nanoporous materials exhibit different pore structures, such as spherical, slit, and cylindrical shapes, which influence the specific capacitance of the supercapacitor. Additionally, the simple and direct proportionality does not hold as the assumption that submicropores are not involved in the formation of EDL is not true. Thus, models in equations 6.2 and 6.3 describe the specific capacitance of microporous and mesoporous carbon electrodes, respectively, where b, a_0 and d represent pore radius, the effective radius of counter ions, and the distance of the approaching ions from the surface of the electrode, correspondingly (Huang, Sumpter, and Meunier 2008b).

$$C = \frac{\varepsilon_r \varepsilon_0}{d} A \tag{6.1}$$

$$C = \frac{\varepsilon_r \varepsilon_0}{b \ln\left[b / (b - d)\right]} A \tag{6.2}$$

$$C = \frac{\varepsilon_r \varepsilon_0}{b \ln\left(b / a_0\right)} A_{zz} \tag{6.3}$$

The EDL structure can be explained using three models, namely the Helmholtz model, the Gouy-Chapman model, and the Stern model (Figure 6.2). In the Helmholtz model, it is assumed that there is formation of an ionic monolayer at the electrode surface so that the differential capacitance is given by $C_l = \varepsilon / 4\pi\sigma$, where σ is the separation of the center of the ionic monolayer and electrode surface.

There is neutralization between a charge of the opposite sign ions and a solid electronic conductor at any distance from the surface to the center of the ions. For simplicity, the model assumes that the charges are rigidly attached to the surface. For the Gouy-Chapman model, it is considered that charges exist in a diffuse layer, which leads to diffuse capacitance. The model suggests that the ions in the solution disperse into the liquid phase to the extent that the offset potential set up by their removal limits the process. The kinetic energy of the ions determines the thickness of the diffuse and the model acquires the Boltzmann distribution. However, the double layer that is highly

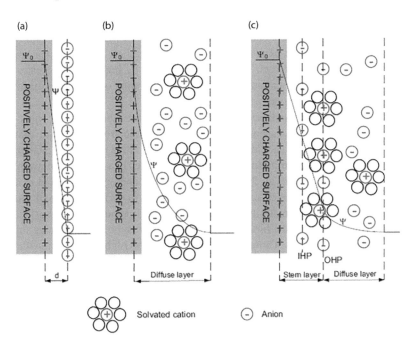

FIGURE 6.2 EDL models: (a) Helmholtz model, (b) Gouy-Chapman model, and (c) Stern model. (Reproduced from González, A. et al., *Renew. Sust. Energ. Rev.*, 58, 1189–1206, 2016.)

charged limits the model (Bagotsky 2005). The constraint of this model is based on the assumption that the surface can be approached by point-charged ions with no limits. A combination of compact layer and diffuse layer is considered in the Stern model, which accounts finite size of ions (see equation 6.4). The compact layer is formed by ions adsorbed by the electrode (the inner Helmholtz plane) and the outer Helmholtz plane is formed by nonspecifically adsorbed counter-ions (González et al. 2016).

$$\frac{1}{c} = \frac{1}{c_l} + \frac{1}{c_d} \tag{6.4}$$

6.1.3.3 Electrochemical Performance of Mesoporous Electrodes for Supercapacitors

For improvement of supercapacitor's performance (e.g., high specific capacitance and high cyclic stability); researchers apply two strategies to increase energy density. The first strategy is to improve the microstructures of the electrode to increase the capacitance of the electrode. The second strategy is to increase the voltage window by studying different electrolytes. The voltage window can be improved by studying electrolytes, while the specific capacitance can be enhanced by synthesizing nanoporous materials or using composite electrode materials. Carbon-based nanomaterials are among the most used nanomaterials for supercapacitor electrode due to their abundance, ease of processing, non-toxicity, tunable pore size for easy transport of electrolyte, and large surface area for providing interaction platform (Zhang and Zhao 2009, Xia et al. 2008, Wang et al. 2012). The major limitation of some carbon materials such as carbon nanotubes and graphene is the high cost of processing and hence their limited availability. Therefore, the exploration of alternative carbon precursors for supercapacitor electrodes is imperative. This can be achieved by improving microstructures and textures of the existing nanomaterials and understanding of ions behavior in small pores and relatively larger pores.

Predominantly, biomass-based organic materials are abundant and have been investigated as source of carbon for supercapacitor electrodes. Several materials such as coffee beans, pine needles, rice straw (He et al. 2013), corn stalk (Chen et al. 2013), sorghum pitch (Senthilkumar et al. 2011), water hyacinth (Kurniawan et al. 2015), human hair (Qian et al. 2014), coconut shells, and egg yolk (Li et al. 2015) have been used to synthesize nanostructured carbon materials. The goal with the aforementioned biomass materials has to be achieved to increase the surface area and other textures so that the biomass-derived carbon fits the application as supercapacitor electrode materials. Though most carbons exhibit high conductivity, they also exhibit low capacitance, thus resulting in low energy density.

As it has been highlighted above, porous carbon plays a crucial role in electrochemical energy storage. The facilitation of energy storage mechanisms depends on the textural properties of the porous carbon. Microporous materials possess a pore diameter less than 2 nm, while the nanomaterials with the pore size ranging between 2 and 50 nm are mesoporous. Microporous nanomaterials exhibit high surface area, thus increasing sites for ion adsorption and high specific capacitance. In contrast, mesoporous nanomaterials exhibit moderate surface area but large pore size. The large pore sizes enhance specific capacitances by increasing the transfer of ions, thus decreasing the transfer resistances. Hence, the present chapter deals with mesoporous electrode for supercapacitors.

6.2 PREPARATION OF MESOPOROUS ELECTRODES FOR SUPERCAPACITORS

Mesoporous nanomaterials can be synthesized using various methods depending on the desired applications of the final products. The commonly used methods are physical or chemical activation of carbon materials: soft templating and hard templating.

6.2.1 ACTIVATION METHOD

Mesoporous carbon is synthesized through single- or double-step carbonization of carbon precursor. For single-step carbonization, the activating agent (KOH, NOH, $ZnCl_2$, and H_3PO_4) is mixed with powder made up of a precursor and then directly carbonized at different temperatures to generate porous carbon network. In double-step carbonization, the precursor is carbonized at temperature, say 400°C, followed by activation at a temperature above 500°C under inert atmosphere. Among other factors, the starting materials of carbon determine the properties of the activated carbons. It has been observed the higher the ratio of carbon to activating agent, the more is the mesopore content (Hou et al. 2014). Physical activation involves two processes: pyrolysis of the carbon precursor at temperatures at 400°C and above under inert atmosphere, followed by gasification at 700°C–1200°C (Bouchelta et al. 2008, Sekirifa et al. 2013, Sun and Chun Jiang 2010). All volatile materials in the carbon precursor are removed through pyrolysis, while closed pores are opened through gasification. The mesopores synthesized through these routes are mainly disordered. Additionally, more active sites can be generated through burning away of more organics using oxidants (Sevilla and Mokaya 2014). However, it has been pointed out that the activated carbons exhibit more micropores than mesopores as most mesopores tend to break at higher temperatures, leading to a decrease in pore diameter (Pagketanang et al. 2015).

6.2.2 SOFT AND HARD TEMPLATING

Soft- and hard-templating routes are very useful in preparing OMCs. The hard-templating route intends to control the size and ordered structure of the synthesized nanomaterials. Silica hard template has been reported to fit in synthesizing mesoporous carbon. MCM-41, SBA-15, and noncommercial mesoporous silica materials have been reported to be an excellent hard template in the synthesis of mesoporous carbon. Soft templating involves the self-assembly of amphiphilic surfactants or block co-polymers to form mesoporous carbon. The removal of the template by etching process generates a porous structure with suitable properties for the fabrication of mesoporous electrode materials (Wang, Xin, and Wang 2014). Primarily, four fundamental requirements are needed for the synthesis of mesoporous carbon materials: (i) both pore-forming component and carbon component, (ii) the self-assembly trend of the precursor and component to yield microstructures, (iii) suitable duration of high temperature for the nanostructures, and (iv) replication of the microstructure starting materials by the highly cross-linked and mechanical strength carbon component (Zhang et al. 2010). It is reported further that the change in preparation methods alter the textural properties of the obtained materials. Self-templated high-surface-area mesoporous carbon material can also be synthesized. For example, in a study, "self-templated synthesis of mesoporous carbon from carbon tetrachloride precursor for supercapacitor electrodes" by Tang et al. (2016), mesoporous carbon was synthesized via reduction of carbon tetrachloride by sodium-potassium alloy. One advantage of using the reduction-generated salt templates is that the template can be easily removed with water.

6.3 TRENDS TOWARD THE PREPARATION OF MESOPOROUS ELECTRODES

Textural properties have been revealed to be crucial in influencing the specific capacitance of energy storage materials. Colossal investigations of various carbon-based mesoporous electrodes are being carried out to establish the synergistic effect between carbon (high power density) and pseudocapacitive materials (high energy density). Primarily, strategies to improve the performance of supercapacitors have expanded from improving power density to enhancing high energy density as well. Apart from forming pure OMC, extensive investigations have been dedicated to improving the supercapacitive performance of the carbon materials. This has exquisitely been accomplished through the incorporation of foreign atoms such as nitrogen, boron, phosphorus, and sulfur (Gao et al. 2016, Song et al. 2016). The unique pore structure and heteroatom functionalities can enhance the specific capacitance of the attained materials. The existence of heteroatoms can promote the electron donor/acceptor mechanisms of the supercapacitors. It has been further delineated that the presence of N and O element improves the carbon surface affinity and contributes to pseudocapacitive properties (Wang et al. 2018, Li et al. 2018). Table 6.1 summarizes the electrochemical performance of some selected carbon-based materials. It can be seen that the electrochemical properties of supercapacitors are highly dependent on the type of materials, preparation method, and structural properties of the final product.

Recently, our research group reported the specific capacitance–pore texture relationship of biogas slurry mesoporous carbon/MnO_2 composite electrodes for supercapacitors (Kibona et al. 2019). The report outlined the relationship between specific capacitance and textural properties of the obtained materials (Amaral et al. 2012, Enock et al. 2017, Enock, King'ondu, and Pogrebnoi 2018). Mesoporous carbon materials with outstanding properties for mesoporous supercapacitors electrodes were obtained by hydrothermal treatment of the biogas slurry (as starting materials). The materials with promising interconnected pores and enhanced electrochemical performances were produced from MnO_2/biogas slurry. It was found that final products with an appealing specific capacitance of 709 Fg^{-1} can be generated, provided that the textural properties of the electrode materials are controlled by the addition of MnO_2 (Kibona et al. 2019).

The main focus in fabricating mesoporous carbon for supercapacitors applications is to understand the existing correlation between microstructures, pore textures, and electrochemical performance of mesoporous carbon-based electrode. The preparation

TABLE 6.1

Comparison of the Electrochemical Performance of the Representative Reports of Selected Carbon-Based Materials Reported in the Literature

Materials Composition	Surface Area (cm^2g^{-1})	Pore Volume (cm^2g^{-1})	Pore Diameter (nm)	Specific Capacitance (Fg^{-1})	Current Density (Ag^{-1})	Electrolyte	Reports
Graphene/nitrogen	1569	1.38	6.4	377	0.2	KOH	Song et al. (2016)
Coconut shell/ sewage sludge	3003	2.04	5–50	420	0.5	KOH	Peng et al. (2018)
N-Phenolic resin	439	0.33	5.0	288	0.1	KOH	Wang et al. (2018)
Carbon/boron	1157	1.2	4.8	267.8	1.0	KOH	Gao et al. (2016)
N-carbon nanofiber	563	0.5	3.6	202.0	1.0	KOH	Chen et al. (2012)
N,S-doped carbon	983	1.1	6.3	180	1.0	KOH	Zhang et al. (2013)
Furfuryl alcohol	—	—	—	700	0.1	Na_2SO_4	Wang, Chen, and Chen (2015)
Carbon/MnO_2	514	0.52	6.2	709	—	KOH	Kibona et al. (2019)
$NiCo_2S_4$/ nitrogen-doped carbon foams			2–8	877	20	—	Shen et al. (2015)

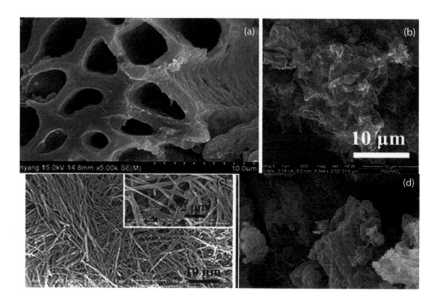

FIGURE 6.3 SEM images of different mesoporous nanomaterials. (a) A mesoporous carbon synthesized through carbonization and activation. (Adapted from Enock, T.K. et al., *Mater. Today Energy*, 5, 126–137, 2017) (b) A mesoporous carbon synthesized through soft templating. (Adapted from Li, W. et al., *Adv. Energy Mater.*, 1, 382–386, 2011) (c) Mesoporous nanofibers. (Adapted from Tong, F. et al., *RSC Adv.*, 9, 6184–6192, 2019) (d) Ordered mesoporous carbon. (Adapted from Sobrinho, R.A.L. et al., *J. Hazard. Mater.*, 362, 53–61, 2019).

method used can improve both textural properties and electrochemical performance of the obtained mesoporous carbon. As shown in Figure 6.3, different synthesis methods yield different morphologies. Furthermore, the type of precursor also determines the type of morphology to be obtained from the mesoporous carbon. For example, Ghosh and coworkers synthesized mesoporous carbon nanofibers through electrospinning and calcination of PAN. It was observed that the electrochemical behavior of porous carbon nanofibers had a direct correlation with a higher volume of mesopores, degree of graphitization, and surface area morphology (Ghosh et al. 2019). Nomura et al. (2019) reported that mesoporous carbon sheet with graphene walls can be synthesized. The synthesized mesoporous sheet exhibited appealing stability under high-voltage and high-temperature environments.

These studies demonstrate the importance of understanding the relationship between pore textures and specific capacitance in forming supercapacitors with improved electrochemical performances. It can further be discerned that the electrochemical properties can also be enhanced through the incorporation of foreign atoms and compounds.

This demonstrates a general paradigm shift toward the formation of mesoporous electrodes with suitable properties for supercapacitors applications. Interestingly, a systematic assessment of the practical reliability of the obtained mesoporous structures has also demonstrated a tremendous improvement of the power density and cycling stability. Hence, designing various methods to fabricate mesoporous electrodes for supercapacitors is very fascinating, commendable, and indisputably in great demand for the sake of boosting large-scale production and commercialization of supercapacitors.

6.4 CONCLUSIONS

The current chapter presented a review of mesoporous electrodes for supercapacitors. It is unquestionable that supercapacitors have several advantages over rechargeable batteries in terms of power density and cycling stability. However, the low energy density of supercapacitors has compelled various researchers to design different strategies to overcome the issue. To date, most of the studies have shown that OMCs are the best candidate for mesoporous electrodes for supercapacitors. High specific capacitances and cycling stability have been witnessed and the designed preparation methods are significantly facile and less expensive. This can eventually boost large-scale production and commercialization of supercapacitors. Interestingly, incorporation of heteroatoms into OMCs has been a common strategy and the generated materials have shown high specific capacitance as well. Blending of supercapacitive metal oxides such as MnO_2, RuO_2, MoO_3, and Fe_3O_4 with OMCs has also received a heightened attention. Generally, the generation of suitable mesoporous electrodes for supercapacitors depends on the pore structure, surface area, nature of the materials, and preparation methods. Therefore, the formation of mesoporous electrodes with desired properties for supercapacitors is very fascinating, and its explicit attraction to scientists and engineers is indisputably incessant.

REFERENCES

Amaral-Labat G., Andrzej Szczurek, Vanessa Fierro, N. Stein, C. Boulanger, A. Pizzi, and Alain Celzard. 2012. "Pore structure and electrochemical performances of tannin-based carbon cryogels." *Biomass and Bioenergy* 39:274–282.

Bagotsky, Vladimir S. 2005. *Fundamentals of Electrochemistry*. Vol. 44. John Wiley & Sons, Hoboken, NJ.

Bouchelta, Chafia, Mohamed Salah Medjram, Odile Bertrand, and Jean Pierre Bellat. 2008. "Preparation and characterization of activated carbon from date stones by physical activation with steam." *Journal of Analytical and Applied Pyrolysis* 82 (1):70–77.

Chen, Li-Feng, Xu-Dong Zhang, Hai-Wei Liang, Mingguang Kong, Qing-Fang Guan, Ping Chen, Zhen-Yu Wu, and Shu-Hong Yu. 2012. "Synthesis of nitrogen-doped porous carbon nanofibers as an efficient electrode material for supercapacitors." *ACS Nano* 6 (8):7092–7102.

Chen, Mingde, Xueya Kang, Tuerdi Wumaier, Junqing Dou, Bo Gao, Ying Han, Guoqing Xu, Zhiqiang Liu, and Lu Zhang. 2013. "Preparation of activated carbon from cotton stalk and its application in supercapacitor." *Journal of Solid State Electrochemistry* 17 (4):1005–1012.

Conway, Brian E. 2013. *Electrochemical Supercapacitors: Scientific Fundamentals and Technological Applications.* Springer Science & Business Media, New York.

Daud, Wan Mohd Ashri Wan, Wan Shabuddin Wan Ali, and Mohd Zaki Sulaiman. 2000. "The effects of carbonization temperature on pore development in palm-shell-based activated carbon." *Carbon* 38 (14):1925–1932.

Divya, KC, and Jacob Østergaard. 2009. "Battery energy storage technology for power systems—An overview." *Electric Power Systems Research* 79 (4):511–520.

Du, Xuan, Cheng-Yang Wang, Ming-Ming Chen, Shuo Zhao, and Jin Wang. 2010. "Effects of carbonization temperature on microstructure and electrochemical performances of phenolic resin-based carbon spheres." *Journal of Physics and Chemistry of Solids* 71 (3):214–218.

Enock, Talam Kibona, Cecil K. King'ondu, Alexander Pogrebnoi, and Yusufu Abeid Chande Jande. 2017. "Biogas-slurry derived mesoporous carbon for supercapacitor applications." *Materials Today Energy* 5 (2017):126–137.

Enock, Talam Kibona, Cecil K. King'ondu, and Alexander Pogrebnoi. 2018. "Effect of biogas-slurry pyrolysis temperature on specific capacitance." *Materials Today: Proceedings* 5 (4):10611–10620.

Gao, Jiao, Xianyou Wang, Youwei Zhang, Jia Liu, Qun Lu, and Min Liu. 2016. "Boron-doped ordered mesoporous carbons for the application of supercapacitors." *Electrochimica Acta* 207:266–274. https://doi.org/10.1016/j.electacta.2016.05.013.

Ghosh, Subrata, Wan Dao Yong, En Mei Jin, Shyamal Rao Polaki, Sang Mun Jeong, and Hangbae Jun. 2019. "Mesoporous carbon nanofiber engineered for improved supercapacitor performance." *Korean Journal of Chemical Engineering* 36 (2):312–320.

González, Ander, Eider Goikolea, Jon Andoni Barrena, and Roman Mysyk. 2016. "Review on supercapacitors: Technologies and materials." *Renewable and Sustainable Energy Reviews* 58:1189–1206.

He, Xiaojun, Pinghua Ling, Moxin Yu, Xiaoting Wang, Xiaoyong Zhang, and Mingdong Zheng. 2013. "Rice husk-derived porous carbons with high capacitance by $ZnCl_2$ activation for supercapacitors." *Electrochimica Acta* 105:635–641.

Hou, Jianhua, Chuanbao Cao, Xilan Ma, Faryal Idrees, Bin Xu, Xin Hao, and Wei Lin. 2014. "From rice bran to high energy density supercapacitors: A new route to control porous structure of 3D carbon." *Scientific Reports* 4:7260.

Huang, Jingsong, Bobby G Sumpter, and Vincent Meunier. 2008a. "Theoretical model for nanoporous carbon supercapacitors." *Angewandte Chemie International Edition* 47 (3):520–524.

Huang, Jingsong, Bobby G Sumpter, and Vincent Meunier. 2008b. "A universal model for nanoporous carbon supercapacitors applicable to diverse pore regimes, carbon materials, and electrolytes." *Chemistry—A European Journal* 14 (22):6614–6626.

Imamura, T., M. Nakamizo, and H. Honda. 1978. "Formation of carbonaceous mesophase at lower temperature." *Carbon.* 16 (6):487–490.

Jayalakshmi, Mandapati, and K. Balasubramanian. 2008. "Simple capacitors to supercapacitors-an overview." *International Journal of Electrochemical Science* 3 (11):1196–1217.

Kibona, Talam E., Godlisten N. Shao, Hee Taik Kim, and Cecil K. King'ondu. 2019. "Specific capacitance–pore texture relationship of biogas slurry mesoporous carbon/MnO_2 composite electrodes for supercapacitors." *Nano-Structures & Nano-Objects* 17:21–33.

Kurniawan, Fredi, Michael Wongso, Aning Ayucitra, Felycia Edi Soetaredjo, Artik Elisa Angkawijaya, Yi-Hsu Ju, and Suryadi Ismadji. 2015. "Carbon microsphere from water hyacinth for supercapacitor electrode." *Journal of the Taiwan Institute of Chemical Engineers* 47 (2015):197–201.

Lanlan, Jiang, Wang Xianyou, Wu Hao, Wu Chun, Zhao Qinglan, and Song Yunfeng. 2013. "Effect of carbonization temperature on structure and electrochemical performance of porous carbon from metal framework." *Journal of Central South University (Science and Technology)* 44 (10):4012–4018.

Lee, Hye-Min, Kay-Hyeok An, and Byung-Joo Kim. 2014. "Effects of carbonization temperature on pore development in polyacrylonitrile-based activated carbon nanofibers." *Carbon Letters* 15 (2):146–150.

Lewis, IC. 1982. "Chemistry of carbonization." *Carbon* 20 (6):519–529.

Li, Jiangfeng, Shuwen Ma, Liya Cheng, and Qingsheng Wu. 2015. "Egg yolk-derived three-dimensional porous carbon for stable electrochemical supercapacitors." *Materials Letters* 139 (2015):429–432.

Li, Jing, Ning Wang, Jiarui Tian, Weizhong Qian, and Wei Chu. 2018. "Cross-coupled macro-mesoporous carbon network toward record high energy-power density supercapacitor at 4 V." *Advanced Functional Materials* 28 (51):1806153.

Li, Wei, Fan Zhang, Yuqian Dou, Zhangxiong Wu, Haijing Liu, Xufang Qian, Dong Gu, Yongyao Xia, Bo Tu, and Dongyuan Zhao. 2011. "A self-template strategy for the synthesis of mesoporous carbon nanofibers as advanced supercapacitor electrodes." *Advanced Energy Materials* 1 (3):382–386.

Nomura, Keita, Hirotomo Nishihara, Naoya Kobayashi, Toshihiro Asada, and Takashi Kyotani. 2019. "4.4 V supercapacitors based on super-stable mesoporous carbon sheet made of edge-free graphene walls." *Energy and Environmental Science* 12 (5):1542–1549.

Oh, SeungáM, and KiáBum Kim. 1999. "Synthesis of a new mesoporous carbon and its application to electrochemical double-layer capacitors." *Chemical Communications* 21:2177–2178.

Pagketanang, Thanchanok, Apichart Artnaseaw, Prasong Wongwicha, and Mallika Thabuot. 2015. "Microporous activated carbon from KOH-activation of rubber seed-shells for application in capacitor electrode." *Energy Procedia* 79 (2015):651–656.

Peng, Lin, Yeru Liang, Hanwu Dong, Hang Hu, Xiao Zhao, Yijing Cai, Yong Xiao, Yingliang Liu, and Mingtao Zheng. 2018. "Super-hierarchical porous carbons derived from mixed biomass wastes by a stepwise removal strategy for high-performance supercapacitors." *Journal of Power Sources* 377:151–160.

Qian, Wenjing, Fengxia Sun, Yanhui Xu, Lihua Qiu, Changhai Liu, Suidong Wang, and Feng Yan. 2014. "Human hair-derived carbon flakes for electrochemical supercapacitors." *Energy and Environmental Science* 7 (1):379–386.

Ryoo, Ryong, Sang Hoon Joo, and Shinae Jun. 1999. "Synthesis of highly ordered carbon molecular sieves via template-mediated structural transformation." *The Journal of Physical Chemistry B* 103 (37):7743–7746.

Sekirifa, Mohamed L, Mahfoud Hadj Mahammed, Stephanie Pallier, Lotfi Baameur, Dominique Richard, and Ammar H Al Dujaili. 2013. "Preparation and characterization of an activated carbon from a date stones variety by physical activation with carbon dioxide." *Journal of Analytical and Applied Pyrolysis* 99 (2013):155–160.

Senthilkumar, ST, B Senthilkumar, S Balaji, C Sanjeeviraja, and R Kalai Selvan. 2011. "Preparation of activated carbon from sorghum pith and its structural and electrochemical properties." *Materials Research Bulletin* 46 (3):413–419.

Sevilla, Marta, and Robert Mokaya. 2014. "Energy storage applications of activated carbons: supercapacitors and hydrogen storage." *Energy and Environmental Science* 7 (4):1250–1280.

Shafizadeh, Fred. 1982. "Introduction to pyrolysis of biomass." *Journal of Analytical and Applied Pyrolysis* 3 (4):283–305.

Shen, Laifa, Jie Wang, Guiyin Xu, Hongsen Li, Hui Dou, and Xiaogang Zhang. 2015. "NiCo$_2$S$_4$ nanosheets grown on nitrogen-doped carbon foams as an advanced electrode for supercapacitors." *Advanced Energy Materials* 5 (3):1400977.

Simon, Patrice, and Gogotsi Yury. 2013. "Capacitive energy storage in nanostructured carbon-electrolyte systems." *Accounts of Chemical Research* 46:1094–1103.

Sobrinho, Raimundo Alves Lima, George Ricardo Santana Andrade, Luiz P. Costa, Marcelo José Barros de Souza, Anne Michelle Garrido Pedrosa de Souza, and Iara F Gimenez. 2019. "Ordered micro-mesoporous carbon from palm oil cooking waste via nanocasting in HZSM-5/SBA-15 composite: Preparation and adsorption studies." *Journal of Hazardous Materials* 362:53–61.

Song, Yanfang, Jun Yang, Ke Wang, Servane Haller, Yonggang Wang, Congxiao Wang, and Yongyao Xia. 2016. "In-situ synthesis of graphene/nitrogen-doped ordered mesoporous carbon nanosheet for supercapacitor application." *Carbon* 96:955–964. https://doi.org/10.1016/j.carbon.2015.10.060.

Sun, Kang, and Jian Chun Jiang. 2010. "Preparation and characterization of activated carbon from rubber-seed shell by physical activation with steam." *Biomass and Bioenergy* 34 (4):539–544.

Tang, Duihai, Shi Hu, Fang Dai, Ran Yi, Mikhail L Gordin, Shuru Chen, Jiangxuan Song, and Donghai Wang. 2016. "Self-templated synthesis of mesoporous carbon from carbon tetrachloride precursor for supercapacitor electrodes." *ACS Applied Materials & Interfaces* 8 (11):6779–6783.

Tong, Fenglian, Wei Jia, Yanliang Pan, Jixi Guo, Lili Ding, Jingjing Chen, and Dianzeng Jia. 2019. "A green approach to prepare hierarchical porous carbon nanofibers from coal for high-performance supercapacitors." *RSC Advances* 9 (11):6184–6192.

Wang, Jian-Gan, Hongzhen Liu, Huanhuan Sun, Wei Hua, Huwei Wang, Xingrui Liu, and Bingqing Wei. 2018. "One-pot synthesis of nitrogen-doped ordered mesoporous carbon spheres for high-rate and long-cycle life supercapacitors." *Carbon* 127:85–92. https://doi.org/10.1016/j.carbon.2017.10.084.

Wang, Jia-Wei, Ya Chen, and Bai-Zhen Chen. 2015. "A synthesis method of MnO$_2$/activated carbon composite for electrochemical supercapacitors." *Journal of the Electrochemical Society* 162 (8):A1654–A1661.

Wang, Jie, Huolin L Xin, and Deli Wang. 2014. "Recent progress on mesoporous carbon materials for advanced energy conversion and storage." *Particle & Particle Systems Characterization* 31 (5):515–539.

Wang, Rutao, Peiyu Wang, Xingbin Yan, Junwei Lang, and Chao Peng, Qunji Xue. 2012. "Promising porous carbon derived from celtuce leaves with outstanding supercapacitance and CO$_2$ capture performance." *ACS Applied Materials & Interfaces* 4 (11): 5800–5806.

Wang, Yonggang, and Yongyao Xia. 2013. "Recent progress in supercapacitors: From materials design to system construction." *Advanced Materials* 25 (37):5336–5342.

Xia, Kaisheng, Qiuming Gao, Jinhua Jiang, and Juan Hu. 2008. "Hierarchical porous carbons with controlled micropores and mesopores for supercapacitor electrode materials." *Carbon* 46 (13):1718–1726.

Xu, Bin, Feng Wu, Shi Chen, Cunzhong Zhang, Gaoping Cao, and Yusheng Yang. 2007. "Activated carbon fiber cloths as electrodes for high performance electric double layer capacitors." *Electrochimica Acta* 52 (13):4595–4598.

Xu, Bin, Yufeng Chen, Gang Wei, Gaoping Cao, Hao Zhang, and Yusheng Yang. 2010. "Activated carbon with high capacitance prepared by NaOH activation for supercapacitors." *Materials Chemistry and Physics* 124 (1):504–509.

Yang, Mei, and Zhen Zhou. 2017. "Recent breakthroughs in supercapacitors boosted by nitrogen-rich porous carbon materials." *Advanced Science* 4 (8):1600408.

Zhang, Deyi, Yuan Hao, Liwen Zheng, Ying Ma, Huixia Feng, and Heming Luo. 2013. "Nitrogen and sulfur co-doped ordered mesoporous carbon with enhanced electrochemical capacitance performance." *Journal of Materials Chemistry A* 1 (26):7584–7591.

Zhang, Li Li, and, and X.S. Zhao. 2009. "Carbon-based materials as supercapacitor electrodes." *Chemical Society Reviews* 38 (9):2520–2531.

Zhang, Weimin, Peter Sherrell, Andrew I Minett, Joselito M Razal, and Jun Chen. 2010. "Carbon nanotube architectures as catalyst supports for proton exchange membrane fuel cells." *Energy & Environmental Science* 3 (9):1286–1293.

7 Honeycomb Nanostructures for Supercapacitors

Huailin Fan

CONTENTS

7.1 INTRODUCTION

Nature is a grand and successful laboratory, which provides largely effective methods to many scientific and technical problems. Honeycomb-like structure closely resembles the bee's nest in nature, from which it gets its name. Honeycomb-like structure can be made by any flat materials and different kinds of honeycomb-like structure have been manufactured in the past. Paper-honeycomb, as an ornament, was first made about 2000 years ago by the Chinese. The basic concept of honeycomb-like structure is that thin, dense, and strong materials are bonded to a thick, lightweight monolithic block. Honeycomb-like structure consists of an array of open cells, formed from very thin sheets of material attached to each other. It is a unique and interesting configuration, which offers many advantages such as excellent mechanical properties, good soundproofing effectiveness, low thermal conductivity coefficients, low diffusion resistances, and large exposed area within the cells.

When the open cells shrink to micrometers/nanometers scale, the honeycomb-like nanostructures will be formed. The nanomaterials can maintain the excellent properties of original macroscopic construction [1]. Moreover, novel honeycomb-like structures at nanoscales have been increasingly found. For example, lots of hexagonal or similar patterns have been discovered in living tissue [2], cell aggregates [3], and molecules [4].

Supercapacitors are a particular kind of capacitors prepared based on charging and discharging on the electrode–electrolyte interface, for instance, carbon materials or certain metallic compounds and so on. They are dominated by the same basic rules as conventional capacitors and are ideally suited for the rapid storage and release of energy. However, supercapacitors' electrode materials own much higher effective surface areas and less thickness of the double layer leading to a tremendous increase in both capacitance and energy by 10,000-fold than those of the regular capacitors. Thus, supercapacitors can be estimated at as high as tens, hundreds, and even thousands of farads. Carbon materials are currently the most widely utilized electrode materials in commercial supercapacitors. Few materials can match carbon materials for the combination of high conductivity and high surface area.

Certain materials utilize rapid and reversible redox reactions at intersurface, which represent a different kind of capacitance compared to electronic double layer capacitor (EDLC), referred to as pseudocapacitance. Pseudocapacitive materials involve fast and reversible redox reactions answerable for charge storage. Compared to ionic adsorption in EDLCs, these faradaic contributions substantially increase the resulting capacitance and energy density of the device. The involved redox reactions have slower kinetics than that of ionic adsorption phenomena as the latter do not involve any charge transfer. Recently, the most studied pseudocapacitive materials were electronic conductive polymers and metal oxides. These materials were either considered as positive or negative electrodes depending on the redox couples involved. For example, manganese dioxide is the pseudocapacitive material of choice. It is naturally abundant, inexpensive, and not too toxic. Its charge storage mechanism has been described as involving very fast Mn^{3+}/Mn^{4+} redox reactions with cation exchanges, between protons and cations of the electrolyte, to balance the change in the oxidation state of manganese.

Generally, the application properties of electrode materials are determined by the structure and surface chemical composition, and much effort has been made to adjust the structure and surface chemical properties. Electrode materials, such as carbon materials and pseudocapacitive materials with various novel morphologies, such as cubes, microtrees, and carbon flower-like composites, have been researched. Thus, carbon materials with honeycomb-like nanostructure or pseudocapacitive materials such as $Ni(OH)_2$ and MnO_2 have been synthesized widely for supercapacitor electrodes.

This chapter is intended to provide a summary of the honeycomb-like nanostructure synthesis and supercapacitor applications. It also aims to discuss and provide a brief summary of recent progress in the preparation and applications of honeycomb-like nanostructures dividing them into honeycomb-like porous carbon and honeycomb-like metallic compounds.

7.2   HONEYCOMB-LIKE POROUS CARBON

Fabricating novel nanostructures is a promising approach to obtain fast diffusion and transport process. Macro/mesopores have been thought to be a transport channel to access the fine porosity and in particular the higher surface-area-to-volume ratio of micropores which makes them effective for preparing high-surface-area materials.

The cell sizes and porous structure of honeycomb-like porous carbon can be controlled by various methods. At the same time, the carbon precursors used for synthesizing honeycomb-like porous carbon are organic small molecules, polymers, biomass, and so on. Furthermore, the chemical composition of honeycomb-like porous carbon can be adjusted by regulating the kinds of precursors, temperatures, and posttreatment processes.

7.2.1 TEMPLATING METHODS

Templating methods are promising techniques to prepare microstructure-controllable carbon materials: the micro-shape, nano-size, and porous structure of carbon materials can be precisely controlled by replicating the original structure of templates. Honeycomb-like porous carbon was prepared from various carbon sources by using gravitationally packed SiO_2, polymethyl methacrylate (PMMA), polystyrene (PS), NaCl, and $CaCO_3$ as template by carbonization.

Porous carbon materials, of various morphologies, replicated from silica templates are widely investigated including nanospheres, nanosheets, and monolithic materials. For example, Lindén et al. prepared monolithic carbon materials embracing wormhole-like mesopores and macropores [5]. A carbon material with co-continuous structure and hierarchical pores was prepared by similar methods [6]. Largely, honeycomb-like carbon materials have been widely synthesized with regular open-cell carbon structure, as shown in Figure 7.1a. For example, hierarchically, honeycomb-like carbon has been prepared using glucose as carbon source and colloidal silica as templates by Liu et al. [7]. The preparation process is illustrated in Figure 7.1b. Firstly, the colloidal silica was synthesized by modified Stöber method with 3-amino-propyltrimethoxysilane. Glucose as carbon source was dissolved in the above mixtures under stirring, which can be coated onto the surface of SiO_2 nanospheres because of the electrostatic interactions between $-NH_2$ on the surface of the colloidal silica spheres and $-OH$ of the glucose. After solvent evaporation and gravitational sedimentation, the obtained composites were carbonized under 800°C for 2 h. Honeycomb-like carbon materials were obtained following

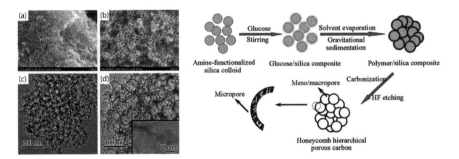

FIGURE 7.1 Left: Typical (a and b) scanning electron microscope (SEM) and (c and d) transmission electron microscopy (TEM) micrographs of the honeycomb-like carbon at different magnifications. The inset in (d) is the high-resolution TEM micrograph of the localized graphitic walls. Right: Schematic illustration of the preparation process of the honeycomb-like carbon. (Reprinted with permission from Han, Y. et al., *J. Power Sources*, 227, 118–122, 2013.)

by etching with hydrofluoric acid to remove silica. The specific capacitances of the honeycomb-like porous carbon were calculated to be 292, 272, and 233 Fg^{-1} at 1, 5, and 10 Ag^{-1}, respectively. The corresponding area capacitances were 44.5, 41.5, and 35.5 mF/cm^2, respectively, which shows highly efficient ion-accessible surface area for charge accumulation. The capacitance maintenance was as high as 80% when the current density increased by 10 times. Honeycomb-like mesoporous nitrogen-doped carbon was also successfully synthesized by Zhen-Bo Wang et al. using polyaniline as carbon and nitrogen precursor, and silica nanoparticles as template, for achieving mesoporous structure [8]. The N contents and distribution proportion of different types of honeycomb-like carbon were closely related to the pyrolysis temperature.

Kim et al. [9] synthesized honeycomb-like carbon using SiO_2 nanosphere templates. Honeycomb-like carbon framework obtained by thermal vapor deposition was activated to enhance the EDLC properties. Honeycomb-like carbon framework/Co_3O_4 composite electrode showed a dramatically improved specific capacitance (456 Fg^{-1} at 1 Ag^{-1}) due to the synergistic roles between the EDLC created by hierarchical ternary pore and the outstanding pseudocapacitance of Co_3O_4. In addition, ferrocene was used as framework to prepare Fe_3O_4 dots (≈ 3 nm) that were embedded in three-dimensional (3D) carbon using SiO_2 nanospheres as template by Zhang et al. [10]. The honeycomb-like carbon/metal oxide composites were prepared in one step in this process. Gan et al. [11] prepared N and O co-doped honeycomb-like carbon skeletons derived from a melamine–formaldehyde precursor showing the surface area of 2379 m^2g^{-1} and high heteroatom N/O dopants (6.90 and 10.17 wt%). Melamine and formaldehyde can crosslink onto the surfaces of SiO_2 nanospheres due to the hydrogen-bond interaction. Increasing the solvent contents led to the significant increase of wall thickness and reduction of pore size (Figure 7.2). Benefiting from such merits, the honeycomb-like porous carbon electrode showed a high

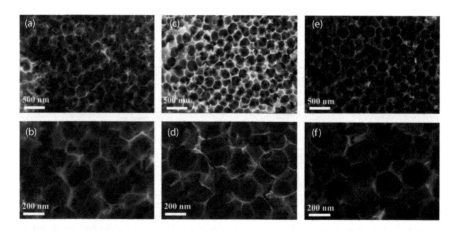

FIGURE 7.2 SEM images of (a and b) honeycomb porous carbon (HPC)-1, (c and d) HPC-2, and (e and f) HPC-3. (Reprinted with permission from Song, Z. et al., *J. Mat. Chem. A*, 7, 816–826, 2019.)

capacitance (533 Fg^{-1} at 0.5 Ag^{-1}) in 6 mol/L KOH electrolyte. The honeycomb-like porous carbon-assembled devices delivered the energy outputs of 12.8 and 26.6 Wh kg^{-1} using KOH and Na_2SO_4, respectively.

Besides, chitosan and SiO_2 nanoparticles were selected as the carbon precursor and mesopores template, and spray drying was used for the preparation of complex microspheres [12]. This approach not only simplified preparation processes but also helped in obtaining various honeycomb-like porous carbon structures and supercapacitors assembled with porous carbon as the electrode exhibited a specific capacitance of 250.5 Fg^{-1} at 0.5 Ag^{-1}. The SiO_2 templates method was often used to prepare porous materials, but removing SiO_2 was difficult and harsh reagents were frequently required. In addition to SiO_2, nano-$CaCO_3$, FeO, MgO, and CaO particles were used as hard template, which can be removed by diluted hydrochloric acid. A nitrogen-doped micron-sized honeycomb-like carbon was developed by Zhang et al. [13]. This novel material was obtained by using $CaCO_3$ as the template and sucrose as the carbon source. With one-step ammonia activation, nitrogen doping and porous structure optimization were realized simultaneously. Zhang et al. [14] also reported similar works in which honeycomb-like carbon materials were prepared using $CaCO_3$ as the hard template. Ordered honeycomb-like carbon nanosheets were prepared by etching self-assembled iron oxide-carbon complex [15]. The two-dimensional (2D) nanosheets had close-packed uniform mesopores of about 20 nm, which resembled the honeycomb structure consisting of an ordered array of hexagonal pores. The excellent electrochemical properties could be attributed to the fact that the uniform mesopores offered sufficient space for sulfur expansion and trapped polysulfides. Honeycomb-like carbon nanoflakes were obtained using MgO templates as a host for SnO_2 nanoparticles [16]. Kwok Feng Chong et al. prepared CaO-impregnated highly porous honeycomb-like carbon from agriculture waste [17]. CaO was obtained from chicken eggshell waste to produce CaO/honeycomb-like carbon, which showed highly porous honeycomb-like structure with CaO nanoparticles (30–50 nm). The prepared materials were evaluated as supercapacitor electrodes, and a specific capacitance of 222 Fg^{-1} at 0.025 Ag^{-1} was achieved which was around three times higher than that of honeycomb-like structure (76 Fg^{-1}).

NaCl was also used as template for the preparation of honeycomb-like porous carbons [18] (Figure 7.3). ZnS nanoparticles with small diameter of 20–40 nm coated with honeycomb-like carbon nanosheets were synthesized via freeze-drying and carbonization methods by using NaCl crystals as the hard template [19]. The porous carbon matrix provided a conductive network and also worked as a buffer to confine ZnS nanoparticle expansion during charge–discharge process. Molten salt-assisted carbonization has been thought of as a high-efficiency method for preparing honeycomb-like porous carbons. For example, wool was converted to N-doped honeycomb-like porous carbons with a surface area of 787.079 m^2g^{-1} and 2.6 wt% nitrogen content by molten salt carbonization method. The carbon showed 318.2 Fg^{-1} at 0.25Ag^{-1} and excellent rate capability [20]. Moreover, the symmetric supercapacitor assembled by the porous carbons showed high specific energy of 20.2 Wh kg^{-1} at 202 W kg^{-1} when operated at a voltage of 1.8 V. Mingbo Wu et al. synthesized 3D porous honeycomb-like carbon materials as supercapacitor electrodes by in situ activation coupled with molten salt [21]. NaCl and KCl were applied

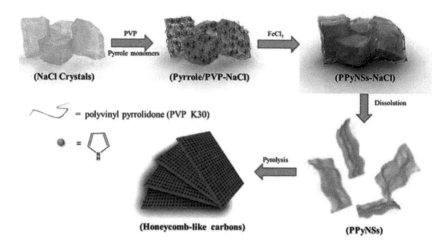

FIGURE 7.3 Schematic illustration of the preparation of nitrogen-doped honeycomb-like porous carbons. (Reprinted with permission from Niu, W.H.et al., *J. Mat. Chem. A.*, 4, 10820–10827, 2016.)

as separators for asphalt, which facilitated the effective touch between asphalt and KOH, and microporous carbons with high surface area were obtained. The interconnected honeycomb structure supplied the channels for ionic and electronic transfer, and showed 264 Fg^{-1} at 0.05 Ag^{-1} in aqueous solution electrolyte, with 83.7% rate performance of 221 Fg^{-1} at 20 Ag^{-1}.

Another sustainable route for synthesizing 3D honeycomb porous carbon using the mixture of starch and Na_2CO_3 followed by KOH activation was developed by Du et al. [22]. During the preparation process, Na_2CO_3 was used as the renewable honeycomb-like macroporous template, which can be easily removed by washing and recovered by recrystallization. The honeycomb carbon with high specific surface area and interconnected pore channels exhibited 249.2 Fg^{-1} at 0.5 Ag^{-1} and good rate capability (the capacitance retained was 90.7% at 18 Ag^{-1}). Moreover, the assembled symmetric supercapacitor delivered 33 Wh kg^{-1} at 100 W kg^{-1} and presented an excellent long-term cycling stability.

With the development of lyophilization technology, ice was also used as template for preparing honeycomb-like carbon foams. Lignin-derived honeycomb-like carbon was fabricated by ice-templating followed by freeze-drying and carbonizing by Zeng et al. [23]. By adjusting the density, good electrical conductivity and mechanical properties could be accomplished. Besides the merged interfaces and good conductivity between carbon and graphene, uniform pores also enhanced electromagnetic interference shielding effectiveness of the foams (Figure 7.4).

KOH is the most commonly used and effective activator for making porous structure, which was also used as template. Honeycomb-like carbons with 3D structure were obtained using polyvinyl alcohol as carbon source and KOH as template by Wu et al. [24]. Honeycomb-like carbons exhibited 367 Fg^{-1} at 0.5 Ag^{-1} and 99.7%

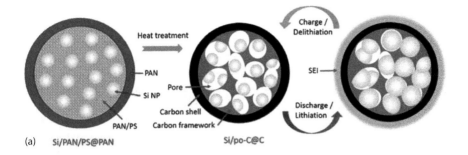

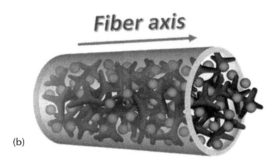

FIGURE 7.4 Schematic illustration of the Si/po-C@C composite fiber: (a) structural formation process; (b) 3D sketch of the overall structure. (Reprinted with permission from Zhang, H.R. et al., *J. Mat. Chem. A*, 3, 7112–7120, 2015.)

maintenance of their initial capacitance after 10,000 cycles resulting from large surface area, honeycomb-like porous structure, and rich oxygen doping. The assembled symmetric device showed 22.4 Wh kg^{-1} in sodium sulfate aqueous electrolyte and excellent electrochemical stability. Thus, the method provided a facile idea to produce 3D honeycomb-like porous carbon on a large scale.

Polymer-based hard templates are removed under high-temperature carbonization without diluted hydrochloric acid/deionized water washing. Honeycomb-like conductive carbon framework-embedded Si nanoparticles were obtained using the coaxial electrospinning technique, and PS nanospheres were used as honeycomb-like template by Li et al. [25]. The carbon skeleton not only buffered the volume expansion of silicon but also improved the fiber conductivity. The carbon shell was able to stop the electrolyte from smoking into the core parts, and therefore, a settled interface can be formed on the fiber surface.

The matrix structure could be effectively duplicated, and various precursors could be applied for the preparation of honeycomb-like carbon frameworks, but the process was tedious; thus, this technique was unfavorable for their large-scale application. A soft-templating method is also proposed for the synthesis of honeycomb-like carbon frameworks. The self-assembly of triblock copolymers and precursor was induced during solvent evaporation, and honeycomb-like carbon frameworks were obtained after carbonization. Yanmei Zhou et al. reported 3D honeycomb-like carbon materials were prepared by a double-soft-templating method [26]. In this preparation strategy, F127 and sodium dodecyl sulfate (SDS) were, respectively, applied as mesoporous and macroporous soft templates. During curing process, the surfactant templates and precursors mixed, which supplied a homogeneous reaction system. At pyrolysis, F127 and SDS were decomposed to prepare the 3D honeycomb-like skeleton, while the micropores were obtained by the activation of SO_2 and NH_3 during pyrolysis. The honeycomb-like porous structure, large specific surface area, and abundant heteroatom doping simultaneously helped to enhance the electrochemical performance of as-obtained porous carbon materials. Yong Liu et al. reported the simple preparation of honeycomb-like nitrogen-doped carbon nanospheres by employing dopamine as carbon and nitrogen precursor and PS-b-PEO diblock copolymer as a soft template [27]. The mesoporous carbon nanospheres had large multilayer adsorption surface areas (554 m^2g^{-1}) and high nitrogen contents (6.9 wt%). Honeycomb-like MnO_2 and carbon composite porous nanofiber structure was prepared via electroblown spinning and in situ hydrothermal synthesis with PVA soft template [28]. Various parameters including the ratios of honeycomb-like carbon nanofiber and $KMnO_4$, hydrothermal process times, and temperatures were explored to optimize reaction conditions. The large specific surface area, proper pore size distribution, and excellent electrical conductivity made it appropriate for the application in supercapacitor. The specific capacity was 421.5 Fg^{-1} at 0.5 Ag^{-1}, and it retained about 81.2% capacitance over 3000 cycles. The honeycomb-like carbons were also synthesized using coal tar pitch and melamine as soft template coupled with KOH activation [29]. The specific surface area of honeycomb-like porous carbons reached 2038 m^2g^{-1}. As electrodes for supercapacitors, the honeycomb-like porous carbons showed 221 Fg^{-1} at 0.05 Ag^{-1} for 6 M KOH electrolyte, excellent rate performance with 81.0% capacitance remaining at 20 Ag^{-1}, and perfect cycle life with 95.3% capacitance retention over 10,000 cycles.

7.2.2   KOH + BIOMASS

Various honeycomb-like porous carbons have been prepared by KOH activation of different types of biomass pyrolysis. Chuanxiang Zhang et al. adopted an immerged method, and the effective pore-forming was contemporaneously achieved by KOH activation using swelling feature of hydrophilic soybean protein gels in alkaline solution (Figure 7.5) [30]. The honeycomb-like porous carbon exhibited advantageous features for EDLC including honeycomb-like carbon morphology, high specific surface area, and governable surface groups, which provided a lot of favorable nano-channels with appropriate surface property for rapid electrolyte diffusion and effective electron transfer in the frameworks. Particularly, lots of macropores ranging from 50 to 60 nm and mesopores with diameter of 2–50 nm disperse equally onto the surface of carbon materials. These mesopores can act as ion storage systems to accommodate more electrolytes and reduce the ion dispersal length to the micropores channel. The electrode presented 330 Fg^{-1} at 0.5 Ag^{-1}, 94.6% capacitance preservation over 10,000 cycles, and excellent rate capability. Besides, the symmetry device delivered 8.3 Wh kg^{-1} energy density. In addition, a new honeycomb-like N and O co-doped carbon has been prepared using soybean residue as the carbon source by

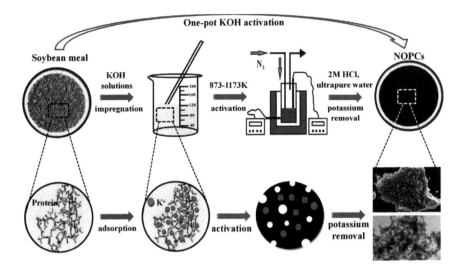

FIGURE 7.5 Schematic illustration of the fabrication processes of N and O co-doped porous carbon (NOPCs). (Reprinted with permission from Zhao, H.H. et al., *J. Alloy Compd.*, 766, 705–715, 2018.)

carbonization and activation [31]. The obtained honeycomb-like porous carbon possessed a unique hierarchical carbon structure, high surface area (2690.3 m^{2}g^{-1}), large pore volume (1.34 cm^3/g), and proper N and O contents.

Honeycomb-like porous carbon structure was obtained using corncob by hydrothermal carbonization with potassium hydroxide templating [32]. The obtained honeycomb-like porous carbon possessed a multiple macroporous and mesoporous structure, and the diameter of cells was around 600 nm. During the synthetic process, the potassium compounds derived from KOH could serve as templates to form macropores, which could serve as ion storeroom for electrolyte, assuring the fast diffusion of electrolyte ions to reduce the diffusion paths. Besides, potassium hydroxide etched the carbon skeleton to form micropores, which significantly enhanced the specific surface area and supplied more efficient active sites. Electrochemical tests showed that the honeycomb-like porous carbon exhibited 452 Fg^{-1} and excellent electrical conductivity.

Three-dimensional interlinked carbon honeycomb was successfully prepared using activated plane tree fluff with the addition of Co^{2+} and potassium [33]. The as-prepared honeycomb-like porous carbon exhibited beneficial configurational characteristics for applying in supercapacitors, including interlinked ways, large specific surface area, high pore volume, and doping heteroatom. The electrode material showed a very good charge storage capacity (493 Fg^{-1}) in a three-electrode system. The energy density of 42.8 Wh kg^{-1} was also assured with 1 M LiPF$_6$ in ethylene carbonate/diethyl carbonate (EC/DEC) as the organic electrolyte at a power density of 299 W kg^{-1}.

The pine cone flowers mainly include 46% glucose, 25% mannose, 24% klason lignin, and bits of galactose/xylose. Cellulosic and lignin derivatives could serve as a good biomass carbon source to prepare the carbon samples. Honeycomb-like carbon frameworks were obtained by the KOH treatment under N_2 atmosphere with pine cone flowers as biomass source. Three-dimensional honeycomb-like carbon skeleton-pillared nickelous hydroxide nanosheets for application as supercapacitor electrode materials were prepared [34]. Since then, the nickelous hydroxide nanosheets were synthesized evenly on the surface of honeycomb-like conducting scaffolds via a solvothermal method. The materials supplied pathways for efficient diffusion of electrolyte ions and fast transportation of electrons when employed as an electrode material. The electrode showed a good specific capacitance of about 916.4 Fg^{-1} at 1 Ag^{-1} with excellent cycling life compared to the pristine honeycomb-like conducting scaffolds and $Ni(OH)_2$ electrodes. Honeycomb-like nitrogen and sulfur co-doped biomass-based carbons were prepared by a green, low-cost, and high-yield method [35]. Firstly, the lotus plumule was carbonized in nitrogen atmosphere and activated with KOH to obtain the honeycomb-like porous carbon powders.

The waste paper was converged by cutting into small pieces, which were disposed with sulfuric acid and used in the hydrothermal system. Then, the filtration powder was activated by KOH followed by carbonization under inert atmosphere leading to the formation of honeycomb-like carbon [36] (Figure 7.6). The present challenges for liquid device in energy storage equipments were closely correlated with the flexibility and portability of devices. The good honeycomb-like and core-shell configuration resulted in large active surface area electrochemically, which exhibited excellent specific capacitance of 227 Fg^{-1} in liquid potassium hydroxide electrolyte. Biomass-based honeycomb-like microstructure carbon with large surface area (1776.9 m^2g^{-1}) was obtained by KOH and urea activation of *Ailanthus altissima* stems at 800°C [37]. The honeycomb-like microstructure carbon showed a large capacitance (300.6 Fg^{-1} at 0.5 Ag^{-1} in 1 M H_2SO_4) and retained 213.4 Fg^{-1} at 20 Ag^{-1} as electrode material.

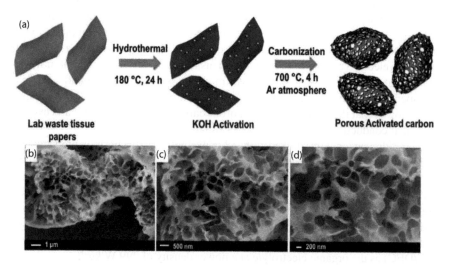

FIGURE 7.6 (a) Schematic representation of the formation of the porous carbon and (b–d) field emission scanning electron microscopy (FE-SEM) images of the carbon. (Reprinted with permission from Veerasubramani, G.K. et al., *J. Mat. Chem. A*, 5, 11100–11113, 2017.)

No distinct capacitance reduction was observed after 5000 cycles demonstrating the robust long-life. The good performance of honeycomb-like microstructure carbon electrode can be ascribed to the internal honeycomb-like porous microstructure, which supplied more surfaces for KOH activation to form more microporous network. The work provided a promising method using the unique microstructure of natural raw materials by simple technologies to obtain sustainable energy development. Porous carbon skeleton with nitrogen doping is current research hot topic for obtaining excellent supercapacitor electrode materials. Three-dimensional honeycomb-like carbon with interlinked porosity and self-doping nitrogen was prepared by one-step KOH activation from waste cotton seed husk [38]. The 3D honeycomb-like carbon had micro-, meso-, and macro-tri-pores, a high specific surface area, and 3D architecture. Due to the obvious characteristics, 3D honeycomb-like porous carbon showed high specific capacitances (238 and 200 Fg^{-1} at 0.5 and 20 Ag^{-1}, respectively), displaying good capacitance maintenance of 84%. The assembled device also showed high specific capacitance of 52 Fg^{-1} at 0.5 Ag^{-1} with 10.4 $Wh\ kg^{-1}$ energy density at 300 $W\ kg^{-1}$ and 91% capacitance retention over 5000 cycles.

The results demonstrated the feasibility of utilizing green and wholesale carbon materials obtained from waste biomass, for example, the pomelo peel as a raw material. Interlinked honeycomb-like microstructures have been triumphantly prepared by using pomelo peel as feedstock [39]. Benefiting from the specific honeycomb-like microstructure and large specific surface area, the honeycomb-like carbon exhibited approvable capacitance characteristics—374 Fg^{-1} at 0.1 Ag^{-1}—and excellent cycling stability—92.5% capacitance maintenance after 5000 cycles. What is more, the assembled symmetric device delivered satisfactory gravimetric and volumetric energy density of 20 $Wh\ kg^{-1}$ and 18.7 $Wh\ L^{-1}$ in solution electrolyte, respectively.

Yahui Wang et al. developed a facile and sustainable method for the preparation of 3D interlinked honeycomb-like carbon derived from sunflowers stem [40]. The optimized sample had large specific surface area with 3D interconnected honeycomb-like porous structure and high oxygen content. Due to its synergistic effect, the material showed a satisfactory specific capacitance of 349 Fg^{-1} at 1 Ag^{-1}, good rate capability (247 Fg^{-1} at 50 Ag^{-1}), and excellent cycling stability (retaining 98.6% after 10,000 cycles) in 6 M KOH aqueous electrolyte. Moreover, the asymmetric supercapacitor showed a high energy density of 58.8 $Wh\ kg^{-1}$ and good electrochemical stability (83.1% of initial capacitance retention after 10,000 cycles). Therefore, these unique properties enable the material to become a promising high-performance electrode material for supercapacitors. Qingli Hao et al. prepared a honeycomb-like enteromorpha-derived nitrogen-doped carbon using one-step pyrolysis under 600°C with a heterogeneous mixture of enteromorpha/NaOH/urea, accomplishing carbonization, activation, and N-doping synchronously, where a simple decolorization treatment of enteromorpha was performed [41]. Green bristlegrass seeds were triumphantly explored as the new biomass raw material to fabricate carbon materials. With the aid of gelatinization, 3D interlinked carbon honeycomb materials with heteroatom doping, large specific surface area, and porous honeycomb structure can be controllably prepared by pyrolysis carbonization [42]. Nitrogen/sulfur/oxygen tri-doped honeycomb carbon was triumphantly prepared controllably from green bristlegrass seeds with one-step carbonization under the assistance of gelatinization

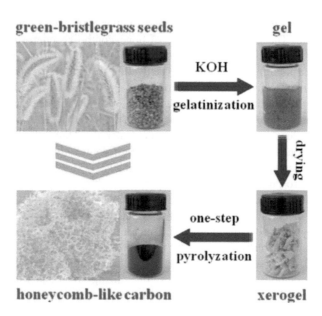

FIGURE 7.7 Flow diagram of the synthesis of 3D interconnected honeycomb-like porous carbon materials derived from green bristlegrass seeds. (Reprinted with permission from Zhou, W. et al., *J. Power Sources*, 402, 203–212, 2018.)

treatment (Figure 7.7). Benefitting from the 3D interlinked hierarchical honeycomb-like architecture coupled with large specific surface area and multi-heteroatom doping, the honeycomb-like carbon materials showed approvable specific capacitance of 391 Fg^{-1} at 0.5 Ag^{-1} with good rate capability—67.8% capacitance retention at 50 Ag^{-1}—and excellent cycling stability—97.2% capacitance retention after 10,000 cycles. Meanwhile, the device in 1 M Na$_2$SO$_4$ aqueous electrolyte delivered a satisfactory energy density of 20.15 Wh kg^{-1} under a power density of 500 W kg^{-1}. Besides, it still maintained 15.56 Wh kg^{-1} even at up to 20 kW kg^{-1}. The supercapacitor device also showed superior cycling life with 94.2% capacitance maintenance after 10,000 cycles. During the calcination process with aid of KOH, the porous carbon was converted from the pine needles and the nanopores were self-assembled into honeycomb carbon interlinked frameworks [43]. Honeycomb carbon exhibited an interlinked structure which was connected with the obvious carbon skeleton. Hongfei Xu et al. reported 3D porous honeycomb structure was fabricated by a straightforward method coupling hydrothermal carbonization and KOH chemical activation [44]. A shape transformation from microsphere to honeycomb-like structure was achieved using biomass-based carbon materials. For working electrode in supercapacitors, a satisfactory capacitance of 264 Fg^{-1} at 0.5 Ag^{-1} was achieved in three-electrode system. Besides, a high retention ratio of about 77.45% at 30 Ag^{-1} and excellent cycling life of 91% capacitance retention over 5000 cycles were also proved. The study provided an effective method to fabricate shape-controlled carbon with specific surface area of 1286 m^{2}g^{-1} that exhibited important potential as new energy devices. Yingliang Liu et al. presented a porous carbon material derived from bagasse by the goal-directed coupling of chemical activation and hydrothermal carbonization [45]. The bagasse-based porous carbon exhibited not only a honeycomb-like texture but also a very satisfactory specific surface area. Benefiting from the association of multiporous structure and well-developed porosity, the prepared carbon electrode showed a high capacitance of 413 Fg^{-1} at 1 Ag^{-1} and a good cycling life of 93.4% capacitance retention over 10,000 cycles in aqueous supercapacitors.

7.2.3 CARBONIZATION OF HONEYCOMB-LIKE STRUCTURE PRECURSORS

Zhihong Zhu et al. reported the fabrication of an interlinked porous carbon material by directly carbonizing mollusc shell [46]. The 3D carbon frameworks consisted of hexangular-arranged channels, which showed efficient electrolyte penetration and enabled the mollusc shell-based carbon material to be an excellent conductive scaffold as supercapacitor electrodes (Figure 7.8). $NiCo_2O_4$/honeycomb-like structure composites acted as supercapacitor electrode, which exhibited anomalously large specific capacitance (1696 Fg^{-1}), good rate discharge performance (58.6% capacity retention at 15 Ag^{-1}), and excellent cycling life (88% retention after 2000 cycles).

Ordered honeycomb-like macroporous carbon from biomass with extremely interlinked hexangular channels was obtained by carbonizing the organic matrix of mollusc shells [47]. The specific structure ensured rapid charge transfer by the whole 3D network, which made the carbon skeleton an ideal substrate as electrode material. Cuboidal Co_3O_4 was embedded with the 3D network by a simple hydrothermal treatment, and the as-obtained Co_3O_4@3D network showed a specific capacitance of 1307 Fg^{-1} at 1 Ag^{-1}, impressive rate performance of 61.0% capacity retention at 20 Ag^{-1}, and 84% of the capacitance maintenance over 3000 cycles.

The green preparation of honeycomb-like carbon microspheres from liquefied larch sawdust by ultrasonic spray pyrolysis has been reported [48]. The size of the

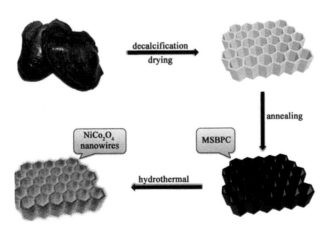

FIGURE 7.8 Preparation process of $NiCo_2O_4$/MSBPC composites. (Reprinted with permission from Xiong, W. et al., *ACS Appl. Mater. Interfaces*, 6, 19416–19423, 2014.)

obtained honeycomb-like carbon microspheres increased from 560 to 890 nm at 5% solution concentration to 0.58–1.4 μm and 0.58–2.1 μm at 11% and 16%, respectively. The optimal carbon spheres prepared with 5% solution concentration possessed a honeycomb-like carbon structure, uniform size, and exact pore size distribution.

Ruqiang Zou et al. [49] showed a novel method to build the nitrogen and sulfur co-doped honeycomb-like porous carbons with Co_9S_8 embedded inside from metal-organic frameworks (MOF). The obtained Co_9S_8@CNS900 showed notably comparable oxygen reduction reaction (ORR) catalytic property and good long-term life compared with 20 wt% Pt/C under KOH electrolyte. The excellent electrochemical performance could be linked with the specific open honeycomb structure with large surface area and pore volume and rich content of the active species. The current results developed a new way for fabricating novel controllable carbon nanostructures. Honeycomb-like carbon nanostructures with N-doping assembled from mesoporous 2D nanosheets have been obtained by MOF via thermal annealing under inert atmosphere [50]. The as-obtained honeycomb-structured nanofilms showed advantageous features including honeycomb-like morphology, 19% N-doping, 3.7 nm mesoporous structure, and abundant defects for electrochemical energy storage (Figure 7.9).

Yusuke Yamauchi et al. triumphantly developed a confined strategy to develop flake-like MOF by using various 2D amphoteric hydroxides as both the precursor and template [51]. It was shown that the MOF-derived porous carbon maintained the 2D morphology and showed specific honeycomb-like structure due to the restricted shrinkage. The whole properties of the honeycomb-like carbon were tunable due to the tailored metal compositions in amphoteric hydroxides, which were desirable for goal-oriented applications in supercapacitors.

Biomass-based porous carbons were developed for their green and wide range of sources. A honeycomb-structured hard carbon material using pine pollen as carbon precursor has been successfully prepared [52]. Pine pollen-based porous carbon showed good electrochemical properties because of its 3D structure and proper layer spacing (about 0.41 nm) with 171.54 m^2g^{-1}.

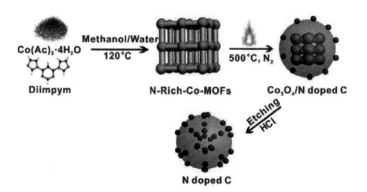

FIGURE 7.9 Schematic illustration of the fabrication of the N-doped C composite by a stepwise pyrolysis and etching process. (Reprinted with permission from Han, X.G. et al., *J. Mat. Chem. A*, 6, 18891–18897, 2018.)

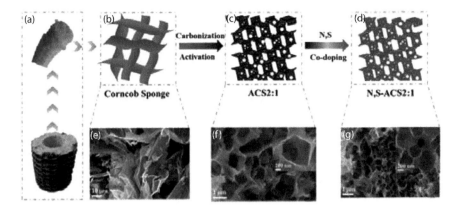

FIGURE 7.10 (a) The images of corncob sponge. (b–d) Schematic of the synthesis for nitrogen and sulfur co-doped activated corncob sponge, and (e–g) the corresponding SEM images of the corncob sponge, activated corncob sponge, and nitrogen and sulfur co-doped activated corncob sponge, respectively 7.17 (a) SiO_2 will be embedded in the assembled lamellar or bulk structure of RuO_2. (c) the RuO_2 self-assembled together layer by layer. (b,d) RuO_2 honeycomb networks (RHCs) and hollow spherical structures (RHSs) are obtained from chemical corrosion of the corresponding precursor respectively.

Corncob sponge is an agricultural secondary product and abundant across the world. A green one-step activation and carbonization method was explored to convert corncob sponge into 3D interlinked honeycomb-like carbon, followed by fabricating N and S double-doped activated corncob sponge [53] (Figure 7.10). The resultant products possessed 1874 m^2g^{-1} accessible surface area induced by the 3D honeycomb-like framework and extremely porous structure for large ion storage and ion transfer. In addition to the EDLC, heteroatom doping evoked faradic contribution. N and S co-doped honeycomb-like porous carbon demonstrated 404 and 253 Fg^{-1} specific capacitance at 0.1 and 10 Ag^{-1} in KOH electrolyte, respectively, with only 1% cycling loss after 10,000 cycles. Besides, the assembled symmetric flexible solid-state devices displayed a satisfactory integrated energy density of 30 Wh kg^{-1} at power density of 8 kW kg^{-1} and 99% capacitance retention over 10,000 cycles.

Bombyx mori possesses an oriented β-sheet crystal structure with a lamellar-like layer and highly porous nonwoven structure. As reported by Zhang Ming [54], after the thermal carbonization process of *Bombyx mori* silk cocoons, the sericin was eliminated and porous and graphene-type honeycomb-structured N-doped and Co-Mn incorporated carbon materials derived from *Bombyx mori* silk cocoons were synthesized via one-step thermal carbonization and graphitization process. Due to the large specific surface area of honeycomb-like structure, the large Co-Mn active sites, the pyridine-N and graphitic-N were exposed onto the prepared 3%Co-Mn/silk cocoon carbon materials. In addition, using natural sericin as the nitrogen source and carbon source, Junkuo Gao et al. [55] synthesized a series of carbon catalysts exhibiting a perfect honeycomb-like porous 3D structure with a high specific surface area (143 m^2g^{-1}), rich macro/mesoporous structure, and Fe_3O_4 doping particles that were little and even.

7.2.4 FOAMING METHOD

The methods for preparing honeycomb-structured carbon with foaming procedures of carbon precursors are divided into two: pyrolysis under pressure and blowing with some chemical additives. The former consists of saturation by decomposition gases from the precursor itself in a closed vessel. Honeycomb-carbon foams with adjustable porous structures were prepared from larch sawdust by Shouxin Liu et al. [56]. The porous structure was controlled by addition of polyethylene glycol (PEG). The PEG molecules were stretched and dispersed in the aqueous phenolic resin system. Some of the PEG molecules self-assembled to form an ordered net and banded structure, and then interacted with the resin to form settled polymers.

Mesoporous honeycomb carbons were obtained from pitch by foaming methods without any templates [57]. Abundant mesopores ranging from 2 to 5 nm and rich honeycomb pore endowed a hierarchical pore-in-porous structure characteristic and an ultrahigh specific surface area (3473 m^2g^{-1}) (Figure 7.11). The bubbles then united and grew under proper conditions leading to generation of mesopore voids or macropores in as-obtained products. With their superior structure, honeycomb-like mesoporous carbons delivered a high capacity of 339 Fg^{-1} at 1 Ag^{-1} in aqueous KOH electrolyte and maintained 85.6% of the capacity even at 50 Ag^{-1} current density. Meanwhile, the cell assembled by two honeycomb-like mesoporous carbon electrodes yielded a conspicuous 34.5 Wh kg^{-1} at 679.4 W kg^{-1}. The work opened a new way for the controllable preparation of advanced carbon materials from pitch and showed further applications in flexible electrochemical supercapacitors.

Xuebo Cao et al. demonstrated the one-pot, chemical-free fabrication of nitrogen and phosphorus co-doped honeycomb-like mesoporous carbons [58]. The material was prepared by the direct thermolysis of popcorn in air environment. For the strategy, N and P groups from proteins were condensed with graphitic matrix, and H_2O, CO_2, and other thermolysis by-products were activated by disordered carbon to form pore channels.

Honeycomb-like nitrogen-doped porous carbons have been successfully synthesized by chemical blowing with melamine using glucose as carbon source and KOH

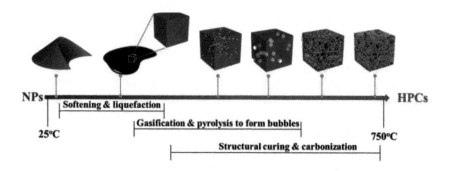

FIGURE 7.11 Schematic diagram displaying the overall evolution from naphthalene-derived pitch (NPs) into HPCs. (Reprinted with permission from Guan, T. et al., *ACS Sustain. Chem. Eng.*, 7, 2116–2126, 2019.)

as an activation reagent [59]. The obtained honeycomb-like nitrogen-doped porous carbon possessed a film-like morphology with ultrathin thickness, micro/mesoporous structure, 1997.5 m^2g^{-1} specific surface area, and 0.94 cm^3g^{-1} pore volume. Moreover, owing to the synergism of EDLC and pseudocapacitance, honeycomb-like nitrogen-doped porous carbon material with 3.06 wt% nitrogen content showed 312 Fg^{-1} specific capacitance at 0.5 Ag^{-1} in 6 M KOH aqueous electrolyte. It also showed excellent rate capability (247 Fg^{-1} at 30 Ag^{-1}) and cycle life (91.3% retention over 4000 cycles). The assembled symmetric devices showed 20.2 Wh kg^{-1} energy density at 448 W kg^{-1} power density under voltage of 1.8 V in 0.5 M Na_2SO_4 aqueous electrolyte. Porous honeycomb-like carbon originated from the gases that were formed during the decomposition of NH_4Cl while blowing the melted sucrose at 160°C [60]. This unique morphology contributed to the higher surface area, which allowed more locations for both sulfur loading and the adsorption of polysulfides.

7.3   HONEYCOMB-LIKE METALLIC COMPOUND

Transition metal-based nanoparticles have attracted widespread attention as supercapacitors electrode material due to their extremely high surface-area-to-volume ratio and intrinsic synergies of individual components; however, they still bear limited interior capacity and cycling life owing to single geometric configuration, low surface activity, and poor structural integrity. The flat utilization rate of active materials has been a key interference for the large-scale ultracapacitors. Three-dimensional honeycomb-like morphology has been proved as an ideal electrode shape in energy storage systems owing to the collaborative combination of the multilevel structure advantages. In this section, the honeycomb-like metallic compounds as hopeful cathode materials for supercapacitors have been discussed.

7.3.1   HONEYCOMB-LIKE NICKEL COMPOUND

Among the supercapacitor materials, nickel compounds such as nickel hydroxide, nickel oxide, nickel sulfide, nickel phosphide, and so on have been considered as a large-scale alternative due to their flat structure with tunable layer spacing relying on intercalated anions.

Honeycomb-like $Ni(OH)_2$ nanosheets were successfully grown in situ onto the surface of nickel foam under mild conditions and low temperature by Weihua Chen et al. [61]. The vibratory growing circumstances maintained the ultrathin nanosheets and ensured tight touch with the nickel foam. At the same time, the nanosheets showed an intersecting structure with 15 nm thickness and offered lots of active surface atoms shortening electron transport distance to the material. Combining the ultrathin $Ni(OH)_2$ nanosheets and ultrashort electron transport, high specific capacitance of 213.55 Fg^{-1} was achieved at 1 Ag^{-1} in a two-electrode system. Meanwhile, the device also achieved 74.94 Wh kg^{-1} energy density at 197.4 W kg^{-1} power density. In addition, size-controllable $Ni(OH)_2$ was electrodeposited on Ti wire to prepare the integrated electrode by Shangbin Sang et al. [62]. The nanosheet $Ni(OH)_2$-forming honeycomb-like structure with a pristine crystal size of 10–15 nm showed 260 mah g^{-1} specific capacity at 5 Ag^{-1} and 150 mah g^{-1} at 100 Ag^{-1}.

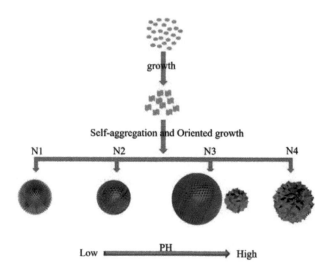

FIGURE 7.12 Illustration of the morphological evolution of samples. (Reprinted with permission from Ren, X.C. et al., *ACS Appl. Mater. Interfaces*, 7, 19930–19940, 2015.)

The mesoporous cathode materials promising for supercapacitors have been obtained using honeycomb-like NiO microspheres by hydrothermal reaction followed by an annealing treatment by Yinghui Wei et al. [63] (Figure 7.12). The electrochemical experiment demonstrated the specific capacitance was 1250 Fg^{-1} at 1 Ag^{-1}, which retained 945 Fg^{-1} at 5 Ag^{-1} and 88.4% retention after 3500 cycles. Besides, the 3D graphene for supercapacitors has been obtained as anode material, which displayed 313 Fg^{-1} at 1 Ag^{-1}. The asymmetric supercapacitor was assembled based on the honeycomb-like NiO and rGO, which achieved 74.4 Fg^{-1} specific capacitance and 23.25 Wh kg^{-1} energy density.

The Mn-doped $Ni(OH)_2$-modified Ni_3S_2 honeycomb-like nanohybrids were synthesized by Lin Ye et al. by a one-step hydrothermal treatment with temperature regulation [64]. The various temperature controls accomplished rapid nanostructure adjustment from thicker to thinner nanosheets. Rich open space and internal ultrathin nanosheets made up the intersected honeycomb-like structure. At the same time, the addition of Ni_3S_2 and Mn-doping into $Ni(OH)_2$ generated collaborative effects. As expected, the composite electrode showed 2518.8 mC/cm^2 areal capacitance at 1.2 mA/cm^2 and still had 1027.7 mC/cm^2 at 120 mA/cm^2. The nanohybrids displayed intersected nano-honeycomb, which was composed of rich open space and ultrathin internal nanosheets. The assembled supercapacitor possessed 51 W kg^{-1} energy density at 425 W kg^{-1}, and maintained 16.4 W kg^{-1} at 8600 W kg^{-1}.

Honeycomb-like $NiMoO_4$ nanosheets with interlinked porous structure, which supplied large electrochemically active sites, facilitated electrolyte immersion and ion diffusion and provided effective channels for electron transport [65]. The as-prepared ultrathin mesoporous $NiMoO_4$ exhibited high capacitance/capacity, excellent rate performance, and good cycling stability, representing a promising difunctional electrode material for energy storage. Therefore, the as-prepared $NiMoO_4$ electrode

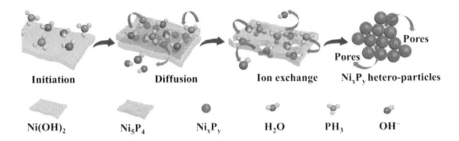

FIGURE 7.13 Schematic illustration of the proposed phosphorization and morphology evolution process at the surface of Ni_xP_y nanosheets. (Reprinted with permission from Liu, S. et al., *ACS Appl. Mater. Interfaces*, 9, 21829–21838, 2017.)

exhibited an excellent rate performance and a high specific capacity and cycling life in supercapacitors. Besides, 43.5 Wh kg^{-1} energy density at 500 W kg^{-1} was obtained for the symmetric supercapacitor consisting of $NiMoO_4$.

A 3D honeycomb-like $NiCo_2S_4$ nanostructure has been successfully prepared by hydrothermal method assisted by sulfuration by Zhiguo Zhang et al. [66]. Over 14 mAh/cm specific capacity at 1 mA/m^2 was achieved and 7.96 mAh/cm^2 still remained even at 50 mA/cm^2. The 3D honeycomb-like $NiCo_2S_4$ retained 99.64% of the original capacity over 1000 cycles. Seong Chan Jun et al. proposed a green strategy to build a complicated architecture consisting of honeycomb-like porous Ni_5P_4-Ni_2P nanosheets with rich interlinked hetero-nanoparticles [67] (Figure 7.13). Benefiting from the collaborative effect of the multicomponent setups, the synthesized Ni_5P_4-Ni_2P delivered 1272 Cg^{-1} specific capacity at 2 Ag^{-1} and retained 90.9% capacity over 5000 cycles. An asymmetric supercapacitor employing as-synthesized honeycomb-like biphasic Ni_5P_4-Ni_2P as the positive electrode and porous carbon as the negative electrode showed 67.2 Wh kg^{-1} at 0.75 kW kg^{-1} and 20.4 Wh kg^{-1} at 15 kW kg^{-1}. The results proved that the biphasic Ni_5P_4-Ni_2P with honeycomb-like innovative nanostructure can be possibly applied in high-performance supercapacitors. The results also proved that the rational assembly and hetero-growth of active materials are promising for high-performance supercapacitors while maintaining a high power density.

Honeycomb-like $Na_3Ni_2SbO_6$ was prepared by single solid-state method, the electrochemical behavior of which was tested as supercapacitor material [68]. The results showed honeycomb-like $Na_3Ni_2SbO_6$ had microcrystal morphology. The results showed that the specific capacitance of $Na_3Ni_2SbO_6$ electrodes was 387 Fg^{-1} at 0.5 Ag^{-1}. The extremely pseudocapacitive property of $Na_3Ni_2SbO_6$ can be ascribed to redox reaction of Ni(II)-Ni(III). Among these sodium-based transition metal oxides, $Na_3Ni_2SbO_6$ showed prominent electrochemical properties when used as an electrode material in Na-ion batteries.

Honeycomb-like $NiCo_2O_4$ sheets consisting of interlinked nanofilms were directly grown on conductive substrates via a H_2O-ethylene glycol-assisted synthesis method [69]. After annealing, the resultant $NiCo_2O_4$ nanoflakes possessed rich pores, which were more helpful for improving electrochemical properties. Particularly, porous structures showed high pseudocapacitance of 920 Fg^{-1} at 40 Ag^{-1} with good capacitance retention at 16 Ag^{-1} after 3000 cycles, which holds great promise for high-performance supercapacitors.

A simple chemical bath deposition strategy to prepare a binder-free electrode material of NiCo$_2$O$_4$ nanoplates adhered to NiMoO$_4$ honeycomb composites on a nickel foam was proposed. Anil Kumar Yedluri et al. reported the synthesis of a honeycomb composite with folded silk-like NiCo$_2$O$_4$ nanoplates adhered to NiMoO$_4$ honeycomb composites onto nickel foam to increase the availability of electrochemically active areas providing additional ways for electron transport and improving the utilization rate of the electrode materials [70]. The as-fabricated honeycomb-like composite electrode showed 2695 Fg^{-1} at 20 A/g. Moreover, the honeycomb-like composite showed 61.2 Wh kg^{-1} energy density and excellent power density of 371.5 W kg^{-1} and 98.9% capacitance retention over 3000 cycles. The good electrochemical performance made the honeycomb-like composite with folded film-like nanostructure a promising candidate for high-performance electrochemical energy storage.

Hee-Je Kim et al. reported a facile two-step fabrication of honeycomb-like NiMoO$_4$@NiWO$_4$ nanocomposites on Ni foam as the electrode for high-performance supercapacitor applications [71]. The electrochemical properties of the as-synthesized honeycomb-like NiMoO$_4$@NiWO$_4$ composites were studied for their application as electrode material of supercapacitors. A maximum specific capacitance of 1290 Fg^{-1} was achieved for honeycomb-like NiMoO$_4$@NiWO$_4$ at a current density of 2 Ag^{-1} in a 3 M KOH electrolyte solution. This electrode exhibited excellent long cycle-life stability with 93.1% specific capacitance retention after 3000 cycles at a current density of 6 Ag^{-1}. Our studies indicated that the as-synthesized honeycomb-like NiMoO$_4$@NiWO$_4$ nanocomposites could be a promising candidate as the electrode material in high-performance electrochemical energy storage applications. Wei-dong Xue et al. reported a rational study showing the synthesis of NiCo$_2$O$_4$@NiMoO$_4$ nano-film core–shell arrays by two-step green hydrothermal treatment, in which Ni foam was directly utilized for supercapacitive investigation [72]. The supercapacitor based on NiCo$_2$O$_4$@NiMoO$_4$ nanofilm core–shell arrays showed good electrochemical performance with 6.29 F/cm^2 at 5 mA/cm^2 and 3.58 F/cm^2 at 100 mA/cm^2. Besides, the composite electrode showed 87% areal capacitance retention over 5000 cycles. Such dramatic electrochemical performance proved that the 3D honeycomb-like structure electrode could be a hopeful candidate for supercapacitors.

Three-dimensional honeycomb-like MnO$_2$/Ni(OH)$_2$ composites for supercapacitor electrode have been produced using microwave irradiation method [73] (Figure 7.14).

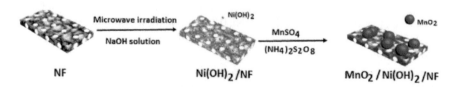

FIGURE 7.14 Schematic illustration of the formation process of 3D MnO$_2$/Ni(OH)$_2$ electrode. (Reprinted with permission from Wang, F. et al., *J. Alloy Compd.* 700: 185-190.

The results revealed that the $MnO_2/Ni(OH)_2$ electrode exhibited 506 Fg^{-1} at 16.7 Ag^{-1} and 96% excellent capacitance retention over 7000 cycles, which suggested the composite has helpful applications in electrochemical capacitors.

Honeycomb-like Co-Ni phosphate with honeycomb-like mesopores was firstly prepared by a facile hydrothermal treatment coupled with low-temperature pyrolysis [74]. Due to the suitable 3D mesoporous nanostructure, providing more active areas for electrochemical reaction and short diffusion paths for electrolyte ions, as well as the collaborative effect of Co and Ni, the CoNiP exhibited high supercapacitive performance. The even honeycomb-like structure offered lots of active sites for chemical reaction and shortened transport ways for the electrolyte ions diffusion. The asymmetric supercapacitor assembly with the CoNiP as positive material and porous carbon as negative material showed 45.8 Wh kg^{-1} energy density at 42.4 W kg^{-1} and even 30.7 Wh kg^{-1} at the power density of 2.8 kW kg^{-1}.

Honeycomb-like Ni-Mn layered double hydroxides (LDH) nanostructures have been bonded on carbon cloth by a simple hydrothermal treatment, which showed good electrochemical performance with 2239 Fg^{-1} at 5 mA/cm^2 [75]. It is obvious that the honeycomb-like nanostructure could supply abundant electrochemically active sites for redox reactions. Moreover, an asymmetric supercapacitor with good flexibility was fabricated based on the Ni-Mn-layered double-hydroxide electrode. Over 110% of the initial capacitance was retained after repeating 10,000 charge–discharge cycles at a current density of 3 mA/cm^2. The honeycomb-like Ni-Mn-layered double-hydroxide flexible electrode with good performance proposed a new pathway for flexible energy storage devices.

7.3.2 HONEYCOMB-LIKE MANGANESE COMPOUND

Manganese oxides were considered as a helpful candidate owing to their low cost, low toxicity, and large theoretical capacitance. Manganese oxides have been drawing burning attention since they were reported for supercapacitor application in 1999. For Mn with five known oxidation states, redox reactions involving ion exchange between manganese oxides and electrolytes occur. Though they have high theoretical capacitance, early studies on manganese oxides-based supercapacitors reported capacitance values of about 200 Fg^{-1} due to low conductivity. Various nanostructures with novel morphologies, improved conductivity, and high surface area have been researched to increase the capacitance.

Porous honeycomb architectures with manganese oxide were in situ formed on Ni foam via electrochemical oxidation from Mn_3O_4 nanosphere electrodes by Cuiyin Liu et al. [76]. The optimized manganese oxide electrodes showed 377 Fg^{-1} specific capacitance at 1 Ag^{-1}, superior rate capability, and 76% retention over 4000 cycles. The work provided helpful methods to develop high-performance manganese oxide electrodes. The manganese oxide electrodes showed porous honeycomb architecture with specific nanosheets attended with exceptional electrochemical performances with high specific capacitance and rate capability.

Honeycomb-like MnO_2 nanospheres were prepared by solution polymerization and in situ PANI polymerization. The MnO_2 microspheres with honeycomb-like structure coated with PANI layers served as a framework for polyaniline (PANI)

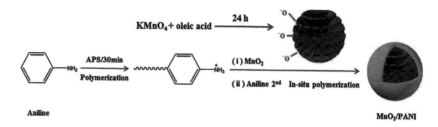

FIGURE 7.15 Schematic illustration of the growth process of MnO$_2$/PANI nanocomposite. (Reprinted with permission from Sun, X.W. et al., *Electrochim. Acta*, 180, 977–982, 2015.)

layer and PANI acted as a shielded coating layer to regulate MnO$_2$ nanospheres from dissolution [77] (Figure 7.15). The nanocomposite exhibited 555 Fg^{-1} (specific capacity of 116 mah g^{-1}) at 0.8 Ag^{-1} and a high specific capacity of 143 mah g^{-1} at the scan rate of 20 mV/s. Besides, the electrode material maintained about 77% of the initial capacitance over 1000 charge–discharge cycles at 8 Ag^{-1}.

Honeycomb-like MnO$_2$ nanofibers have been fabricated by solution reaction between KMnO$_4$ and carbon fibers [78]. The honeycomb porous MnO$_2$ nanofibers were composed of radially grown MnO$_2$ nanosheets with thickness of about 3–7 nm, interlinked with each other, forming the honeycomb pores. Very low KMnO$_4$ concentrations and sufficiently long reaction time were needed for the formation of this unique structure (Figure 7.16). The MnO$_2$ nanosheets constituting the honeycomb porous MnO$_2$ nanofibers were well separated with the sheet edges oriented on the surface, leading to excellent supercapacitive performance. Symmetric aqueous supercapacitors were assembled using the honeycomb porous MnO$_2$ nanofibers and 1 M Na$_2$SO$_4$ electrolyte, which exhibited a working voltage as high as 2.2 V and high energy density of 41.1 Wh kg^{-1} at the power density of 3.3 kW kg^{-1}. The supercapacitor capacity was retained at about 76% of its initial value after 3500 cycles, which is acceptable due to its high energy density. The results proved that the honeycomb porous MnO$_2$ nanofibers were of high help in developing advanced supercapacitors with high working voltage and energy density for practical applications.

The MnO$_2$ nanosheets were self-assembled to honeycomb-like microstructure by controlling the electroplating time [79]. The honeycomb-like microstructure showed good electrochemical performance in Na$_2$SO$_4$ electrolyte with 1110.85 Fg^{-1} at 1 Ag^{-1} based on the MnO$_2$ mass and 77.44% capacity retention from 1 Ag^{-1} to 10 Ag^{-1}. A flexible asymmetric device with good energy density was built using carbonized silk fabrics as flexible substrate decorated with carbon nanotube, and MnO$_2$ nanosheets as the positive and porous carbon powders as the negative electrodes, respectively. The assembled asymmetric device delivered a maximum energy density of 43.84 Wh kg^{-1} and a maximum power density of 6.62 kW kg^{-1}.

The 3D honeycomb-like porous structure was assembled from interlinked sheet-architectural MnO$_2$ on CNT [80]. For the specific MnO$_2$ architectures integrated

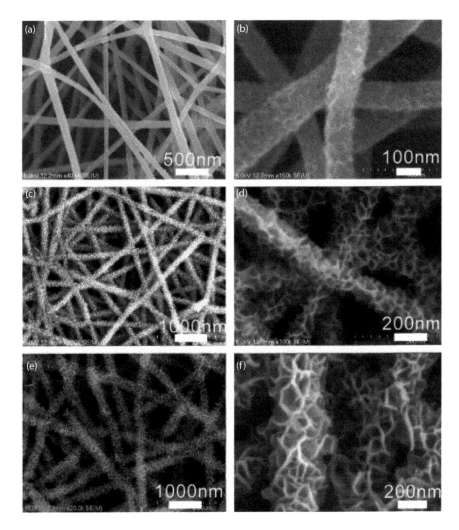

FIGURE 7.16 SEM images of the (a) pristine CNFs and the samples obtained in 2 mM KMnO$_4$ solution after reaction for (b) 12 h, (c and d) 48 h, and (e and f) 168 h. (Reprinted with permission from Zhao, L. et al., *Nano Energy*, 4, 39–48, 2014.)

onto CNT flexible films, excellent electrochemical performance was achieved with 324 Fg^{-1} at 0.5 Ag^{-1}. A maximum energy density of 7.2 Wh kg^{-1} and a power density as high as 3.3 kW kg^{-1} were achieved by the honeycomb MnO$_2$/CNT network device, which was comparable to the performance of other carbon-based and metal oxide/ carbon-based solid-state supercapacitor devices. Specifically, the long cycling life of the material was excellent, without loss of its original capacitance and with excellent coulombic efficiency of 82% over 5000 cycles.

7.3.3 HONEYCOMB-LIKE OTHER METAL COMPOUNDS

The large family of transition metal oxides has been researched widely for super-capacitors electrode materials. Starting from RuO_2 as a supercapacitor electrode in aqueous H_2SO_4 electrolyte in 1971, hollow spherical RuO_2 with honeycomb-like skeletons was rationally designed and synthesized with modified-SiO_2 templating via two hydrothermal methods by Zhongyi Liu et al. [81] (Figure 7.17). The work

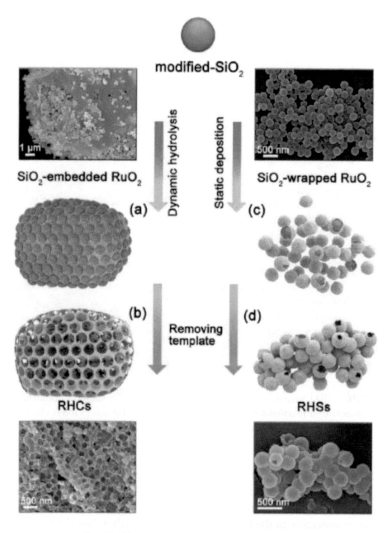

FIGURE 7.17 Schematic illustration of the fabrication procedure to produce the RuO_2 3D macroporous architecture. (Reprinted with permission from Peng, Z.K. et al., *ACS Appl. Mater. Interfaces*, 9, 4577–4586, 2017.)

proved the practicability of the SiO_2-templating method in preparing hollow nano-structured functional materials offering great benefits in electrochemical energy storage applications. The RuO_2 honeycomb-like frameworks exhibited the specific capacitance as high as 628 Fg^{-1} at 20 Ag^{-1}; this was about 81% of the capacitance retention at 0.5 Ag^{-1}. The honeycomb networks still retained 86% of initial specific capacitances over 4000 cycles. The honeycomb network frameworks had a well-defined pathway that allowed the transmission/diffusion of electrolyte and surface redox reactions. As a result, they exhibited good supercapacitor performance in both acid and alkaline electrolytes. Honeycomb networks with rough surfaces, void inner space, or interconnected porous structure increased the effective utilization of active materials. This led to excellent electrochemical energy storage performance.

Honeycomb-like Co_3S_4 nanosheet-decorated electrode was successfully prepared by using highly dispersive Co_3O_4 nanowires with Ni foam as template [82]. The nanosheets contained lots of 3–5 nm micropores assembled into 3D honeycomb-like structure for improving specific surface area of electrode. The enhanced electrochemical performance was achieved, including an excellent cyclability of 10,000 cycles at 10 Ag^{-1} and large specific capacitances of 2425 and 1252 Fg^{-1} at 1 and 20 Ag^{-1}, respectively. As expected, ultrathin 2D Co_3S_4 nanosheets with high mass loading were successfully fabricated on 3D nanofibre (NF) by using the mentioned ultrafine Co_3O_4 nanowires as precursors. The mass loading of Co_3S_4 nanosheets reached 6 mg/cm^2. The possible formation process of ultrathin nanosheets was attributed to the template function of the nearly monodisperse Co_3O_4 nanowires. Even so, the mass-specific capacitance (2415 Fg^{-1}) and areal capacitance (13.31 F/cm^2) were more than two times that of the Co_3O_4 nanowire precursors at the current density of 1 Ag^{-1}. Furthermore, a Co_3S_4 asymmetric supercapacitor was fabricated using the Co_3S_4/NF electrode and activated carbon electrode as the positive and negative electrode, respectively. The asymmetric supercapacitor exhibited high energy density, power density, and good long-term electrochemical stability. Honeycomb-like Co_3O_4 with high specific surface area was successfully prepared on Ni foam by the facile hydrothermal method followed by an annealing treatment, which was used as high-performance supercapacitor electrode by Jun Li et al. [83]. The specific capacitance of the electrode was 743.00 Fg^{-1} at 1 Ag^{-1} in the galvanostatic charge-discharge (GCD) test. Besides, the electrode also showed an excellent cyclic life, in which 96% of the initial specific capacitance remained at 1 Ag^{-1} for 500 cycles in the GCD test. This excellent electrochemical performance was ascribed to high specific surface area of Co_3O_4 nanosheets that provided added channels and space for the ions transportation.

7.4 CONCLUSION

An open-ended nanostructure is also highly needed for electrode materials to improve the accessible surface area, ion transport, or diffusion. Much attempts for constructing 3D honeycomb-like scaffold frameworks to enhance the structural stability, prevent stacking-induced loss of active surface area, and achieve fast ion diffusion have been made. Particularly, combined with interlinked macropores and rich micropores, honeycomb-like porous structure allows ions to fully penetrate electrode material and the pore walls provide continuous conductive pathway for charges, which are promising to obtain high energy densities and excellent power densities simultaneously. The obtained honeycomb-like metal compounds possessed a unique 3D hierarchical porous structure with interlinked meso/macropores, which facilitated the transport of electrolyte ions and offered a good environment for charge accumulation, thus showing good performance as electrode materials for supercapacitors.

REFERENCES

1. A.D. Marmorstein, V.L. Bonilha, S. Chiflet, J.M. Neill, E. Rodriguez-Boulan. 1996. The polarity of the plasma membrane protein RET-PE2 in retinal pigment epithelium is developmentally regulated. *Journal of Cell Science.* 109: 3025–3034.
2. Y. Uraki, Y. Tamai, T. Hirai, K. Koda, H. Yabu, M. Shimomura. 2011. Fabrication of honeycomb-patterned cellulose material that mimics wood cell wall formation processes. *Materials Science and Engineering: C.* 31: 1201–1208.
3. T. Lecuit, P.-F. Lenne. 2007. Cell surface mechanics and the control of cell shape, tissue patterns and morphogenesis. *Nature Reviews Molecular Cell Biology.* 8: 633.
4. G. Pawin, K.L. Wong, K.-Y. Kwon, L. Bartels. 2006. A homomolecular porous network at a Cu (111) surface. *Science.* 313: 961–962.
5. A.-H. Lu, J.-H. Smått, S. Backlund, M. Lindén. 2004. Easy and flexible preparation of nanocasted carbon monoliths exhibiting a multimodal hierarchical porosity. *Microporous and Mesoporous Materials.* 72: 59–65.
6. Z. Shi, Y. Feng, L. Xu, S. Da, M. Zhang. 2003. Synthesis of a carbon monolith with trimodal pores. *Carbon.* 41: 2677–2679.
7. Y. Han, X.T. Dong, C. Zhang, S.X. Liu. 2013. Easy synthesis of honeycomb hierarchical porous carbon and its capacitive performance. *Journal of Power Sources.* 227: 118–122.
8. L.M. Zhang, Z.B. Wang, J.J. Zhang, X.L. Sui, L. Zhao, D.M. Gu. 2015. Honeycomb-like mesoporous nitrogen-doped carbon supported Pt catalyst for methanol electrooxidation. *Carbon.* 93: 1050–1058.
9. M. Kim, I. Oh, H. Ju, J. Kim. 2016. Introduction of Co_3O_4 into activated honeycomb-like carbon for the fabrication of high performance electrode materials for supercapacitors. *Physical Chemistry Chemical Physics.* 18: 9124–9132.
10. C.F. Li, L.Q. Li, Z.P. Li, W.H. Zhong, Z.H. Li, X.Q. Yang, G.Q. Zhang, H.Y. Zhang. 2017. Fabrication of Fe_3O_4 dots embedded in 3D honeycomb-like carbon based on metallo-organic molecule with superior lithium storage performance. *Small.* 13: 1701351.
11. Z. Song, L. Li, D. Zhu, L. Miao, H. Duan, Z. Wang, W. Xiong, Y. Lv, M. Liu, L. Gan. 2019. Synergistic design of a N, O co-doped honeycomb carbon electrode and an ionogel electrolyte enabling all-solid-state supercapacitors with an ultrahigh energy density. *Journal of Materials Chemistry A.* 7: 816–826.
12. X.W. Liu, X.H. Liu, B.F. Sun, H.L. Zhou, A.P. Fu, Y.Q. Wang, Y.G. Guo, P.Z. Guo, H.L. Li. 2018. Carbon materials with hierarchical porosity: Effect of template removal strategy and study on their electrochemical properties. *Carbon.* 130: 680–691.
13. J. Li, Y.N. Zhang, W. Zhou, H.J. Nie, H.M. Zhang. 2014. A hierarchically honeycomb-like carbon via one-step surface and pore adjustment with superior capacity for lithium-oxygen batteries. *Journal of Power Sources.* 262: 29–35.
14. J. Li, H.M. Zhang, Y.N. Zhang, M.R. Wang, F.X. Zhang, H.J. Nie. 2013. A hierarchical porous electrode using a micron-sized honeycomb-like carbon material for high capacity lithium-oxygen batteries. *Nanoscale.* 5: 4647–4651.
15. S.K. Park, J. Lee, T. Hwang, B. Jang, Y. Piao. 2017. Scalable synthesis of honeycomb-like ordered mesoporous carbon nanosheets and their application in lithium-sulfur batteries. *ACS Applied Materials & Interfaces.* 9: 2430–2438.

16. H.K. Wang, J.K. Wang, D.X. Cao, H.Y. Gu, B.B. Li, X. Lu, X.G. Han, A.L. Rogach, C.M. Niu. 2017. Honeycomb-like carbon nanoflakes as a host for SnO_2 nanoparticles allowing enhanced lithium storage performance. *Journal of Materials Chemistry A.* 5: 6817–6824.

17. G.A.M. Ali, O.A. Habeeb, H. Algarni, K.F. Chong. 2019. CaO impregnated highly porous honeycomb activated carbon from agriculture waste: Symmetrical supercapacitor study. *Journal of Materials Science.* 54: 683–692.

18. W.H. Niu, L.G. Li, N. Wang, S.B. Zeng, J. Liu, D.K. Zhao, S.W. Chen. 2016. Volatilizable template-assisted scalable preparation of honeycomb-like porous carbons for efficient oxygen electroreduction. *Journal of Materials Chemistry A.* 4: 10820–10827.

19. H.Z. Chen, B. Zhang, Y. Cao, X. Wang, Y.Y. Yao, W.J. Yu, J.C. Zheng, J.F. Zhang, H. Tong. 2018. ZnS nanoparticles embedded in porous honeycomb-like carbon nanosheets as high performance anode material for lithium ion batteries. *Ceramics International.* 44: 13706–13711.

20. D. Zeng, Y.P. Dou, M. Li, M. Zhou, H.M. Li, K. Jiang, F. Yang, J.J. Peng. 2018. Wool fiber-derived nitrogen-doped porous carbon prepared from molten salt carbonization method for supercapacitor application. *Journal of Materials Science.* 53: 8372–8384.

21. L. Pan, X.X. Li, Y.X. Wang, J.L. Liu, W. Tian, H. Ning, M.B. Wu. 2018. 3D interconnected honeycomb-like and high rate performance porous carbons from petroleum asphalt for supercapacitors. *Applied Surface Science.* 444: 739–746.

22. G.X. Du, Q.X. Bian, J.B. Zhang, X.H. Yang. 2017. Facile fabrication of hierarchical porous carbon for a high-performance electrochemical capacitor. *RSC Advances.* 7: 46329–46335.

23. Z.H. Zeng, Y.F. Zhang, X.Y.D. Ma, S.I.S. Shahabadi, B.Y. Che, P.Y. Wang, X.H. Lu. 2018. Biomass-based honeycomb-like architectures for preparation of robust carbon foams with high electromagnetic interference shielding performance. *Carbon.* 140: 227–236.

24. X. Hu, X.H. Xu, R.Q. Zhong, L.J. Shang, H.T. Ma, X.L. Wu, P.Y. Jia. 2018. Facile synthesis of microporous carbons with three-dimensional honeycomb-like porous structure for high performance supercapacitors. *Journal of Electroanalytical Chemistry.* 823: 54–60.

25. H.R. Zhang, X.Y. Qin, J.X. Wu, Y.B. He, H.D. Du, B.H. Li, F.Y. Kang. 2015. Electrospun core-shell silicon/carbon fibers with an internal honeycomb-like conductive carbon framework as an anode for lithium ion batteries. *Journal of Materials Chemistry A.* 3: 7112–7120.

26. L. Sun, Y. Zhou, L. Li, H. Zhou, X. Liu, Q. Zhang, B. Gao, Z. Meng, D. Zhou, Y. Ma. 2019. Facile and green synthesis of 3D honeycomb-like N/S-codoped hierarchically porous carbon materials from bio-protic salt for flexible, temperature-resistant supercapacitors. *Applied Surface Science.* 467–468: 382–390.

27. Y.M. Zhang, Y. Liu, W.H. Liu, X.Y. Li, L.Q. Mao. 2017. Synthesis of honeycomb-like mesoporous nitrogen-doped carbon nanospheres as Pt catalyst supports for methanol oxidation in alkaline media. *Applied Surface Science.* 407: 64–71.

28. J.G. Ju, H.J. Zhao, W.M. Kang, N.N. Tian, N.P. Deng, B.W. Cheng. 2017. Designing MnO_2 & carbon composite porous nanofiber structure for supercapacitor applications. *Electrochimica Acta.* 258: 116–123.

29. X. He, H. Yu, L. Fan, M. Yu, M. Zheng. 2017. Honeycomb-like porous carbons synthesized by a soft template strategy for supercapacitors. *Materials Letters.* 195: 31–33.

30. H.H. Zhao, B.L. Xing, C. Zhang, G.X. Huang, Q.R. Liu, G.Y. Yi, J.B. Jia, M.J. Ma, Z.F. Chen, C.X. Zhang. 2018. Efficient synthesis of nitrogen and oxygen co-doped hierarchical porous carbons derived from soybean meal for high-performance supercapacitors. *Journal of Alloys and Compounds.* 766: 705–715.

31. F. Chen, J. Yang, T. Bai, B. Long, X.Y. Zhou. 2016. Biomass waste-derived honeycomb-like nitrogen and oxygen dual-doped porous carbon for high performance lithium-sulfur batteries. *Electrochimica Acta.* 192: 99–109.

32. X.F. Zhang, B. Wang, J. Yu, X.N. Wu, Y.H. Zang, H.C. Gao, P.C. Su, S.Q. Hao. 2018. Three-dimensional honeycomb-like porous carbon derived from corncob for the removal of heavy metals from water by capacitive deionization. *RSC Advances.* 8: 1159–1167.

33. Y.Q. Zhang, X. Liu, S.L. Wang, S.X. Dou, L. Li. 2016. Interconnected honeycomb-like porous carbon derived from plane tree fluff for high performance supercapacitors. *Journal of Materials Chemistry A.* 4: 10869–10877.

34. G. Nagaraju, S.M. Cha, J.S. Yu. 2017. Ultrathin nickel hydroxide nanosheet arrays grafted biomass-derived honeycomb-like porous carbon with improved electrochemical performance as a supercapacitive material. *Scientific Reports.* 7: 45201.

35. M. Chen, S. Jiang, C. Huang, X. Wang, S. Cai, K. Xiang, Y. Zhang, J. Xue. 2017. Honeycomb-like nitrogen and sulfur dual-doped hierarchical porous biomass-derived carbon for lithium-sulfur batteries. *ChemSusChem.* 10: 1803–1812.

36. G.K. Veerasubramani, A. Chandrasekhar, M.S.P. Sudhakaran, Y.S. Mok, S.J. Kim. 2017. Liquid electrolyte mediated flexible pouch-type hybrid supercapacitor based on binderless core-shell nanostructures assembled with honeycomb-like porous carbon. *Journal of Materials Chemistry A.* 5: 11100–11113.

37. M. Zhu, J.L. Lan, X. Zhang, G. Sui, X.P. Yang. 2017. Porous carbon derived from Ailanthus altissima with unique honeycomb-like microstructure for high-performance supercapacitors. *New Journal of Chemistry.* 41: 4281–4285.

38. H. Chen, G. Wang, L. Chen, B. Dai, F. Yu. 2018. Three-dimensional honeycomb-like porous carbon with both interconnected hierarchical porosity and nitrogen self-doping from cotton seed husk for supercapacitor electrode. *Nanomaterials.* 8: 412.

39. J.Y. Liu, H.P. Li, H.S. Zhang, Q. Liu, R.M. Li, B. Li, J. Wang. 2018. Three-dimensional hierarchical and interconnected honeycomb-like porous carbon derived from pomelo peel for high performance supercapacitors. *Journal of Solid State Chemistry.* 257: 64–71.

40. Y.H. Wang, Z.Y. Zhao, W.W. Song, Z.C. Wang, X.L. Wu. 2019. From biological waste to honeycomb-like porous carbon for high energy density supercapacitor. *Journal of Materials Science.* 54: 4917–4927.

41. L. Lu, X.Y. Jiao, J.W. Fan, W. Lei, Y. Ouyang, X.F. Xia, Z.X. Xue, Q.L. Hao. 2019. Cobalt ferrite on honeycomb-like algae-derived nitrogen-doped carbon for electrocatalytic oxygen reduction and ultra-cycle-stable lithium storage. *Electrochimica Acta.* 295: 461–471.

42. W. Zhou, S.J. Lei, S.Q. Sun, X.L. Ou, Q. Fu, Y.L. Xu, Y.H. Xiao, B.C. Cheng. 2018. From weed to multi-heteroatom-doped honeycomb-like porous carbon for advanced supercapacitors: A gelatinization-controlled one-step carbonization. *Journal of Power Sources.* 402: 203–212.

43. L.L. Xing, X. Wu, K.J. Huang. 2018. High-performance supercapacitor based on three-dimensional flower-shaped $Li_4Ti_5O_{12}$-graphene hybrid and pine needles derived honeycomb carbon. *Journal of Colloid and Interface Science.* 529: 171–179.

44. Z. Wang, S. Yun, X. Wang, C. Wang, Y. Si, Y. Zhang, H. Xu. 2018. Aloe peel-derived honeycomb-like bio-based carbon with controllable morphology and its superior electrochemical properties for new energy devices. *Ceramics International.* 45: 4208–4218.

45. P. Yu, Y. Liang, H. Dong, H. Hu, S. Liu, L. Peng, M. Zheng, Y. Xiao, Y. Liu. 2018. Rational synthesis of highly porous carbon from waste bagasse for advanced supercapacitor application. *ACS Sustainable Chemistry & Engineering.* 6: 15325–15332.

46. W. Xiong, Y.S. Gao, X. Wu, X. Hu, D.N. Lan, Y.Y. Chen, X.L. Pu, Y. Zeng, J. Su, Z.H. Zhu. 2014. Composite of macroporous carbon with honeycomb-like structure from mollusc shell and $NiCo_2O_4$ nanowires for high-performance supercapacitor. *ACS Applied Materials & Interfaces*. 6: 19416–19423.

47. Y.H. Liu, W. Yu, L. Hou, G.H. He, Z.H. Zhu. 2015. Co_3O_4@Highly ordered macroporous carbon derived from a mollusc shell for supercapacitors. *RSC Advances*. 5: 75105–75110.

48. X. Zhao, W. Li, S.S. Zhang, L.H. Liu, S.X. Liu. 2015. Facile fabrication of hollow and honeycomb-like carbon spheres from liquefied larch sawdust via ultrasonic spray pyrolysis. *Materials Letters*. 157: 135–138.

49. Q.-L. Zhu, W. Xia, T. Akita, R. Zou, Q. Xu. 2016. Metal-organic framework-derived honeycomb-like open porous nanostructures as precious-metal-free catalysts for highly efficient oxygen electroreduction. *Advanced Materials*. 28: 6391–6398.

50. X.G. Han, L.M. Sun, F. Wang, D. Sun. 2018. MOF-derived honeycomb-like N-doped carbon structures assembled from mesoporous nanosheets with superior performance in lithium ion batteries. *Journal of Materials Chemistry A*. 6: 18891–18897.

51. J. Wang, J. Tang, B. Ding, Z. Chang, X. Hao, T. Takei, N. Kobayashi, Y. Bando, X. Zhang, Y. Yamauchi. 2018. Self-template-directed metal-organic frameworks network and the derived honeycomb-like carbon flakes via confinement pyrolysis. *Small*. 14: e1704461.

52. Y.J. Zhang, X. Li, P. Dong, G. Wu, J. Xiao, X.Y. Zeng, Y.J. Zhang, X.L. Sun. 2018. Honeycomb-like hard carbon derived from pine pollen as high-performance anode material for sodium-ion batteries. *ACS Applied Materials & Interfaces*. 10: 42796–42803.

53. Y. Liu, Z. Xiao, Y. Liu, L.-Z. Fan. 2018. Biowaste-derived 3D honeycomb-like porous carbon with binary-heteroatom doping for high-performance flexible solid-state supercapacitors. *Journal of Materials Chemistry A*. 6: 160–166.

54. M. Zhang, T. Li, J. Wang, Y. Pan, S.J. Ma, H. Zhu, M.L. Du. 2018. Honeycomb-like structured and Co-Mn incorporated carbon materials derived from Bombyx mori cocoons act as a bifunctional catalyst for water splitting. *Chinese Journal of Inorganic Chemistry*. 34: 942–950.

55. C.Q. Li, T. Zhao, S.Y.H. Abdalkarim, Y.H. Wu, M.T. Lu, Y.W. Li, J.K. Gao, J.M. Yao. 2018. Fe_2O_3-N-doped honeycomb-like porous carbon derived from nature silk sericin as electrocatalysts for oxygen evolution reaction. *Zeitschrift Fur Anorganische Und Allgemeine Chemie*. 644: 1103–1107.

56. W. Li, Z. Huang, Y. Wu, X. Zhao, S. Liu. 2015. Honeycomb carbon foams with tunable pore structures prepared from liquefied larch sawdust by self-foaming. *Industrial Crops and Products*. 64: 215–223.

57. T. Guan, J. Zhao, G. Zhang, J. Wang, D. Zhang, K. Li. 2019. Template-free synthesis of honeycomblike porous carbon rich in specific 2–5 nm mesopores from a pitch-based polymer for a high-performance supercapacitor. *ACS Sustainable Chemistry & Engineering*. 7: 2116–2126.

58. L.W. Zhu, J. Wu, Q. Zhang, X.K. Li, Y.M. Li, X.B. Cao. 2018. Chemical-free fabrication of N, P dual-doped honeycomb-like carbon as an efficient electrocatalyst for oxygen reduction. *Journal of Colloid and Interface Science*. 510: 32–38.

59. Y. Wang, H. Xuan, G. Lin, F. Wang, Z. Chen, X. Dong. 2016. A melamine-assisted chemical blowing synthesis of N-doped activated carbon sheets for supercapacitor application. *Journal of Power Sources*. 319: 262–270.

60. W.X. Su, W.J. Feng, Y. Cao, L.J. Chen, M.M. Li, C.K. Song. 2018. Porous honeycomb-like carbon prepared by a facile sugar-blowing method for high-performance lithium-sulfur batteries. *International Journal of Electrochemical Science*. 13: 6005–6014.

61. S. Zhou, S. Cui, W. Wei, W. Chen, L. Mi. 2018. Development of high-utilization honeycomb-like α-Ni(OH)$_2$ for asymmetric supercapacitors with excellent capacitance. *RSC Advances.* 8: 37129–37135.
62. Y. Wu, S.B. Sang, W.J. Zhong, F. Li, K.Y. Liu, H.T. Liu, Z.G. Lu, Q.M. Wu. 2018. The nanoscale effects on the morphology, microstructure and electrochemical performances of the cathodic deposited alpha-Ni(OH)(2). *Electrochimica Acta.* 261: 58–65.
63. X.C. Ren, C.L. Guo, L.Q. Xu, T.T. Li, L.F. Hou, Y.H. Wei. 2015. Facile synthesis of hierarchical mesoporous honeycomb-like NiO for aqueous asymmetric supercapacitors. *ACS Applied Materials & Interfaces.* 7: 19930–19940.
64. Y.N. Zou, Y.H. Cui, Z.G. Zhou, P. Zan, Z.X. Guo, M. Zhao, L. Ye, L.J. Zhao. 2018. Formation of honeycomb-like Mn-doping nickel hydroxide/Ni$_3$S$_2$ nanohybrid for efficient supercapacitive storage. *Journal of Solid State Chemistry.* 267: 53–62.
65. K. Xiao, L. Xia, G.X. Liu, S.Q. Wang, L.X. Ding, H.H. Wang. 2015. Honeycomb-like NiMoO$_4$ ultrathin nanosheet arrays for high-performance electrochemical energy storage. *Journal of Materials Chemistry A.* 3: 6128–6135.
66. Z.G. Zhang, X. Huang, H. Li, Y.Y. Zhao, T.L. Ma. 2017. 3-D honeycomb NiCo$_2$S$_4$ with high electrochemical performance used for supercapacitor electrodes. *Applied Surface Science.* 400: 238–244.
67. S. Liu, K.V. Sankar, A. Kundu, M. Ma, J.Y. Kwon, S.C. Jun. 2017. Honeycomb-like interconnected network of nickel phosphide heteronanoparticles with superior electrochemical performance for supercapacitors. *ACS Applied Materials & Interfaces.* 9: 21829–21838.
68. D. Gyabeng, D.A. Anang, J.I. Han. 2017. Honeycomb layered oxide Na$_3$Ni$_2$SbO$_6$ for high performance pseudocapacitor. *Journal of Alloys and Compounds.* 704: 734–741.
69. Y.B. Zhang, Z.G. Guo. 2016. Honeycomb-like NiCo$_2$O$_4$ films assembled from interconnected porous nanoflakes for supercapacitor. *Materials Chemistry and Physics.* 171: 208–215.
70. A.K. Yedluri, H.J. Kim. 2018. Wearable super-high specific performance supercapacitors using a honeycomb with folded silk-like composite of NiCo$_2$O$_4$ nanoplates decorated with NiMoO$_4$ honeycombs on nickel foam. *Dalton Transactions* 47: 15545–15554.
71. A.E. Reddy, T. Anitha, C. Gopi, S.S. Rao, H.J. Kim. 2018. NiMoO$_4$@NiWO$_4$ honeycombs as a high performance electrode material for supercapacitor applications. *Dalton Transactions.* 47: 9057–9063.
72. W.D. Xue, W.J. Wang, Y.F. Fu, D.X. He, F.Y. Zeng, R. Zhao. 2017. Rational synthesis of honeycomb-like NiCo$_2$O$_4$@NiMoO$_4$ core/shell nanofilm arrays on Ni foam for high-performance supercapacitors. *Materials Letters.* 186: 34–37.
73. F. Wang, Q. Zhou, G. Li, Q. Wang. 2017. Microwave preparation of 3D flower-like MnO$_2$/Ni(OH)$_2$/nickel foam composite for high-performance supercapacitors. *Journal of Alloys and Compounds.* 700: 185–190.
74. Y.F. Tang, Z.Y. Liu, W.F. Guo, T. Chen, Y.Q. Qiao, S.C. Mu, Y.F. Zhao, F.M. Gao. 2016. Honeycomb-like mesoporous cobalt nickel phosphate nanospheres as novel materials for high performance supercapacitor. *Electrochimica Acta.* 190: 118–125.
75. L. Shi, P. Sun, L.H. Du, R.P. Xu, H.J. He, S.Z. Tan, C.X. Zhao, L.H. Huang, W.J. Mai. 2016. Flexible honeycomb-like NiMn layered double hydroxide/carbon cloth architecture for electrochemical energy storage. *Materials Letters.* 175: 275–278.
76. C.Y. Liu, Y.F. Chen, W.X. Huang, S.T. Yue, H. Huang. 2018. Birnessite manganese oxide nanosheets assembled on Ni foam as high-performance pseudocapacitor electrodes: Electrochemical oxidation driven porous honeycomb architecture formation. *Applied Surface Science.* 458: 10–17.

77. X.W. Sun, M.Y. Gan, L. Ma, H.H. Wang, T. Zhou, S.Y. Wang, W.Q. Dai, H.N. Wang. 2015. Fabrication of PANI-coated honeycomb-like MnO_2 nanospheres with enhanced electrochemical performance for energy storage. *Electrochimica Acta*. 180: 977–982.

78. L. Zhao, J. Yu, W.J. Li, S.G. Wang, C.L. Dai, J.W. Wu, X.D. Bai, C.Y. Zhi. 2014. Honeycomb porous MnO_2 nanofibers assembled from radially grown nanosheets for aqueous supercapacitors with high working voltage and energy density. *Nano Energy*. 4: 39–48.

79. Y.L. Chen, C. Chen, R.T. Lv, W.C. Shen, F.Y. Kang, N.H. Tai, Z.H. Huang. 2018. Flexible asymmetric supercapacitor based on MnO_2 honeycomb structure. *Chinese Chemical Letters*. 29: 616–619.

80. W.Y. Ko, Y.F. Chen, K.M. Lu, K.J. Lin. 2016. Porous honeycomb structures formed from interconnected MnO_2 sheets on CNT-coated substrates for flexible all-solid-state supercapacitors. *Scientific Reports*. 6: 18887.

81. Z.K. Peng, X. Liu, H. Meng, Z.J. Li, B.J. Li, Z.Y. Liu, S.C. Liu. 2017. Design and tailoring of the 3D macroporous hydrous RuO_2 hierarchical architectures with a hard-template method for high-performance supercapacitors. *ACS Applied Materials & Interfaces*. 9: 4577–4586.

82. B.B. Xin, Y.Q. Zhao, C.L. Xu. 2016. A high mass loading electrode based on ultra-thin Co_3S_4 nanosheets for high performance supercapacitor. *Journal of Solid State Electrochemistry*. 20: 2197–2205.

83. W.L. Jia, J. Li, Z.J. Lu, Y.F. Juan, Y.Q. Jiang. 2018. Synthesis of Honeycomb-like Co_3O_4 nanosheets with excellent supercapacitive performance by morphological controlling derived from the alkaline source ratio. *Materials*. 11: 1560.

8 Chemical Synthesis of Hybrid Nanoparticles Based on Metal–Metal Oxide Systems

Vivek Ramakrishnan and Neena S. John

CONTENTS

8.1   INTRODUCTION: NANOSTRUCTURES AND ENERGY

Metal oxide nanoparticles symbolize an important class of materials attracting considerable interest since they possess fascinating optical and electrical properties suitable for applications in sensing, information technology, catalysis, energy storage, medicine, electronics, optics, and photonics (Fernández-García and Rodriguez 2011). Their diverse applications and needfulness have driven research toward synthetic pathways for obtaining various types of tailored nanostructures and have resulted in a wide variety of morphologies such as nanorods, nanotubes, nanoporous structures, nanowires, nanorings, nanobelts, nanospheres, and so on (Kolmakov and Moskovits 2004, Sun, Liu, et al. 2012, Wang and Zhou 2012). New-generation nanodevices with high performance can be attained by tuning the properties of metal oxides by proper doping (Franke, Koplin, and Simon 2006, Keller et al. 2010). Researchers have tried improving the inherent characteristics of metal oxides with metals, metal ions, layered materials, noble metals, etc. (Wang et al. 2009, Zhi et al. 2013, Jiang et al. 2012).

Among the doped systems and composites, metal–metal oxide (MMO) systems have attained significant interest for decades till date and will continue to gain attention with much higher needfulness (Ng et al. 2013). MMOs can play appreciable roles in multitude areas of chemical, physical, and biological applications. The overall architecture of nanocomposites formed by combining metal with metal oxide will dictate the generation of novel and unique nature in addition to the improvement in physical, chemical, electronic, and magnetic properties. Metal oxides in combination with metals result in a plethora of nanocomposites including metal-incorporated metal oxide nanoparticles, nanoarrays, nanocages, decorated nanoflowers and nanorods, core/shell, yolk/shell, Janus nanostructures, and so on (Ray and Pal 2017,

Liu et al. 2017). The regulated and tailored syntheses of nanohybrids composed of metals and metal oxides have received significant attention for various applications in catalysis, photocatalysis, solar cells, drug delivery, photo and electrochemical water splitting, field emitting diodes and transistors, sensors, etc. (Liu et al. 2014, Sreeprasad et al. 2011, Kochuveedu, Jang, and Kim 2013).

Energy storage accomplished by rechargeable lithium-ion battery technology has taken a great tremendous turn since its introduction over the last two decades. Nanomaterial-based technology has got potential and scope in the field of energy storage, particularly in supercapacitors. By tuning the size, shape, structure, electronic properties, etc., the scientific community will be able to engineer the materials and improve the efficiency toward energy storage. So, it is crucial that these nanomaterials should have desirable properties while designing and preparation, and hence, synthesis plays a significant role.

Synthesis of MMO nanocomposites by a facile, simple, low-cost, eco-friendly, feasible, and less time-consuming method is still essential for usage in various applications. There are numerous methods reported for the syntheses of metal oxide systems such as hydrothermal, co-precipitation, sol–gel, sonochemical, electrodeposition, seeding, photochemical, microwave, impregnation, and so on (Guo et al. 2015, Lu, Chang, and Fan 2006). To prepare MMOs, such methods could be applied as such, or a combination of the methods can be used to prepare hetero-nanostructure efficiently with desirable properties (Figure 8.1).

In this chapter, we focus on the significant advances in traditional approaches to prepare MMO hetero-nanostructures with emphasis on potential applications. Coupling highly potent metals and metal oxides could allow for the energy-efficient preparation of a vast variety of hybrid nanoparticles whose resulting improved properties can be exploited in a whole range of technologically important areas of science and industry, rendering importance to highly tunable MMO nanohybrid materials.

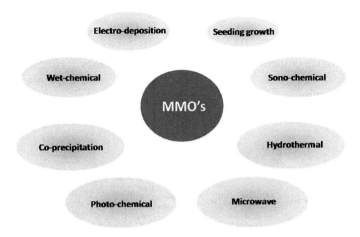

FIGURE 8.1 Various methods used for the synthesis of hybrid metal–metal oxide nanostructures.

8.2 HYDROTHERMAL SYNTHESIS

8.2.1 PREFACE

Due to its simplicity, low cost, and lower synthesis temperatures, the hydrothermal method is one of the most extensively employed techniques (Whittingham 1996). It is a solution-based chemical method used for the preparation of a host of nanoscale systems with varied dimensions and morphologies including single crystals, thin films, and nanoparticles (Yuan et al. 2014). As a subset of solvothermal methods, hydrothermal synthesis has been utilized for the preparation of a bunch of materials, *viz.* zeolites, coordination complexes, metal chalcogenides, halides, tungstates, and mixed metal oxides (Titirici, Antonietti, and Thomas 2006, Chi et al. 2012). Hydrothermal synthesis has particularly been implemented for the synthesis of nanoscale semiconducting metal oxides. Recently, hydrothermal method has been employed to grow molybdenum dioxide nanostructures on fluorine doped tin (FTO) substrates by a simple reaction involving ammonium heptamolybdate (AHM) and citric acid (CA) in water (Figure 8.2). These MoO_2 nanostructures are shown to be excellent electrocatalysts for energy conversion (Ramakrishnan et al. 2018). There are numerous noteworthy factors that affect the hydrothermal formulation of different nanostructures, *viz.* reaction temperature, nature of precursors, surfactants, pH, crystallography, substrate nature, and seed layer (Guo et al. 2015, Lin and Ding 2013).

For performing the hydrothermal synthesis, a particular type of equipment is used, *viz.* autoclave in which in-built pressure is generated and sustenance is supplied in an appropriate solvent. The system is heated at the required temperature, and as the reaction proceeds, nucleation of the crystal structure is initiated. An in situ temperature gradient is generated, giving rise to the formation of two temperature regions wherein the reactants get dissolved in the lower-high temperature region and get carried to the upper-low temperature region. Due to supersaturation with lowering of temperature, the nucleates get deposited as seeds that further grow to form the product (Byrappa and Yoshimura 2012).

Hydrothermal method is marked with following supremacy over other methods:

- Usage of lower temperature, relatively less time consuming, and environmentally friendly
- Preparation of thermodynamically lesser-stable phases of nanostructures
- Growth of crystals of compounds with high melting points at lower temperatures

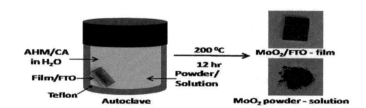

FIGURE 8.2 Schematic diagram of a typical hydrothermal synthesis. (copyright 2018 WILEY-VCH Verlag GmbH & Co. KGaA, Weinheim.)

- Preparation of nanostructures with high crystallinity in good yield with suitable control over their composition
- Preparation of nanostructures on flexible templates at low reaction temperature (Byrappa and Adschiri 2007, Komarneni et al. 2010)

Among various metal oxides, TiO_2- and ZnO-based metal oxides, MMOs, and metal oxide–metal oxide-based systems have been studied and used majorly in various fields such as solar cells, sensors, cosmetics, ceramics, photocatalysts, diodes, transistors, electrocatalysts, and so on (Considine and Kulik 2005, Yu and Yu 2008, Byrappa et al. 2008, Kanjwal et al. 2010). There are numerous reports of TiO_2-based nanostructured systems prepared via hydrothermal synthesis for various applications as listed in Table 8.1.

8.2.2 Metal–TiO₂-Based Systems

Doping techniques have been applied to TiO_2 systems for various reasons to overcome the limitations such as wide bandgap, ineffectiveness of photocatalysis under sunlight, and thermal instability (Tan, Wong, and Mohamed 2011). Owing to its large bandgap (3.2 eV), TiO_2 can only harness 6% of the entire solar irradiation, but doping techniques shift the activity from UV to the visible light region. There are numerous reports on the doping of TiO_2 for superior properties using metals as well as nonmetals (Table 8.2).

TABLE 8.1
TiO₂-Based Systems Prepared by Hydrothermal Route

No.	System	Method	Application	References
1	Phase pureTiO_2	Hydrothermal	Photocatalysis	H. Yin et al. (2001)
2	Mesoporous TiO_2	Hydrothermal	Photocatalysis	J. Yu et al. (2007)
3	Colloidal TiO_2	Hydrothermal	Solar cell	C.-Y. Huang et al. (2006)
4	TiO_2	Hydrothermal	Surface Properties	G.J. Wilson et al. (2006)
5	TiO_2 microemulsions	Hydrothermal	Photocatalytic Wet oxidation	M. Andersson et al. (2002)
6	TiO_2 nanosheets	Hydrothermal	Lithium storage	L. Kavan et al. (2004)

TABLE 8.2
Effect of Doping of Metal Ions in TiO₂ Nanostructures Prepared by Hydrothermal Route

No.	System	Method	Application	References Year
1	$Ce^{2+/3+}/N$-TiO_2	Hydrothermal	Photocatalysis	M. Nasir et al. (2014)
2	$Fe^{2+/3+}$-TiO_2	Hydrothermal	Photocatalysis	J. Zhu et al. (2004)
3	M^{x+}-TiO_2	Hydrothermal	Photocatalysis	K. Lee et al. (2006)
4	Fe^{3+}-TiO_2	Hydrothermal	Photocatalysis	V.N. Nguyen et al. (2011)
5	Sr-TiO_2	Hydrothermal	Solarcell	H.F. Mehnane et al. (2017)

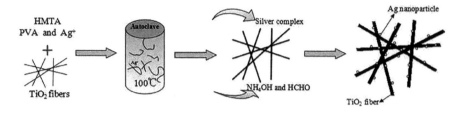

FIGURE 8.3   The proposed schematic synthesis process to form Ag/TiO_2 nanoheterostructures by the hydrothermal reducing reaction in HMTA solvent. (© 2015 Royal Society of Chemistry.)

F. Zhang et al. reported the fabrication of heterostructured Ag-doped TiO_2 via a facile two-step preparation involving electrospun technique followed by hydrothermal process (2016). An illustrative diagram of the preparation of the nanostructure is depicted in Figure 8.3. Primarily, TiO_2 nanofibers were fabricated through electrospinning technique using tetrabutyl titanate as the precursor, which was added to thoroughly mixed polyvinylpyrrolidone, methyl alcohol, and acetic acid medium. The electrospun composite nanofibers were fabricated by applying 12 kV direct voltage followed by annealing at 520°C for 6 hours to form pure TiO_2 fibers. The final $Ag@TiO_2$ composite was prepared by subsequent hydrothermal reaction of as-prepared TiO_2 fibers in a mixture of $AgNO_3$, polyvinyl alcohol (PVA), and hexamethylenetetramine (HMTA) by keeping at 90°C for 4 hours where HMTA served as alkali and reducing agent. Ag nanoparticles anchored on the periphery of TiO_2 fiber with varying size and load could be controlled with concentrations of reactants during the hydrothermal process. By anchoring Ag, visible light-induced photocatalytic activity on the TiO_2 fibers was found to be enhanced significantly. Catalytic activity of composite was established by the higher degradation efficiency than the pure TiO_2 fiber. There are some other notable works on the metal–metal oxide systems of TiO_2 obtained by hydrothermal synthesis and briefed in Table 8.3.

TABLE 8.3
Effect of Metal Deposition on TiO_2 Nanostructures Prepared by Hydrothermal Route

No.	System	Method	Application	References
1	Pd, Cr, or Ag-TiO_2	Hydrothermal synthesis and impregnation	Photocatalysis	C.-G. Wu et al. (2004)
2	Urchin-like ternary TiO_2	Hydrothermal	Photocatalysis	Y. Liu et al. (2016)

8.2.3   Metal–ZnO Systems

For ZnO nanostructures, it is well established that lifetime of electrons (e⁻) and holes (h⁺) could be efficiently enhanced by blending with noble metal nanoparticles and carbon nanomaterials. K. Bramhaiah et al. have reported comparative photocatalytic activity of binary and ternary hybrid systems of ZnO possessing nanostructured morphology with reduced graphene oxide (rGO) or with rGO and Au nanoparticles (2016). The hybrids were prepared via a simple solution route and hydrothermal method. At the outset, rGO and rGO-Au were prepared by the reduction of GO with hydrazine hydrate or in the presence of HAuCl₄. rGO-ZnO and rGO-Au-ZnO composites were prepared by addition of zinc acetylacetonate and KOH to the respective as-prepared dispersion by keeping at 120°C for 8 h in an oil bath. The conventional synthesis produced nanoparticle composites. rGO-ZnO nanorods were prepared via hydrothermal route (120°C/8 h) using the same precursor and decorated with Au nanoparticles. Among all the systems under study, rGO-Au-ZnO nanoparticles showed the uppermost reaction kinetics for photodegradation activity, five times faster than bare ZnO. The next highest photodegradation activity was found for the binary system, rGO-ZnO nanoparticles, and finally nanorods. Figure 8.4 illustrates the feasible charge-transfer (C-T) processes occurring in rGO-Au-ZnO nanoparticle hybrid systems for dye photodegradation.

Following hydrothermal pathway for the preparation of metal-doped ZnO nanostructures could be propitious. In addition, engineering of ZnO nanostructures with vivid 1D and 2D structures on wide varieties of substrates can be accomplished on a bulk scale, and their material characteristics may be tuned by controlling the growth factors (Janisch, Gopal, and Spaldin 2005). Some of the research works are briefed in Table 8.4.

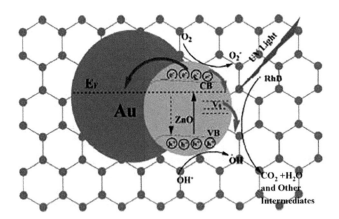

FIGURE 8.4   A schematic of the possible (C-T) processes in rGO-Au-ZnO nanoparticle hybrid during dye photodegradation. (© 2015 Royal Society of Chemistry.)

TABLE 8.4

Effect of Doping of Transition Metals in ZnO Nanostructures Prepared by Hydrothermal Route

No.	System	Method	Application	References
1	Fe-ZnO nanorods	Hydrothermal	UV photodetector	C.O. Chey et al. (2014)
2	Mn-ZnO nanorods	Hydrothermal	Room temperature ferromagnetism	J. Panda et al. (2016)
3	Co-ZnO nanorods	Hydrothermal	Visible light activity	Y. Febrianti et al. (2017)
4	(Cr, Mn, Ni)-ZnO	Hydrothermal	Magnetism	M. Zhong et al. (2015)
5	Ce-ZnO nanorods	Hydrothermal	Photocatalysis	N. Aisah et al. (2017)

8.2.4 METAL–NiO SYSTEMS

Majumder et al. reported direct Ag coating on NiO flake-like structures and its effect on the optical and catalytic properties. Ag nanoparticles on NiO were anchored by a two-step process for the preparation of hetero-nanostructures. NiO nanoflakes were first prepared by hydrothermal synthesis using $Ni(NO_3)_2$, urea, and HMTA in aqueous medium at 90°C for 6 hours followed by calcination at 450°C for 2 h. In the subsequent step, a solution-phase chemical method, involving in situ reduction of $AgNO_3$ by oleylamine in toluene by heating at 85°C for 8 hours, was adopted. Finally, Ag/NiO composite was obtained with highly dense Ag nanoparticles distributed over NiO flakes. The strong coupling and interaction between MMOs led to the high photocatalytic activity.

In literature, there are other reports of metal–NiO nanostructures prepared via hydrothermal synthesis. S. Hasan et al. reported the hydrothermal-mediated two-step synthesis of $NiO@Ni_xMn_{1-x}O$ magnetic nanoparticles for magnetic recording applications (2017). H. Xu et al. reported Co-doped hierarchical nanosheets of NiO with a flower-like morphology for high-performance lithium-ion battery (2015).

8.2.5 METAL–MnO₂ SYSTEMS

R. Liu et al. reported the synthesis of elliptical-shaped porous MnO_2/Co nanostructures through a facile three-step hydrothermal-assisted decomposition method. In the first step (2015), co-embedded manganese and cobalt oxalate were obtained by hydrothermal method using potassium oxalate and chloride salts of corresponding metal precursors at 100°C for 10 h. Consequently, the products were annealed at 500°C for 5 h utilizing specifically controlled decomposition method in a muffle furnace to yield Mn_2O_3/Co nanocomposites. Retaining their original ellipsoidal nanofibrous micro-framework from the parental components, MnO_2/Co hybrids were formed by further hydrothermal reaction of annealed product with $(NH_4)_2S_2O_8$ solution at 60°C for 10 h. Hybrid nanostructure of MnO_2 with a doping amount of 1.83 mol% of Co produced the upper-edge reactivity for electrochemical capacitive nature (initial cycle with specific capacity of 107.5 Fg^{-1}), and 93.5% of the capacitive behavior was retained even after 500 cycles. It was concluded that Co doping imparts its influence in directing the uniformity and structural formation as well as enhances electrochemical activity of MnO_2.

There are numerous studies established to improve the intrinsic properties of Mn-based oxides by doping various metals. M. Sun et al. reported the Sn-doped MnO_2 obtained via one-step hydrothermal synthesis for achieving better supercapacitive property (187 Fg^{-1}) (2012). M. Sun et al. also studied the effect of boron doping of MnO_2 nanostructures on decolorization performance (2014).

8.2.6 METAL–CuO SYSTEMS

In past years, enhancing the electronic as well as magnetic property of CuO has been of significant interest, and a series of efforts have been made by metal doping (Pasha et al. 2009, Shaikh et al. 2011, Filippetti and Fiorentini 2006). CuO is a promising semiconducting material as it possesses high photocatalytic activity, physiochemical durability, exorbitant specific surface area, etc. (Xu and Sun 2009, Liu et al. 2011, Han et al. 2015). Albeit having such superior properties and advantages over other semiconductors, it is reported to have the drawback of fast recombination rate of photo-generated (e^-) and (h^+), which can be suppressed by doping with transition metals such as Zn, In, Mn, Ni, and Fe to enhance the photocatalytic property (Sonia et al. 2015, Yildiz et al. 2014, Sharma, Gaur, and Kotnala 2015, Basith et al. 2014, 2013).

T. Jiang et al. reported the synthesis of novel Zn-doped CuO nanostructures (2016). The low-cost and simple hydrothermal method has been adopted for the synthesis of Zn-doped CuO nanostructures using homogeneous aqueous solution of copper (II) acetate and sodium hydroxide followed by the addition of an appropriate amount of zinc acetate with different concentrations. A series of MMO nanocomposites were prepared by varying the amount of dopant (Zn) during the hydrothermal reaction at 110°C for 2 h. A black powder of nanocomposite having pine-needle-like morphology was obtained and was composed of smaller nanosheets. Zn was found to be present in the lattice of CuO system, and an improved photocatalytic performance was observed for the hybrid nanostructure. This surfactant-free method paves the way for the synthesis of metal-doped metal oxide nanostructures. In literature, there are numerous reports on metal–copper oxide nanoparticles engineered for various applications and are listed in Table 8.5.

8.2.7 OTHER MMO SYSTEMS

In addition to the abovementioned major metal oxide supports, including TiO_2, ZnO, NiO, CuO, MnO_2, and so on, there are other reports with Co_3O_4, WO_3, MoO_3, etc. as useful materials. M.M. Rahman et al. detailed the large-scale hydrothermal

TABLE 8.5

Metal-CuO Nanostructures Prepared by Hydrothermal Route

No.	System	Method	Application	References
1	Zn-CuO	Hydrothermal	Antibacterial activity	J. Iqbal et al. (2015)
2	Zn-CuO	Hydrothermal	Supercapacitor	C. Karunakaran et al. (2013)
3	Sn-CuO	Hydrothermal	Structural and optical properties	S. Mohebbi et al. (2013)
4	Ni-CuO	Hydrothermal	Catalysis	K. Lakshmi et al. (2017)
5	Fe-CuO	Hydrothermal	Ferromagnetism	S. Manna and De (2010)
6	Fe-CuO	Hydrothermal	Synthesis and properties	K. Kannaki et al. (2016)
7	Cd-CuO	Hydrothermal	Synthesis and properties	S. Rejitha et al. (2013)

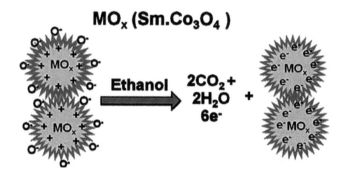

FIGURE 8.5 Schematic diagram showing the Sm-doped Co_3O_4. (© 2011 American Chemical Society.)

synthesis of Sm-doped Co_3O_4 crystalline nanokernels showing high performance in ethanol sensing with respect to excellent response time, detection range, robustness, and high sensitivity (Figure 8.5) (2011). W. Mu et al. portrayed the synthesis and enhanced photocatalytic performance of Nb-doped WO_3 nanowires using a low-temperature hydrothermal method (2014). Z. Li et al. outlined the synthesis of Ce-doped nanobelts of MoO_3 by in situ hydrothermal reaction with enhanced gas-sensing properties (2017).

8.3 SOL–GEL ROUTE

8.3.1 PREFACE

Sol–gel method is a subset of wet chemical method apposite for both glass-type and ceramic-type materials, as it provides unusual compositions even for lower temperatures than that required by the conventional melting techniques (Chiang et al. 2003, Hou et al. 2009, Jia et al. 2002). Because of its simplicity, reliability, repeatability, cost-effectiveness, and requirement of low temperature for processing, this method has been used for the preparation of laser dyes, enzymes, nanometer-sized semiconductors, and metal nanoparticles dispersed in SiO_2 glasses (Katoch et al. 2012, Koelsch et al. 2002, Larsen et al. 2003, Lee et al. 2010, Li et al. 2013, 2004, Mackenzie and Bescher 2007, Moncada, Quijada, and Retuert 2007, Nagarale et al. 2004). Sol–gel method can be further subclassified into ultrasonic-assisted sol–gel method, aerogel method, photo-reductive decomposition, precipitation, two-step wet chemical method, and extremely low-temperature precipitation (Schmidt et al. 2000, Sudarsan et al. 2005, Thota and Kumar 2007, Vafaee and Ghamsari 2007, Wen and Wilkes 1996, Woo et al. 2003).

Sol–gel method has been employed via nonaqueous or popular aqueous solution route. The resultant nanoparticles prepared via sol–gel method show good optical properties by good control of the morphology and size of the particles. One of the crucial factors is excellent adsorption of the surfactant onto the particle surface aiding the size control in a sol–gel process by inhibiting the growth of nuclei (Yang et al. 2005, Yang and Wang 2006).

The sol–gel process involves the transformation of solution, i.e., sol, slowly toward a phase incorporating both liquid and solid states having coagulated-network-like nature. In this process, precursors get converted to colloidal form followed by hydrolysis and polycondensation. The well-known precursors are metal-derived chlorides and alkoxides. The solid phase may have a versatile morphology ranging from infinitesimal gel-like grains to fibrous polymer criss cross structure (Schmidt et al. 2000, Sudarsan et al. 2005).

In a typical sol–gel method, the precursor solution employed contains metal compounds (metal alkoxide, acetylacetonate, carboxylate, etc.) and soluble inorganic species of the dopant along with the hydrolysis agent (water), alcohol (solvent), and an acid or base (catalyst). Initially, a sol is formed in which polymers or colloidal particles are dispersed without precipitation by the hydrolysis and polycondensation of metal precursor. As reaction proceeds, sol gets converted to wet gel (Figure 8.6),

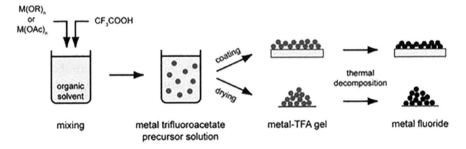

FIGURE 8.6 Schematic representation of sol–gel process of synthesis of nanomaterials. © 2015 The Royal Society of Chemistry.)

and further vaporization of solvent formulates a dry gel (Kemnitz and Noack 2015). Finally, glasses and ceramics are formed by removal of organic constituents and residues when heated at elevated temperatures (Lee et al. 2010, Li et al. 2013).

8.3.2 METAL–TiO₂-BASED SYSTEMS

Metal–TiO₂ hybrid nanoparticles prepared via sol–gel method route have been widely investigated. Typical examples are sol–gel-derived Cr(III)-doped TiO₂ composites (Lu et al. 2007), Pt-deposited TiO₂ thin films (Sanchez et al. 1996), surface-anchored Fe(III)/TiO₂ particles (Piera et al. 2003), vanadium/N-doped TiO₂ (Higashimoto et al. 2008), Al/TiO₂ (Lee et al. 2003), La-TiO₂ (Zhang et al. 2006), etc.

W.-C. Hung et al. reported the synthesis of Fe/TiO₂ photocatalyst (2007). A typical two-step sol–gel process was adopted for the preparation of these photocatalysts. At first, Ti-precursor sol was prepared by mixing titanium isopropoxide with water in a 1:51 ratio followed by the addition of 0.25 M HNO_3 in a 1:1 weight ratio to Ti-precursor. The dehydrated sample is further processed for Fe doping by its nitrate salts. The as-prepared samples were then coated on pyrex cylinder glass for thin-film preparation. They have varied Fe doping from 5 mol% to 0.005 mol%. Fe doping was found to show significant improvement in the visible absorption efficiency as evidenced by optical spectroscopy. An Fe-doped TiO₂ system with 0.005 mol% dopant was found to exhibit optimum activity for photocatalysis.

8.3.3 METAL–ZnO-BASED SYSTEMS

Physiochemical properties of ZnO have been tremendously varied by the doping of transition metals. Chakrabarti et al. (2008) and Xu et al. (2009) reported the effect of doping by Mn/Fe and Cu/Co on the magnetic properties of ZnO, respectively.

G.J. Naz et al. reported sol–gel-mediated preparation of Mn/ZnO and Fe/ZnO sols starting from $Zn(OOCCH_3)_2$, iron nitrate, and $Mn(OOCCH_3)_2$ (2015). In the two-step process, ZnO sol was initially prepared by heating Zn precursor in deionized water with the addition of isopropyl alcohol and triethylamine. Mn or Fe source in triethylamine was added to as-prepared ZnO sol to obtain metal-incorporated hybrid ZnO. The doped sols were spin-coated onto Cu films and annealed under magnetic field. Interestingly, dopant amount was found to have an inverse relation with the size of the crystallite. The incorporation of dopants was not found to disturb the crystal behavior of ZnO retaining the hexagonal wurtzite structure. Ferromagnetic activity (~34.63 emu/cm³) was found to be amplified, as evidenced by vibrating sample magnetometer analysis.

8.3.4 METAL–NiO-BASED SYSTEMS

NiO-based bimetallic mixed metal oxides are well studied for applications such as in the regeneration of alkanes (anticoking behavior), catalytic oxidation of carbon monoxide, purification of halogenated organic mixtures, decomposition of H_2O_2, and so on. N. Bayal et al. reported the synthesis of NiO-ZnO, NiO-CuO, and NiO-MgO mixed metal oxides by simple sol–gel method (Bayal and Jeevanandam 2012). The respective precursors of Cu, Zn, and Mg along with different concentrations of nickel acetate were mixed with toluene and ethanol, and the suspension was stirred vigorously to obtain a homogenous mixture followed by the addition of 1 mL of water, which on further stirring and drying produced mixed metal oxide gel. Further grounding and annealing in ambient conditions at 500°C yielded the final product in the form of powder. The bandgap of NiO calculated for all loading levels in mixed metal oxide nanoparticles showed different values owing to intercalated amount and size effect. Such kind of synthetic pathways holds the potential for the bulk production of all varieties of mixed metal oxide nanoparticles.

8.3.5 Metal–CuO-Based Systems

A. Soria et al. reported the utilization of sol–gel method in the expulsion and recuperation of Cu by ozone treatment of remaining cyanide solution to prepare Cu-CuO (2015). A sol–gel technique was implemented for the synthesis of metal nanoparticles using 2-hydroxypropane-1,2,3-tricarboxylic acid and ethane-1,2-diol, following the ozone oxidation and separation (solid–liquid). Subsequently, residual aqueous contents were removed by thermal evaporation. As-prepared gel product in translucent state was subjected to annealing at various temperatures (400°C, 600°C, and 800°C for 0.25 h) in tubular incinerator yielding the final product of Cu-CuO hybrid. This study portrays an efficient, but cost-effective auxiliary tool for regenerating nanoparticles from industrially cast-off waste solution, especially noble metals.

8.3.6 Metal@Metal Oxide Core–Shell Systems

Metal oxide@metal oxide core–shell nanostructures have been widely used for gas-sensing application and are being replaced by metal@metal oxide core–shell nanoparticles owing to merging of core–metal with shell–metal oxide leading to the enhanced response (Cheng et al. 2008, Yu and Dutta 2011, Chung, Wu, and Cheng 2014). A. Mirzaei et al. originally reported gas detection by utilizing the nanohybrid of α-Fe_2O_3 (shell) with Ag (core) adopting a sol–gel approach. The aqueous solutions of PVP and $AgNO_3$ in water were mixed and thoroughly stirred to form Ag^+-PVP solution to which $NaBH_4$ solution in water was added dropwise to convert Ag^+ ions to metallic silver. Temperature of bath (ice) and the rate of addition of $NaBH_4$ had great control over as-prepared particle size/shape. A sol of colloidal Fe (III) was prepared by special sol–gel reaction (Pechini approach) with $Fe_2(SO_4)_3$ (iron donor), 2-hydroxypropane-1,2,3-tricarboxylic acid (chelation ligand), and poly(ethylene oxide) (esterifying moiety) as the precursors. The addition of the as-prepared iron sol to the so-formed silver nanoparticles yielded an eventual product, $Ag@\alpha$-Fe_2O_3 core–shell nanostructure. Figure 8.7 shows a simple schematic diagram of the synthesis procedure (Mirzaei et al. 2015).

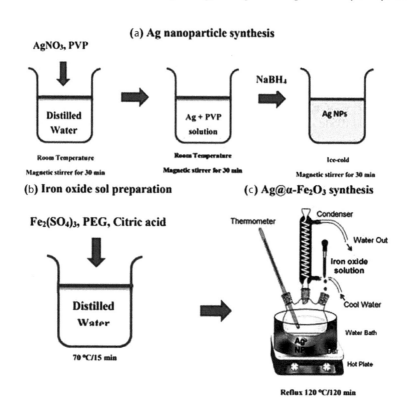

FIGURE 8.7 Schematic representation of synthesis procedure of Ag@Fe$_2$O$_3$ core–shell nanoparticles. (a) Ag nanoparticle synthesis, (b) iron oxide sol preparation, and (c) Ag@Fe$_2$O$_3$ synthesis. (© 2015 Multidisciplinary Digital Publishing Institute.)

8.4 PHOTOCHEMICAL SYNTHESIS

8.4.1 PREFACE

Among the many approaches used for the production of metal nanoparticles knitted with metal oxide nanostructures, photo-deposition has received significant attention by virtue of its simplified execution either in single step or in multiple steps. Photochemical approach comprises preparation of noble metal nanoparticles with subsequent anchoring of the same to an appropriate backbone. Photo-deposition is a phenomenon which involves the illumination of an aqueous solution of metal salt resulting in in situ imprinting of those target materials (metal or metal oxide nanoparticles) on supporting materials. So, photo-deposition is nothing but light-induced electrochemistry. For a feasible photo-deposition, there should be close matching of the band positions and bandgap (redox potential) of the target materials (metal/metal oxide) to that of support system (semiconductor/metal oxide). Photochemically prepared nanostructures are found to have applications mainly in, but not limited to, photocatalytic water splitting, solar cells, gas sensors, wastewater treatment, and air purification (Wenderich and Mul 2016, Wang et al. 2012).

Even though there are numerous approaches for the synthesis of well-defined co-catalyst nanoparticles integrated with semiconductor nanostructures, *viz.* electrodeposition, physical and chemical vapor deposition, sputtering (Maeda, Abe, and Domen 2011), hydrothermal reduction (Kang and Sohn 2012, Ramakrishnan et al. 2016, Murata et al. 2012), and simple physical blending (Bamwenda et al. 1995), photo-deposition is an attractive method owing to the nonrequirement of external high energy such as heat (temperature) or electrical (current or potential) energy (Tada, Fujishima, and Kobayashi 2011). The size and valence state of the as-deposited nanoparticulated materials on the surface of supporting semiconducting nanostructures can be governed in a desirable manner. Moreover, facile simultaneous measurement of gas evolution can be done when applied for photocatalytic water splitting (Peled 1995, Wilson and Houle 1985, Lee et al. 2018).

8.4.2 METAL–TiO$_2$ SYSTEMS

S.F. Chen et al. reported the large-scale photochemical synthesis of noble metals (Ag, Pd, Au, and Pt) anchored via a facile photochemical pathway, hardly requiring any additives or surfactants, for antibacterial applications (2010). H. Li et al. reported the preparation of Ag-decorated TiO$_2$ nanotubes by photochemical-assisted hydrothermal pathway for photocatalytic studies (2008). S. Ko et al. have given an account of the detailed preparation of titanium dioxide nanostructures incorporated with silver particles by photochemical synthesis. Further, they are tested for the catalytic activity under visible light radiation (Ko, Banerjee, and Sankar 2011). L. Sun et al. reported a novel photochemical synthesis coupled with ultrasonication for preparing TiO$_2$ nanotube array photocatalysts loaded with highly dispersed Ag nanoparticles (2009).

It was primarily shown by P.D. Cozzoli et al. that heterojunctions of semiconductors with that of metals could be prepared with colloidal suspension-mediated route, in which Ag/TiO$_2$ nanocomposites were prepared by a two-step process.

1. Chemical synthesis of organic moiety-capped TiO$_2$ nanostructures
2. Subsequent anchoring of Ag on TiO$_2$ to produce hybrids by photochemical approach

Initially, tetrabutylammonium hydroxide was hydrolyzed in alkaline medium using oleic acid as a surfactant to form titanium dioxide (anatase phase) nanoparticles protected with organic moiety. The temperature was maintained around boiling point of the solvent used (aqueous solution) and with varying amount of precursor. As-prepared nanoparticles of TiO$_2$ were tuned for morphological aspect with difference in hydration content and nature of organic bases (trimethylamino-N-oxide dehydrate, trimethylamine, tetramethylammonium hydroxide, tetrabutylammonium hydroxide, etc.) in the reaction mixture. Introduction of ethyl alcohol at ambient conditions resulted in the brisk precipitation and growth of rod-like and spherical titania nanocrystals.

Ag nanoparticles were anchored onto TiO$_2$ by two different methods. In the first method, utilizing a photocatalytic pathway, readily synthesized titanium dioxide nanostructures were mixed with AgNO$_3$ in CHCl$_3$:ethanol mixtures.

The deaerated solution was subjected to ultraviolet irradiation with the aid of an Hg (high-pressure) light source (200 W) and constant stirring. The stability of the reaction was controlled by adjusting the concentration of precursor and light intensity inhibiting premature precipitation of Ag nanoparticles and photo-oxidation of TiO_2. In the second method, deaerated solutions of sodium borohydride in ethyl alcohol and TiO_2 nanocrystals with $AgNO_3$ in $CHCl_3$ were mixed with vigorous stirring. The proposed and implemented photoreaction can be labeled as a model sequence for producing Ag-decorated titanium dioxide nanocomposites in different sizes and morphological regimes as a function of the irradiation time enabled by light-induced photo-fragmentation and ripening, hardly using any kind of surfactants or additives.

In another work, S. Ren and Liu (2016) reported the preparation of Pd/Au-alloyed nanoparticles on TiO_2 nanowires employing one-step photochemical approach for ultrasensitive H_2 detection.

8.4.3 METAL–ZNO SYSTEMS

M. Wu et al. reported a novel matchstick-like hetero-nanostructure of gold nanoparticles-embedded ZnO exhibiting a plasmonic-enhanced photo-electrochemical performance for light-induced hydrogen evolution (2014). The matchstick-like heterostructures of zinc oxide nanorods uniformly decorated with gold nanoparticles on the tips (Figure 8.8a) were synthesized on a substrate (Zn foil) by a two-step process involving hydrothermal synthesis and photoreduction. In the first step, uniform vertically aligned ZnO nanorods on Zn foil (Figure 8.8b–d) were prepared by hydrothermal synthesis with the aid of aqueous 1,6-hexanediamine at a reaction temperature of 180°C maintained for a time period of 5 h. In the second step, Au nanostructures with tunable Au contents were deposited onto ZnO nanorod/Zn foil by irradiation in aqueous gold chloride solution using a light source (300 W xenon lamp, 4 h) (Figure 8.8e–g). The hetero-nanostructure of Au/ZnO was finally obtained by the conversion of Au^{3+} to Au^0. More interestingly, morphology of the decorated Au nanoparticles could be tailored by controlling the light irradiation time. Matchstick-like ZnO/Au heterostructure showed a photocurrent density of ~9 mA/cm^2 at a bias of 1.0 V (with respect to Ag/AgCl reference electrode), which was 30 times larger than bare ZnO nanorods, and the enhanced properties were attributed to the plasmonic-assisted improved light assimilation and e$^-$/h$^+$ separation. This study paves the way for other researchers to explore designing of the semiconductor/metal heterostructures-based photo-anodes with visible light-responsive plasmonic activity for harvesting the solar spectrum.

8.4.4 METAL–SNO₂ SYSTEMS

C.G. Carvajal et al. reported highly oriented SnO_2 nanorods on Si/SiO_2 substrates decorated with Au, Ag, and NiO nanoparticles for preparing improved hybrid gas sensors and photo-detectors (2015). The hybrid nanostructures were prepared by a two-step process. Initially, metal oxide nanorods were architectured on Si/SiO_2 substrates by vapor–liquid–solid technique employing anhydrous zinc chloride and stannous chloride ($SnCl_2$) powder as precursors. $ZnCl_2$ plays a very important

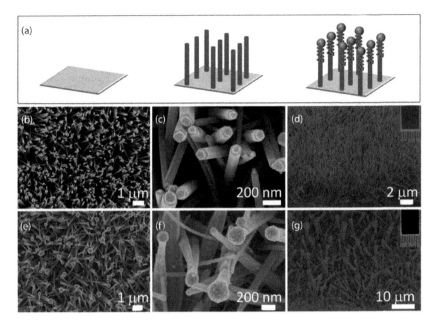

FIGURE 8.8 (a) Schematic illustration of the synthesis of a matchstick-like ZnO/Au hetero-structure. (b and c) Top views. (d) Cross-sectional-view SEM images of ZnO nanorod arrays. (e and f) Top views. (g) Cross-sectional-view SEM images of the typical matchstick-like ZnO/Au heterostructure. The insets of parts (d) and (g) show substrate photographs of pristine ZnO nanorod arrays and the ZnO/Au heterostructure, respectively. (© 2014 American Chemical Society.)

role in the nanorod synthesis, and once the nanorods of SnO_2 were formed, the remaining traces of ZnO produced during the vapor reaction were removed by a chemical bath etching in hydrochloric acid. In the second step, SnO_2 nanorods were decorated with Au, Ag, and NiO particles by photo-illumination of the as-prepared SnO_2/substrate samples in respective metal ion precursor solution with a dual wavelength of 185 and 254 nm simultaneously at 25°C for 5 minutes. For gold, $HAuCl_4.3H_2O$ (2 mM aqueous solution) was used, whereas for silver, 2 mM aqueous solution of $AgNO_3$, and for nickel oxide, 2 mM aqueous solution of nickel (II) nitrate hexahydrate, was used. Gas-sensing studies revealed a significant improvement in selectivity to ethanol and acetone vapors for SnO_2-Au and SnO_2-NiO hybrid devices. Photoconductivity experiments showed a significant increase in photocurrent density by Au and Ag decoration. The study reported here suggests the selective tuning of gas-sensing and photosensitivity properties of SnO_2-based hybrid devices.

In addition, photochemical loading of Pd and Pt on SnO_2 thin films and its potential application to gas sensing were reported by G.E. Buono-core et al. (2006). M. Yuasa et al. reported a photochemical method for the preparation of a stable suspension of Pd-loaded SnO_2 nanocrystals for highly sensitive semiconductor gas sensors (Yuasa, Kida, and Shimanoe 2012).

8.4.5 Metal–Cr$_2$O$_3$ Systems

K. Maeda et al. reported the utilization of noble metal/Cr$_2$O$_3$ core/shell nanoparticles as a co-catalyst for photocatalytic water splitting (2006). Rh/Cr$_2$O$_3$ (core/shell)-loaded (Ga$_{1-x}$Zn$_x$) (N$_{1-x}$O$_x$) nanoparticles were prepared by an in situ two-step photo-deposition method from Na$_3$RhCl$_6$·2H$_2$O and K$_2$CrO$_4$ as precursors. Rh was loaded onto (Ga$_{1-x}$Zn$_x$) (N$_{1-x}$O$_x$) by a typical experiment in which the reaction mixture was irradiated with Hg–Xe lamp operated at 1600 W under N$_2$ stream for ~4 h at 55°C. This was followed by metal oxide shell construction employing a second photodecomposition process wherein Rh-loaded sample in aqueous K$_2$CrO$_4$ solution was exposed to visible light irradiation. The co-catalyst outlined by Maeda et al. is reported to have extensive advantages such as selective introduction of activation sites on the photocatalyst directly influencing the overall water splitting and introduction of various noble metals for capturing photo-generated e⁻ from the bulk. The present method also offers a varied strategy for the removal of Cr(VI) ions from the industrial effluents by photoreduction with simultaneous production of hydrogen.

8.4.6 Metal–CeO$_2$ Systems

Raudonyte-Svirbutaviciene et al. (2018) reported the fabrication of Ag-incorporated CeO$_2$ hetero-nanostructures by a simple and cost-effective sustainable pathway via photochemical deposition method devoid of any capping agents or stabilizers.

Starting from CeCl$_3$.7H$_2$O or Ce(NO$_3$)$_3$.6H$_2$O as the cerium source, MMO nanostructures have been prepared by reaction with NaN$_3$ through a photochemical two-stage synthetic pathway. The thoroughly mixed reaction mixture was then subjected to light irradiation (40 W) for 300 minutes. Agglomeration of the so-formed CeO$_2$ nanoparticles could be controlled effectively by lowering the precursor concentration. Finally, the hybrids were prepared by exposing ceria with the silver source in water under a dark and inert atmosphere to ultraviolet light for various durations.

Other ceria-based MMO nanostructures such as Pt/CeO$_2$ were also prepared by simple photochemical synthesis of highly dispersed Pt nanoparticles on porous CeO$_2$ nanofibers for the water–gas shift reaction as reported by P. Lu et al. (2015).

8.5 CO-PRECIPITATION METHOD

8.5.1 Outline

In chemistry, co-precipitation is the process in which salting out of a substance, which is normally soluble under the conditions employed, occurs (Cocero et al. 2009). This simple method is most effective for binary precipitation from the reaction mixture. As a result, it can be used to prepare multicomponent MMO or metal oxide–metal oxide catalysts. Co-precipitation method is reported to be useful for engineering well-known metal oxide systems such as Al$_2$O$_3$, SiO$_2$, TiO$_2$, ZrO$_2$, and so on in bulk quantities. A simple schematic diagram outlining the co-precipitation-based technique is shown in Figure 8.9 (Theiss, Ayoko, and Frost 2016).

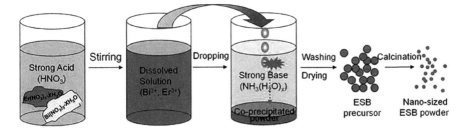

FIGURE 8.9   Schematic diagram showing the co-precipitation-assisted nanoparticles formation. (© 2013 Royal Society of Chemistry.)

8.5.2   Metal–TiO$_2$-Based Systems

M. Bellardita et al. reported the preparation of TiO$_2$ nanoparticles doped with metal ions, particularly rare earths (Ce, Co, Eu, Sm, W, and Yb), by co-precipitation to enhance the photocatalytic activity under visible light irradiation in both powder and film form (Bellardita et al. 2007). For powder sample, hydrolysis reaction of aqueous TiCl$_4$ was carried out at room temperature to get milky white TiO$_2$ dispersion. For metal ion loading, the salt or the oxide of the metal (CeO$_2$, CoCl$_2$·6H$_2$O, Na$_2$WO$_4$·2H$_2$O, Sm$_2$O$_3$, Yb$_2$O$_3$, and Eu$_2$O$_3$) desired was added to the as-prepared solution of TiO$_2$ and boiled for 2 h. In order to compare the photocatalytic activity, films of TiO$_2$ with metal ions were prepared by a simple two-step dip-coating method using the same precursors as those used for powder samples. The samples loaded with W and Sm showed better photocatalytic activity under UV illumination than bare TiO$_2$. With visible light, Co-doped system showed superior activity among all.

Many other research groups have investigated metal ion-doped TiO$_2$ nanostructures for various applications by co-precipitation method. Y. Li et al. reported the preparation of TiO$_2$ photocatalysts doped with alkaline-earth metal ions (Be^{2+}, Mg^{2+}, Ca^{2+}, Sr^{2+}, and Ba^{2+}) by combining the impregnation and co-precipitation methods for photocatalytic hydrogen production (Li et al. 2007). T.J. Kemp et al. reported the influence of transition metal (V, Mo, and W)-doped titanium (IV) dioxide prepared by co-precipitation method on the photo-degradation of polystyrene and poly(vinyl chloride) (Kemp and McIntyre 2006a, Kemp and McIntyre 2006b). M. Haruta et al. have done numerous works on metal-doped metal oxide systems, specifically on TiO$_2$, for various applications. In one such work, they have shown that gold can be highly dispersed on a variety of metal oxides by co-precipitation and deposition–precipitation followed by calcination for use as catalysts for CO oxidation (1993). They have also shown that Au displays very high catalytic activity, comparable with that of a conventional Cu/ZnO/Al$_2$O$_3$ catalyst. This was found to be true for both forward and reverse water–gas shift reactions at a lower temperature (Sakurai et al. 1997). In another pioneering work, they have compared the catalysis of MMOs with a variety of metal oxides (Haruta 1997). J. Moon et al. have explored the possibility of applying Sb-doped TiO$_2$ nanostructures prepared by co-precipitation method for photocatalysis (2001). K. Ranjit et al. reported the use of Fe-doped TiO_2 catalysts prepared via co-precipitation for photocatalysis (Ranjit and Viswanathan 1997).

8.5.3 METAL–ZnO SYSTEMS

V. Gandhi et al. reported the effect of Co doping on structural, optical, and magnetic properties of ZnO nanoparticles prepared via co-precipitation method (2014). ZnO nanoparticles were prepared initially by alkaline hydrolysis of zinc acetate (Figure 8.10a). For the preparation of Co-doped ZnO, the precipitation was done from a mixture of cobalt acetate and aqueous Zn precursor. A series of Co-doped ZnO nanoparticles were synthesized with different Co loading levels ($Zn_{1-x}Co_xO$, where $x =$ 0.05, 0.10, and 0.15) (Figure 8.10b–d). Diamagnetic behavior of pure ZnO was changed to ferromagnetic nature in Co-doped ZnO nanoparticles with significant changes in M–H loop, which was attributed to the oxygen vacancies and zinc interstitials.

In addition, across the scientific world, people have put much effort into coupling various metal ions with ZnO systems for achieving desired properties and applications. M. Mukhtar et al. reported the co-precipitation synthesis and characterization of copper-doped nanocrystalline ZnO nanoparticles (Mukhtar, Munisa, and Saleh 2012). T. Thangeeswari et al. accounted co-precipitation method for

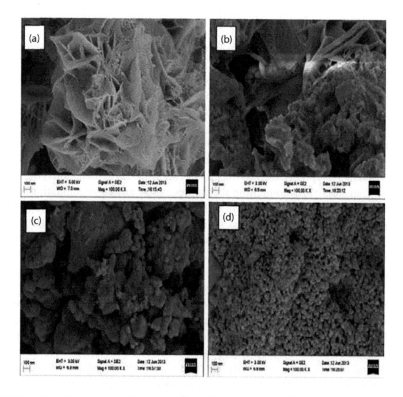

FIGURE 8.10 Morphology of (a) ZnO, (b) $Zn_{0.95}Co_{0.05}O$, (c) $Zn_{0.90}Co_{0.10}O$, and (d) $Zn_{0.85}Co_{0.15}O$ nanoparticles. (© 2014 American Chemical Society.)

obtaining Ni, Pb, and Cu-doped ZnO nanoparticles with emphasis on the structural and luminescence properties. P.G. Devi et al. reported the optical properties of ZnO and Co-doped ZnO nanoparticles prepared by the co-precipitation method (Thangeeswari, Velmurugan, and Priya 2014). Structural, optical, and magnetic properties of Ni-doped ZnO synthesized by co-precipitation method were reported by A.A. Gadalla et al. (2017).

8.5.4   Metal–MgO-Based Systems

The co-precipitation route was employed for the preparation of Ni-MgO hybrid MMO catalysts, and Y.-G. Chen et al. studied their catalytic performance for CO_2 reforming of methane. Ni ion loaded MgO solid solution catalysts ($Ni_xMg_{1-x}O$) were prepared by co-precipitating nickel acetate and magnesium nitrate aqueous solutions with the aid of potassium carbonate. They have prepared a series of catalysts with varying amounts of Ni (5, 10, and 15 wt%). For comparative study, MgO-supported Ni metal catalysts (Ni/MgO) were prepared by impregnation. $Ni_xMg_{1-x}O$ samples showed higher resistance to carbon formation and excellent anticoking performance than Ni/MgO.

Al-Fatesh et al. investigated methane decomposition on Fe/MgO-based catalyst prepared by co-precipitation with Ni, Co, and Mn as additives (2017).

8.5.5   Metal–WO₃-Based Systems

S. Upadhyay et al. (2014) prepared nanostructures of WO_3 blended with Sn via co-precipitation. The doped hybrids synthesized from chloride salt of Sn and Na_2WO_4 were employed for sensor applications. At the outset, stannic tetrachloride was added to metallic tungsten in dilute HNO_3. The aforementioned mixture was stirred at room temperature for 48 h followed by ageing for 20 h. The alcohol response characteristics of the samples were found to have a linear relationship with the amount of Sn in the hybrid. Sensing tests were extended for a series of alcohols to understand their range and activity (propan-2-ol, ethanol, methanol, etc.).

W. Dedsuksophon et al. reported the synthesis of a hybrid catalyst (Pd/WO_3-ZrO_2) employing co-precipitation technique and utilized it for a series of catalytic organic reactions on lignocellulosic biomass and its derivatives which include hydrolyzation, dehydration, aldol condensation, and hydrogenation (2011). In another study using metal–WO_3-based systems prepared by co-precipitation, C. Pang et al. reported the effect of V-doped WO_3 nanoparticles on the inhibition of tungsten particle growth during hydrogen reduction (2010). S. Vives et al. reported the sol–gel-coupled co-precipitation method for the preparation and characterization of zirconia–tungsten composite powders (1999). T. Kodama et al. studied the thermochemical reforming of methane using a reactive WO_3/W redox system prepared by the co-precipitation method (2000).

8.5.6   Metal–CeO₂-Based Systems

I. Luisetto et al. reported the surfactant-assisted co-precipitation synthesis of Co, Ni, or Co/Ni-incorporated CeO_2 hybrid materials for catalytic applications (Luisetto, Tuti, and Di Bartolomeo 2012). The catalysts were prepared by mixing the nitrate salts of Ce, Co, and Ni with ionic surfactant (cetyl trimethyl ammonium bromide) at an elevated temperature of 100°C while maintaining the pH at 10 with the addition of triethylamine. The hydroxide precipitate thus formed was kept for ageing for 20 h followed by thorough washing, drying, and annealing. The loading amount was varied by varying the nature and concentration of doped samples. The bimetallic Co-Ni/CeO$_2$ catalyst showed higher CH$_4$ conversion and sensing activity in comparison with other monometallic systems at the temperature range of 600°C–800°C.

D.J. Guo et al. reported a novel co-precipitation method for the preparation of Pt-doped CeO$_2$ composites on multiwalled carbon nanotubes for direct methanol fuel cell applications (Guo and Jing 2010). W. Luhui et al. reported the effect of precipitants on Ni-CeO$_2$ catalysts for the reverse water–gas shift reaction (2013). Y. Li et al. described the hydrogen production from methane decomposition over Ni-CeO$_2$ catalysts prepared by co-precipitation coupled with impregnation method (2006).

In addition, there are other MMO-based systems for various applications synthesized by co-precipitation pathway. M.A. Marselin et al. reported the co-precipitation synthesis of Co-doped NiO (Marselin and Jaya 2014). J. Liu et al. reported Pd/Fe$_3$O$_4$ catalyst prepared by co-precipitation method for aqueous-phase reforming of ethylene glycol to hydrogen (2010).

8.6 SONOCHEMICAL SYNTHESIS

8.6.1 Preface

Sonochemistry is the application of ultrasound to chemical reactions. In this method, agitation of materials in a medium is achieved by sound energy in terms of ultrasonic frequencies, and it can be used in various fields depending on its requirement. Acoustic cavitation is the phenomenon that causes the sonochemical effects in liquids. A number of studies have used the ultrasound-assisted physical and chemical effects generated by acoustic cavitation for synthesizing nanomaterials and for degradation of organic pollutants in last decade. Figure 8.11 demonstrates the nanoparticle formation by sonochemical method typically for an Sn-C system (Kumar et al. 2016).

8.6.2 Metal–TiO$_2$-Based Systems

K. Cheng et al. reported selective growth of gold nanoparticles by sonochemical pathway on single crystals of titanium dioxide in anatase phase (Figure 8.12) and their resulting photocatalytic performance (2013). At the outset, TiO$_2$ nanoparticles with different facets were synthesized using the precursors TiOSO$_4$ and Hydrofluoric acid (HF) by hydrothermal technique followed by annealing. Both the TiO$_2$ samples with (101) and (001) orientation were subjected to sonochemical deposition of gold precursor for ~0.33 h to prepare hybrid samples at ambient conditions. The crystal plane orientation was found to have a crucial role in the plasmon-assisted photocatalytic activity of the hybrid samples, and among the two catalysts, Au/TiO$_2$ (001) is shown to have better activity toward the degradation of 2,4-dichlorophenol.

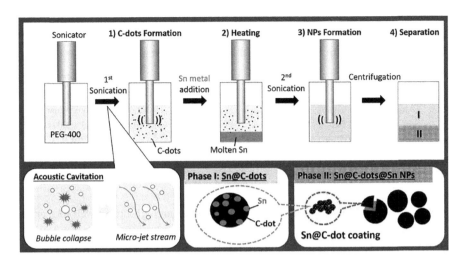

FIGURE 8.11   Schematic diagram showing the formation of Sn@C nanoparticles. (© 2016 Royal Society of Chemistry.)

FIGURE 8.12   FESEM images of (a) Au/TiO$_2$ (101) and (b) Au/TiO$_2$ (001) and their individual crystals (right). (© 2013 American Chemical Society.)

TABLE 8.6

Metal-TiO$_2$ Nanostructures Prepared by Sonochemical Route

No.	System	Method	Application	References
1	Ru-TiO$_2$	Sonochemical	Catalysis	N. Perkas et al. (2005, 2006)
2	Au-TiO$_2$	Sonochemical	Photocatalysis	Anandan and Ashokkumar (2009)
3	(Pt, Au and Pd)-TiO$_2$	Sonochemical	Photocatalysis	Y. Mizukoshi et al. (2007)
4	Fe and N-TiO$_2$ TiO$_2$	Sonochemical	Photocatalysis	T.-H. Kim et al. (2013)
5	Au-TiO$_2$	Sonochemical	Photocatalysis	D. Yang et al. (2009)

Metal-doped TiO$_2$ has been reported by other research groups as well for various applications accounted in Table 8.6.

8.6.3 METAL–ZnO-BASED SYSTEMS

O. Yayapao et al. reported the synthesis, concentration effect, morphology, optical properties, and photocatalytic activities of Dy-doped ZnO nanostructures by a sonochemical method (2013b). For the preparation of hybrid sample in a one-step process with varying amounts of doping (0%–3%), zinc nitrate hexahydrate and dysprosium nitrate hexahydrate were mixed vigorously in water followed by the addition of ammonium hydroxide until the pH was 9.5. The colorless solutions obtained were processed in 35 kHz ultrasonic bath at 80°C for 5 h. In the end, the precipitate collected of hybrid nanorods was found to have a hexagonal wurtzite structure. The photocatalytic activities of the as-synthesized products were tested for the degradation of methylene blue, and 3% Dy-doped ZnO showed the highest photocatalytic activity.

L.B. Arruda et al. reported the sonochemical synthesis and magnetism of Co-doped ZnO nanoparticles (2013). O. Yayapao et al. reported the sonochemical synthesis, photocatalysis, and photonic properties of 3% Ce-doped ZnO nano-needles. (2013a). X. Lu et al. studied ZnO nanoparticles incorporated with magnesium prepared by sonochemical pathway and found that hybrid nanoparticles exhibited excellent photocatalytic behavior in comparison with bare ZnO nanoparticles (2011). In another exciting work, H.M. Xiong et al. presented the sonochemical synthesis of highly luminescent zinc oxide nanoparticles doped with Mg (2009).

8.6.4 METAL–ZrO$_2$-BASED SYSTEMS

Y. Wang et al. reported the layered and hexagonal mesostructures of yttrium–zirconium oxides synthesized via sonochemical method (2001a). In a typical synthesis of hybrid nanoparticles, Y$_2$O$_3$ was dissolved in HNO$_3$ and heated until dry followed by the addition of water, ZrO(NO$_3$)$_2$, sodium dodecyl sulfate (surfactant templating agent), and urea (precipitant agent), keeping the pH at about 4.5, and then this mixture was kept for sonication at room temperature for 9.5, 3, and 6 h using a high-intensity ultrasonic probe. The surface areas of the hexagonal mesostructures were found to be 98 and 245 m^2g^{-1} before and after extraction with sodium acetate. In another report, they have explained the synthesis of yttrium–zirconium oxides via sonochemical method using different precursors (Wang et al. 2001b). The possibility of development of Pt (or Pd, sonochemically synthesized nanoparticles)-impregnated zirconia nano-aggregates (20–45 nm) was investigated by O. Vasylkiv et al. (2004).

8.6.5   OTHER MMO SYSTEMS

In addition to these systems, several other metal oxide systems have been explored for anchoring with metals for tuning their properties toward various applications. In a typical work, Y. Yao et al. reported the AgNP-sensitized WO_3 hollow nano-spheres fabricated via a simple sonochemical synthesis route for localized surface plasmon-enhanced gas sensors (2016). N. Perkas et al. reported a Pt/CeO_2 hybrid prepared via sonochemical route and its application as a catalyst in ethyl acetate combustion (2006). E. Malka et al. reported the simultaneous synthesis and deposition of Zn-doped CuO nanocomposite on cotton fabric using ultrasound irradiation for the eradication of multidrug-resistant bacteria (2013).

8.7   MICROWAVE SYNTHESIS

8.7.1   OUTLINE

Microwave-assisted synthesis is a novel method and a very rapidly developing area of research (Zhu and Chen 2014). The microwave-synthesized materials include a wide variety of solids that are of industrial and technological significance (Bilecka and Niederberger 2010). Microwave method is attractive due to advantages such as swiftness, cleanliness, and cost-effectiveness in comparison with the existing well-known techniques. Resonance and relaxation are the two probable pathways of energy conveyance happening during microwave irradiation of the materials causing uniform rise in temperature/heating. Only a narrow frequency window, centered at 900 MHz and 2.45 GHz, is allowed for microwave heating purposes. Synthesis via microwave irradiation also results in a shortening of reaction interval, decreased particle volume, smaller window for size classification, and product formation with high selectivity (Rao et al. 1999, Singh et al. 2017).

8.7.2   METAL–TiO$_2$-BASED SYSTEMS

K. Esquivel et al. (2013) have reported the microwave-mediated synthesis of a series of titanium dioxide-based systems. These TiO_2 photocatalysts were employed for the photodegradation of dyes (methyl red) at ambient conditions with the usage of the ultraviolet component of light source. For the preparation of bare titanium dioxide nanostructures, tetraisopropyl titanate precursor in propan-2-ol was used. The iron precursor, $FeSO_4 \cdot 7H_2O$, was used for obtaining Fe-modified TiO_2 samples, and thiourea was used for obtaining S-loaded TiO_2 samples. Sols obtained after the dissolution of respective precursors in aqueous medium were heated using a microwave reaction system. Crystalline, chemical, and electronic characteristics of TiO_2 were found to be heavily dependent on the dopant type, Fe and S ions, and also the temperature and duration of reaction under microwave irradiation.

Nanostructured metals (Sn, Cu, and Ni)-doped TiO_2 were synthesized by microwave irradiation method as reported by J. Maragatha et al. (2017). $TiOCl_2$ was added to the transition metal precursors (stannous chloride dihydrate, copper acetate monohydrate, nickel (II) acetate tetrahydrate), and TiO_2 and metal-doped TiO_2 powders were formed when microwave irradiation was passed through the solution with the frequency of 2.45 GHz at 160 W. Further, decomposition of the methylene blue dye under UV light irradiation was studied using the prepared pure TiO_2 and metal (Sn, Cu, and Ni)-doped TiO_2 photocatalyst.

Q. Xiang et al. reported a template-free synthesis of Ag on TiO_2 hollow spheres under microwave–hydrothermal conditions that induce self-transformation chemically (2010).

8.7.3 METAL–ZnO-BASED SYSTEMS

A one-pot route for the synthesis of Au-ZnO nanopyramids using microwave irradiation involving stepwise homogeneous and heterogeneous nucleation was reported by N.P. Herring et al. (2011) Two methods were used for the preparation of Au-ZnO nanoparticles. In the first method, $HAuCl_4$ was added to anhydrous zinc acetate in a mixture of oleic acid and oleylamine, whereas in the second method, preformed gold nanoparticles were added to the above reaction mixture. The reaction mixtures were microwave-treated continuously for different reaction times. Nanopyramids of hexagonal ZnO were formed upon rapid decomposition of zinc acetate precursor when subjected to microwave exposure assisted by the surfactant mixtures mentioned above. In both cases of ZnO nanopyramid formation, in situ-present Au nanocrystals governed the growth of former by acting as nuclei growth sites.

One-step, room-temperature microwave synthesis method for the preparation of M-doped ZnO, where M = Co, Cr, Fe, Mn, and Ni, was reported by G. Glaspell et al. (2005). Microwave synthesis was performed using zinc nitrate solution as Zn precursor and nitrates of desired transition metal as dopant precursors. Magnetic studies of the as-synthesized samples were performed, which revealed paramagnetic nature (Glaspell, Dutta, and Manivannan 2005).

C. Karunakaran et al. reported the synthesis of nanocrystalline ZnO and Ag-ZnO via microwave method and studied the electronic, optical, photocatalytic, and bactericidal properties of the samples (Karunakaran, Rajeswari, and Gomathisankar 2011). One-step microwave synthesis of Ag-ZnO nanocomposite was also reported in the work by F. Sun et al., in which photocatalytic performance of the Ag-ZnO nanocomposites with different Ag contents was systematically studied (2012).

8.7.4 METAL–WO₃-BASED SYSTEMS

The preparation of manganese (Mn)-doped tungsten oxide (WO_3) nanoparticles by a simple, one-step microwave irradiation method was reported by M. Karthik et al. for the first time (2017). H_2WO_4 and $MnCl_2 \cdot 4H_2O$ were used as precursors, and nanoparticles were prepared by microwave irradiation method without any postannealing treatment. The paper demonstrates for the first time the Mn-doped WO_3 nanoparticles as excellent candidates for supercapacitor applications. The authors discuss the synergetic effect of Mn-doped WO_3 nanoparticles and their crucial role in the enhancement of electrochemical behavior and also the possibility of use of these materials in supercapacitor and other energy storage devices.

J. Ye et al. reported pyrolysis using microwave for synthesis of WO_3-C hybrid materials (2010). Ammonium tungstate was used as a precursor, and Pt nanoparticles were interfaced with WO_3-C by microwave-assisted polyol process. The authors attribute improved synergistic properties of the composite as the reason for superior electrochemical performance of Pt/WO_3-C catalyst aided by heat treatment.

C. Yang et al. (2012) briefed a microwave-governed microemulsion-mediated growth of platinum on WO_3 nanoparticles supported on a carbon framework. After microwave deposition of tungsten oxide nanoparticles on carbon, Pt decoration was performed utilizing chloroplatinic acid in a microemulsion. The enhanced catalytic behavior of the composite in carbon monoxide oxidation as well as H_2 spillover and its mechanistic aspects were discussed in detail stressing the crucial role of physiochemical refinement of WO_3 nanoparticles observed by anchoring of Pt.

8.7.5 METAL–AL$_2$O$_3$-BASED SYSTEMS

W. Wong et al. reported an innovative microwave-assisted rapid sintering technique for the synthesis of magnesium composites reinforced with Al_2O_3 particulates using powder metallurgy (Wong, Karthik, and Gupta 2005b, a). Pure magnesium powder and submicron Al_2O_3 powder were uniaxially compacted to billets, which were further sintered using microwave-assisted rapid sintering technique. SiC was used as the microwave susceptor material, which absorbed microwave energy readily and thus, assisted rapid external heating of the billet while the compacted billet was also heated from within by absorption of microwave. The authors mentioned the improvement in mechanical properties as a result of the presence of alumina particulates in the composite formulations thus obtained.

8.7.6 METAL–NiO-BASED SYSTEMS

Parada and Morán (2006) have reported synthesis of Ni/NiO nanoparticles from two different Ni sources (nickel (II) di-acetate and nickel (II) di-formate) by subjecting to microwave exposure. Nickel salts and carbon black were ground hard to obtain a fine powder, and uniform pellets were fabricated and further processed under microwave for different reaction times. The acetate led to the formation of core–shell composites, whereas the formate salt of Ni produced reverse nanostructure to that of the former. Electronic structure and bonding nature influenced the formation of contrastingly behaving nanostructures. Magnetic behavior of the two precursors had significant influence, as the acetate counterpart gave rise to antiferromagnetic oxide leading to the formation of Ni with ferromagnetic nature while formate expressed ferromagnetic nature permanently.

Synthesis of Ni/NiO composite via microwave was also reported by O. Palchik et al. (1999), wherein amorphous Ni nanoparticles were used as the precursor and oxygen as the oxidizing agent. A glass reactor containing the alumina boat was used to place the nickel precursors, and subsequently, microwave irradiation was carried out with a continuous flow of gas. Nickel nanostructures in amorphous phase were formed by microwave method coupled with sonochemical reaction, whereas NiO nanostructures were grown when the reaction was carried out in ambient condition/ gaseous O_2. The above study can be adopted as a general method for preparing core–shell nanostructures. More importantly, both parts (core and shell) can be tuned without affecting the size of nanostructure.

8.8 CONCLUDING REMARKS

In the above narration, major chemical approaches toward the synthesis of MMO hybrid nanostructures are covered. Synthesis of desired morphology of metal oxides by any of the abovementioned methods followed by in situ anchoring of metal nanoparticles is the most popular route for achieving a good physical interface between the two constituents and subsequent improvement in properties. Simultaneous incorporation of both components is preferred when metal ion doping has to be achieved in controlled concentrations. Among the various methods described, hydrothermal approach is widely used for synthesizing metal oxides and MMO hybrids as the higher temperatures and pressures achieved in the sealed vessel can generate highly crystalline and kinetically controlled or thermodynamically stable forms by a simple tailoring using surfactants or other reaction parameters, and advantageously, requires only lower input energy. Lately, microwave method is fast becoming popular for synthesis of various nanoarchitectures of even complex oxides with multicomponents due to the extremely rapid reaction times and larger yield. An understanding of the role of various reagents under microwave irradiation may aid toward better control of the nanostructure morphology and growth.

REFERENCES

Aisah, N., D. Gustiono, V. Fauzia, I. Sugihartono, and R. Nuryadi. 2017. "Synthesis and enhanced photocatalytic activity of Ce-Doped zinc oxide nanorods by hydrothermal method." *IOP Conference Series: Materials Science and Engineering* no. 172 (1):012037.

Al-Fatesh, Ahmed-S, Siham Barama, Ahmed-A Ibrahim, Akila Barama, Wasim-Ullah Khan, and Anis Fakeeha. 2017. "Study of methane decomposition on Fe/MgO-based catalyst modified by Ni, Co, and Mn additives." *Chemical Engineering Communications* no. 204 (7):739–749.

Anandan, Sambandam, and Muthupandian Ashokkumar. 2009. "Sonochemical synthesis of Au–TiO$_2$ nanoparticles for the sonophotocatalytic degradation of organic pollutants in aqueous environment." *Ultrasonics Sonochemistry* no. 16 (3):316–320.

Arruda, Larisa B., Douglas M.G. Leite, Marcelo O. Orlandi, Wilson A. Ortiz, and Paulo Noronha Lisboa-Filho. 2013. "Sonochemical synthesis and magnetism in Co-doped ZnO nanoparticles." *Journal of Superconductivity and Novel Magnetism* no. 26 (7):2515–2519.

Bamwenda, Gratian R., Susumu Tsubota, Toshiko Nakamura, and Masatake Haruta. 1995. "Photoassisted hydrogen production from a water-ethanol solution: A comparison of activities of Au·TiO$_2$ and Pt·TiO$_2$." *Journal of Photochemistry and Photobiology A: Chemistry* no. 89 (2):177–189.

Basith, N. Mohamed, J. Judith Vijaya, L. John Kennedy, and M. Bououdina. 2013. "Structural, optical and room-temperature ferromagnetic properties of Fe-doped CuO nanostructures." *Physica E: Low-dimensional Systems and Nanostructures* no. 53:193–199.

Basith, N. Mohamed, J. Judith Vijaya, L. John Kennedy, and M. Bououdina. 2014. "Structural, morphological, optical, and magnetic properties of Ni-doped CuO nanostructures prepared by a rapid microwave combustion method." *Materials Science in Semiconductor Processing* no. 17:110–118.

Bayal, Nisha, and P. Jeevanandam. 2012. "Synthesis of NiO based bimetallic mixed metal oxide nanoparticles by sol-gel method." *Advanced Materials Research* no. 585:164–168.

Bellardita, Marianna, Maurizio Addamo, Agatino Di Paola, and Leonardo Palmisano. 2007. "Photocatalytic behaviour of metal-loaded TiO$_2$ aqueous dispersions and films." *Chemical Physics* no. 339 (1–3):94–103.

Bilecka, Idalia, and Markus Niederberger. 2010. "Microwave chemistry for inorganic nanomaterials synthesis." *Nanoscale* no. 2 (8):1358–1374.

Bramhaiah, Kommula, Vidya N. Singh, and Neena S. John. 2016. "Hybrid materials of ZnO nanostructures with reduced graphene oxide and gold nanoparticles: Enhanced photodegradation rates in relation to their composition and morphology." *Physical Chemistry Chemical Physics* no. 18 (3):1478–1486.

Buono-Core, Gonzalo E., Gerardo A. Cabello, H Espinoza, A. Hugo Klahn, Marisol Tejos, and Rose H. Hill. 2006. "Photochemical deposition of Pd-loaded and Pt-loaded tin oxide thin films." *Journal of the Chilean Chemical Society* no. 51 (3):950–956.

Byrappa, Kullaiah, and Masahiro Yoshimura. 2012. *Handbook of Hydrothermal Technology*: William Andrew Amsterdam: Elsevier.

Byrappa, Kullaiah, and Tadafumi Adschiri. 2007. "Hydrothermal technology for nanotechnology." *Progress in Crystal Growth and Characterization of Materials* no. 53 (2):117–166.

Byrappa, Kullaiah, Alaloor S. Dayananda, C. Ponnappa Sajan, B. Basavalingu, M.B. Shayan, Kohei Soga, and Masahiro Yoshimura. 2008. "Hydrothermal preparation of ZnO: CNT and TiO$_2$: CNT composites and their photocatalytic applications." *Journal of Materials Science* no. 43 (7):2348–2355.

Carvajal, Christian G., Killani Kadri, Gugu N. Rutherford, Robin Mundle, and Aswini K. Pradhan. 2015. "Photochemical decoration of metal and metal-oxide nanoparticles on highly oriented SnO$_2$ nanorod films for improved hybrid gas sensors and photo-detectors." *ECS Journal of Solid State Science and Technology* no. 4 (10):S3038–S3043.

Chakrabarti, Mahuya, Siddhartha Dechoudhury, D Sanyal, Tapatee Kundu Roy, Debasis Bhowmick, and Alok Chakrabarti. 2008. "Observation of room temperature ferromagnetism in Mn–Fe doped ZnO." *Journal of Physics D: Applied Physics* no. 41 (13):135006.

Chen, Shao Feng, Jian Ping Li, Kun Qian, Wei Ping Xu, Yang Lu, Wei Xin Huang, and Shu Hong Yu. 2010. "Large scale photochemical synthesis of M@TiO$_2$ nanocomposites (M = Ag, Pd, Au, Pt) and their optical properties, CO oxidation performance, and antibacterial effect." *Nano Research* no. 3 (4):244–255.

Cheng, Kun, Wenbin Sun, Hai-Ying Jiang, Jingjing Liu, and Jun Lin. 2013. "Sonochemical deposition of Au nanoparticles on different facets-dominated anatase TiO$_2$ single crystals and resulting photocatalytic performance." *The Journal of Physical Chemistry C* no. 117 (28):14600–14607.

Cheng, Zhipeng, Fengsheng Li, Yi Yang, Yi Wang, and Weifan Chen. 2008. "A facile and novel synthetic route to core-shell Al/Co nanocomposites." *Materials Letters* no. 62 (12–13):2003–2005.

Cheng-yu Chi, Hong-yao Yu, Jian-xin Dong, Wen-qing Liu, Shi-chang Cheng, Zheng-dong Liu, Xi-shan Xie. 2012. "The precipitation strengthening behavior of Cu-rich phase in Nb contained advanced Fe–Cr–Ni type austenitic heat resistant steel for USC power plant application."*Progress in Natural Science: Materials International* no. 22 (3):175–185.

Chey, Chan Oeurn, Ansar Masood, Anastasiia Riazanova, Xianjie Liu, K Venkat Rao, Omer Nur, and Magnus Willander. 2014. "Synthesis of Fe-doped ZnO nanorods by rapid mixing hydrothermal method and its application for high performance UV photodetector." *Journal of Nanomaterials* no. 2014:222.

Chiang, Chin-Lung, Chen-Chi M Ma, Dai-Lin Wu, and Hsu-Chiang Kuan. 2003. "Preparation, characterization, and properties of novolac-type phenolic/SiO_2 hybrid organic–inorganic nanocomposite materials by sol–gel method." *Journal of Polymer Science Part A: Polymer Chemistry* no. 41 (7):905–913.

Chung, Feng-Chao, Ren-Jang Wu, and Fu-Chou Cheng. 2014. "Fabrication of a Au@SnO_2 core–shell structure for gaseous formaldehyde sensing at room temperature." *Sensors and Actuators B: Chemical* no. 190:1–7.

Cocero, María José, Ángel Martín, Facundo Mattea, and Salima Varona. 2009. "Encapsulation and co-precipitation processes with supercritical fluids: Fundamentals and applications." *The Journal of Supercritical Fluids* no. 47 (3):546–555.

Considine, Glenn D., and Peter H. Kulik. 2005. "Hydrothermal processing." In *Van Nostrand's Scientific Encyclopedia*. Hoboken, NJ: John Wiley & Sons.

Dedsuksophon, Wassana, Kajornsak Faungnawakij, Verawat Champreda, and Navadol Laosiripojana. 2011. "Hydrolysis/dehydration/aldol-condensation/hydrogenation of lignocellulosic biomass and biomass-derived carbohydrates in the presence of Pd/WO_3–ZrO_2 in a single reactor." *Bioresource Technology* no. 102 (2):2040–2046.

Devi, P. Geetha, and A. Sakthi Velu. 2016. "Synthesis, structural and optical properties of pure ZnO and Co doped ZnO nanoparticles prepared by the co-precipitation method." *Journal of Theoretical and Applied Physics* no. 10 (3):233–240.

Esquivel, Karen, Rufino Nava, Alma Zamudio-Méndez, Marta Vega González, Oscar E. Jaime-Acuña, Luis Escobar-Alarcón, Juan M. Peralta-Hernández, Barbera Pawelec, and José L. G. Fierro. 2013. "Microwave-assisted synthesis of (S) Fe/TiO2 systems: Effects of synthesis conditions and dopant concentration on photoactivity." *Applied Catalysis B: Environmental* no. 140–141:213–224.

Febrianti, Yosi, Nanda Novia Putri, Iwan Sugihartono, Vivi Fauzia, and Djati Handoko. 2017. "Synthesis and characterization of Co-doped zinc oxide nanorods prepared by ultrasonic spray pyrolysis and hydrothermal methods". *AIP Conference Proceedings* no. 1862 (1):030056.

Fernández-García, Marcos, and José A. Rodriguez. 2011. "Metal oxide nanoparticles." *Encyclopedia of Inorganic and Bioinorganic Chemistry*. Hoboken, NJ: John Wiley and Sons.

Filippetti, Alessio, and Vincenzo Fiorentini. 2006. "Double-exchange driven ferromagnetic metal-paramagnetic insulator transition in Mn-doped CuO." *Physical Review B* no. 74 (22):220401.

Franke, Marion E., Tobias J. Koplin, and Ulrich Simon. 2006. "Metal and metal oxide nanoparticles in chemiresistors: Does the nanoscale matter?" *Small* no. 2 (1):36–50.

Gadalla, A. A., I. Abood, and M. M. Elokr. 2017. "Structural, optical and magnetic properties of Ni-Doped ZnO synthesized by Co-precipitation method." *Journal of Nanotechnology and Materials Science* no. 4 (1):19–26. doi: 10.15436/2377-1372.17.1374.

Gandhi, Vijayaprasath, Ravi Ganesan, Haja Hameed Abdulrahman Syedahamed, and Mahalingam Thaiyan. 2014. "Effect of cobalt doping on structural, optical, and magnetic properties of ZnO nanoparticles synthesized by coprecipitation method." *The Journal of Physical Chemistry C* no. 118 (18):9715–9725.

Glaspell, Garry, Prasanta Dutta, and A Manivannan. 2005. "A room-temperature and microwave synthesis of M-doped ZnO (M= Co, Cr, Fe, Mn & Ni)." *Journal of Cluster Science* no. 16 (4):523–536.

Guo, Dao-Jun, and Zhi-Hong Jing. 2010. "A novel co-precipitation method for preparation of Pt-CeO$_2$ composites on multi-walled carbon nanotubes for direct methanol fuel cells." *Journal of Power Sources* no. 195 (12):3802–3805.

Guo, Ting, Ming-Shui Yao, Yuan-Hua Lin, and Ce-Wen Nan. 2015. "A comprehensive review on synthesis methods for transition-metal oxide nanostructures." *CrystEngComm* no. 17 (19):3551–3585.

Han, Jingfeng, Xu Zong, Xin Zhou, and Can Li. 2015. "Cu$_2$O/CuO photocathode with improved stability for photoelectrochemical water reduction." *RSC Advances* no. 5 (14):10790–10794.

Haruta, Masatake. 1997. "Novel catalysis of gold deposited on metal oxides." *Catalysis Surveys from Asia* no. 1 (1):61–73.

Haruta, Masatake, Susumu Tsubota, Tetsuhiko Kobayashi, Hiroyuki Kageyama, Michel J. Genet, and Bernard Delmon. 1993. "Low-temperature oxidation of CO over gold supported on TiO$_2$, α-Fe$_2$O$_3$, and Co$_3$O$_4$." *Journal of Catalysis* no. 144 (1):175–192.

Hasan, Samiul, RA Mayanovic, and Mourad Benamara. 2017. "Synthesis and characterization of novel inverted NiO@Ni$_x$Mn$_{1-x}$O core-shell nanoparticles." *MRS Advances* no. 2 (56):3465–3470.

Herring, Natalie P., Khaled AbouZeid, Mona B. Mohamed, John Pinsk, and M. Samy El-Shall. 2011. "Formation mechanisms of gold–zinc oxide hexagonal nanopyramids by heterogeneous nucleation using microwave synthesis." *Langmuir* no. 27 (24):15146–15154.

Higashimoto, Shinya, Wataru Tanihata, Yoshinori Nakagawa, Masashi Azuma, Hiroyoshi Ohue, and Yoshihisa Sakata. 2008. "Effective photocatalytic decomposition of VOC under visible-light irradiation on N-doped TiO$_2$ modified by vanadium species." *Applied Catalysis A: General* no. 340 (1):98–104.

Hou, Xing-Gang, Mei-Dong Huang, Xiao-Ling Wu, and An-Dong Liu. 2009. "Preparation and studies of photocatalytic silver-loaded TiO$_2$ films by hybrid sol–gel method." *Chemical Engineering Journal* no. 146 (1):42–48.

Hung, Wen-Chi, Ssu-Han Fu, Jeou-Jen Tseng, Hsin Chu, and Tzu-Hsing Ko. 2007. "Study on photocatalytic degradation of gaseous dichloromethane using pure and iron ion-doped TiO$_2$ prepared by the sol–gel method." *Chemosphere* no. 66 (11):2142–2151.

Iqbal, Javed, Tariq Jan, Sibt Ul-Hassan, Ishaq Ahmed, Qaisar Mansoor, M Umair Ali, Fazal Abbas, and Muhammad Ismail. 2015. "Facile synthesis of Zn doped CuO hierarchical nanostructures: Structural, optical and antibacterial properties." *AIP Advances* no. 5 (12):127112.

Janisch, Rebecca, Priya Gopal, and Nicola A Spaldin. 2005. "Transition metal-doped TiO$_2$ and ZnO—present status of the field." *Journal of Physics: Condensed Matter* no. 17 (27):R657.

Jia, Jianbo, Bingquan Wang, Aiguo Wu, Guangjin Cheng, Zhuang Li, and Shaojun Dong. 2002. "A method to construct a third-generation horseradish peroxidase biosensor: Self-assembling gold nanoparticles to three-dimensional sol–gel network." *Analytical Chemistry* no. 74 (9):2217–2223.

Jiang, Jian, Yuanyuan Li, Jinping Liu, Xintang Huang, Changzhou Yuan, and Xiong Wen David Lou. 2012. "Recent advances in metal oxide-based electrode architecture design for electrochemical energy storage." *Advanced materials* no. 24 (38):5166–5180.

Jiang, Tingting, Yongqian Wang, Dawei Meng, and Dagui Wang. 2016. "One-step hydrothermal synthesis and enhanced photocatalytic performance of pine-needle-like Zn-doped CuO nanostructures." *Journal of Materials Science: Materials in Electronics* no. 27 (12):12884–12890.

Kang, Jun-Gill, and Youngku Sohn. 2012. "Interfacial nature of Ag nanoparticles supported on TiO_2 photocatalysts." *Journal of Materials Science* no. 47 (2):824–832.

Kanjwal, Muzafar A., Nasser A.M. Barakat, Faheem A. Sheikh, Soo Jin Park, and Hak Yong Kim. 2010. "Photocatalytic activity of ZnO-TiO_2 hierarchical nanostructure prepared by combined electrospinning and hydrothermal techniques." *Macromolecular Research* no. 18 (3):233–240.

Kannaki, K, Pugalenthi S. Ramesh, and Devarajan Geetha. 2016. "Facile hydrothermal synthesis and structural, optical, and morphological investigations on PVP assisted Fe doped CuO nanocomposites." *Materials Today: Proceedings* no. 3 (6):2329-2338.

Karthik Maithili, Mathivanan Parthibavarman, Subramaniyan Prabhakaran, Venkatesan Hariharan, Ramaswamy Poonguzhali, Sekar Sathishkumar 2017. "One-step microwave synthesis of pure and Mn doped WO_3 nanoparticles and its structural, optical and electrochemical properties." *Journal of Materials Science: Materials in Electronics* no. 28 (9):6635–6642.

Karunakaran, Chockalingam, and Govindasamy Manikandan. 2013. "Synthesis and characterization of Zn-doped CuO nanomaterials for electrochemical supercapacitor applications." In National Conference on Recent Advances in Surface Science (pp. 93–95).

Karunakaran, Chockalingam, Velayutham Rajeswari, and Paramasavian Gomathisankar. 2011. "Optical, electrical, photocatalytic, and bactericidal properties of microwave synthesized nanocrystalline Ag–ZnO and ZnO." *Solid State Sciences* no. 13 (5):923–928.

Katoch, Akash, Markus Burkhart, Taejin Hwang, and Sang Sub Kim. 2012. "Synthesis of polyaniline/TiO_2 hybrid nanoplates via a sol–gel chemical method." *Chemical Engineering Journal* no. 192:262–268.

Keller, Arturo A., Hongtao Wang, Dongxu Zhou, Hunter S Lenihan, Gary Cherr, Bradley J. Cardinale, Robert Miller, and Zhaoxia Ji. 2010. "Stability and aggregation of metal oxide nanoparticles in natural aqueous matrices." *Environmental Science & Technology* no. 44 (6):1962–1967.

Kemnitz, Erhard. and Noack, Johannes. 2015. "The non-aqueous fluorolytic sol-gel synthesis of nanoscaled metal fluorides." *Dalton Transaction* no. 44:19411–19431.

Kemp, Terence J., and Robin A. McIntyre. 2006a. "Influence of transition metal-doped titanium (IV) dioxide on the photodegradation of polystyrene." *Polymer Degradation and Stability* no. 91 (12):3010–3019.

Kemp, Terence J., and Robin A. McIntyre. 2006b. "Transition metal-doped titanium (IV) dioxide: Characterisation and influence on photodegradation of poly (vinyl chloride)." *Polymer Degradation and Stability* no. 91 (1):165–194.

Kim, Tae-Ho, Vicent Rodríguez-González, Gobinda Gyawali, Sung-Hun Cho, Tohru Sekino, and Soo-Wohn Lee. 2013. "Synthesis of solar light responsive Fe, N co-doped TiO_2 photocatalyst by sonochemical method." *Catalysis Today* no. 212:75–80.

Ko, Seonghyuk, Chandra K. Banerjee, and Jagannathan Sankar. 2011. "Photochemical synthesis and photocatalytic activity in simulated solar light of nanosized Ag doped TiO_2 nanoparticle composite." *Composites Part B: Engineering* no. 42 (3):579–583.

Kochuveedu, Saji Thomas, Yoon Hee Jang, and Dong Ha Kim. 2013. "A study on the mechanism for the interaction of light with noble metal-metal oxide semiconductor nanostructures for various photophysical applications." *Chemical Society Reviews* no. 42 (21):8467–8493.

Kodama, Tatsuya, Hiro Ohtake, Sohkichi Matsumoto, Akira Aoki, Tadaaki Shimizu, and Yoshie Kitayama. 2000. "Thermochemical methane reforming using a reactive WO_3/W redox system." *Energy* no. 25 (5):411–425.

Koelsch, M., Sophie Cassaignon, Jean-François Guillemoles, and Jean P Jolivet. 2002. "Comparison of optical and electrochemical properties of anatase and brookite TiO_2 synthesized by the sol–gel method." *Thin Solid Films* no. 403:312–319.

Kolmakov, Andrei, and Martin Moskovits. 2004. "Chemical sensing and catalysis by one-dimensional metal-oxide nanostructures." *Annual Review of Materials Research* no. 34:151–180.

Komarneni, Sridhar, Young Dong Noh, Joo Young Kim, Seok Han Kim, and Hiroaki Katsuki. 2010. "Solvothermal/hydrothermal synthesis of metal oxides and metal powders with and without microwaves." *Zeitschrift für Naturforschung B* no. 65 (8):1033–1037.

Kumar, Vijay B., Jialiang Tang, Kay Jangweon Lee, Vilas G. Pol, and Aharon Gedanken. 2016. "In situ sonochemical synthesis of luminescent Sn@C-dots and a hybrid Sn@C-dots@Sn anode for lithium-ion batteries." *RSC Advances* no. 6: 66256–66265.

Lakshmi, Krishnasamy, K Kadirvelu, and Palathurai Subramaniam Mohan. 2017. "Catalytic reduction of hazardous compound (Triethylphosphate) using Ni doped CuO nanoparticles." *Defence Life Science Journal* no. 2 (4):458–462.

Larsen, Gustavo, Raffet Velarde-Ortiz, Kevin Minchow, Antonio Barrero, and Ignacio G Loscertales. 2003. "A method for making inorganic and hybrid (organic/inorganic) fibers and vesicles with diameters in the submicrometer and micrometer range via sol–gel chemistry and electrically forced liquid jets." *Journal of the American Chemical Society* no. 125 (5):1154–1155.

Lee, Byung-Yong, Sang-Hyuk Park, Misook Kang, Sung-Chul Lee, and Suk-Jin Choung. 2003. "Preparation of Al/TiO_2 nanometer photo-catalyst film and the effect of H_2O addition on photo-catalytic performance for benzene removal." *Applied Catalysis A: General* no. 253 (2):371–380.

Lee, Sangwook, In-Sun Cho, Ji Hae Lee, Dong Hoe Kim, Dong Wook Kim, Jin Young Kim, Hyunho Shin, Jung-Kun Lee, Hyun Suk Jung, and Nam-Gyu Park. 2010. "Two-step sol–gel method-based TiO_2 nanoparticles with uniform morphology and size for efficient photo-energy conversion devices." *Chemistry of Materials* no. 22 (6):1958–1965.

Lee, Yoonkyung, Eunpa Kim, Yunjeong Park, Jangho Kim, WonHyoung Ryu, Junsuk Rho, and Kyunghoon Kim. 2018. "Photodeposited metal-semiconductor nanocomposites and their applications." *Journal of Materiomics* 4 (2):83–94.

Li, Haibin, Xuechen Duan, Guocong Liu, and Xiaoqi Liu. 2008. "Photochemical synthesis and characterization of Ag/ TiO_2 nanotube composites." *Journal of Materials Science* no. 43 (5):1669–1676.

Li, Wei, Fei Wang, Shanshan Feng, Jinxiu Wang, Zhenkun Sun, Bin Li, Yuhui Li, Jianping Yang, Ahmed A Elzatahry, and Yongyao Xia. 2013. "Sol–gel design strategy for ultra-dispersed TiO_2 nanoparticles on graphene for high-performance lithium ion batteries." *Journal of the American Chemical Society* no. 135 (49):18300–18303.

Li, Yong, Baocai Zhang, Xiaolan Tang, Yide Xu, and Wenjie Shen. 2006. "Hydrogen production from methane decomposition over Ni/CeO_2 catalysts." *Catalysis Communications* no. 7 (6):380–386.

Li, Yuexiang, Shaoqin Peng, Fengyi Jiang, Gongxuan Lu, and Shuben Li. 2007. "Effect of doping TiO_2 with alkaline-earth metal ions on its photocatalytic activity." *Journal of the Serbian Chemical Society* no. 72 (4):393–402.

Li, Zhuoqi, Weijie Wang, Zhicheng Zhao, Xinrong Liu, and Peng Song. 2017. "One-step hydrothermal preparation of Ce-doped MoO_3 nanobelts with enhanced gas sensing properties." *RSC Advances* no. 7 (45):28366–28372.

Li, Zhu-Zhu, Li-Xiong Wen, Lei Shao, and Jian-Feng Chen. 2004. "Fabrication of porous hollow silica nanoparticles and their applications in drug release control." *Journal of Controlled Release* no. 98 (2):245–254.

Lin, Ronghe, and Yunjie Ding. 2013. "A review on the synthesis and applications of meso-structured transition metal phosphates." *Materials* no. 6 (1):217–243.

Liu, Jie, Wei Wang, Tong Shen, Zhiwei Zhao, Hui Feng, and Fuyi Cui. 2014. "One-step synthesis of noble metal/oxide nanocomposites with tunable size of noble metal particles and their size-dependent catalytic activity." *RSC Advances* no. 4 (58):30624–30629.

Liu, Jun, Bo Sun, Jiye Hu, Yan Pei, Hexing Li, and Minghua Qiao. 2010. "Aqueous-phase reforming of ethylene glycol to hydrogen on Pd/Fe$_3$O$_4$ catalyst prepared by co-precipitation: Metal–support interaction and excellent intrinsic activity." *Journal of Catalysis* no. 274 (2):287–295.

Liu, Ruirui, Dalai Jin, and Linhai Yue. 2015. "Synthesis and electrochemical properties of Co doped MnO$_2$ framework with nanofibrous structure." *International Journal of Applied Ceramic Technology* no. 12 (S2):E59–E64.

Liu, Xueqin, James Iocozzia, Yang Wang, Xun Cui, Yihuang Chen, Shiqiang Zhao, Zhen Li, and Zhiqun Lin. 2017. "Noble metal–metal oxide nanohybrids with tailored nanostructures for efficient solar energy conversion, photocatalysis and environmental remediation." *Energy & Environmental Science* no. 10 (2):402–434.

Liu, Yuhuan, Yi Zhou, Luyue Yang, Yutang Wang, Yiwei Wu, Chaocheng Li, and Jun Lu. 2016. "Hydrothermal synthesis of 3D urchin-like Ag/TiO$_2$/reduced graphene oxide composites and its enhanced photocatalytic performance." *Journal of Nanoparticle Research* no. 18 (9):283.

Liu, Zhaoyang, Hongwei Bai, Shiping Xu, and Darren Delai Sun. 2011. "Hierarchical CuO/ZnO 'corn-like' architecture for photocatalytic hydrogen generation." *International Journal of Hydrogen Energy* no. 36 (21):13473–13480.

Lu, Jia Grace, Paichun Chang, and Zhiyong Fan. 2006. "Quasi-one-dimensional metal oxide materials—Synthesis, properties and applications." *Materials Science and Engineering: R: Reports* no. 52 (1–3):49–91.

Lu, Ping, Botao Qiao, Ning Lu, Dong Choon Hyun, Jinguo Wang, Moon J Kim, Jingyue Liu, and Younan Xia. 2015. "Photochemical deposition of highly dispersed Pt nanoparticles on porous CeO$_2$ nanofibers for the water-gas shift reaction." *Advanced Functional Materials* no. 25 (26):4153–4162.

Lu, Xianyong, Zhaoyue Liu, Ying Zhu, and Lei Jiang. 2011. "Sonochemical synthesis and photocatalytic property of zinc oxide nanoparticles doped with magnesium (II)." *Materials Research Bulletin* no. 46 (10):1638–1641.

Lu, Xiaojuan, Duowang Fan, Duojin Fan, and Hongzhong Liu. 2007. Synthesis and photocatalytic capability of Cr-TiO$_2$ film. *International Nano-Optoelectronics Workshop*, IEEE, pp. 276–277.

Luhui, Wang, Liu Hui, Liu Yuan, CHEN Ying, and YANG Shuqing. 2013. "Effect of pre-cipitants on Ni- CeO$_2$ catalysts prepared by a co-precipitation method for the reverse water-gas shift reaction." *Journal of Rare Earths* no. 31 (10):969–974.

Luisetto, Igor, Simonetta Tuti, and Elisabetta Di Bartolomeo. 2012. "Co and Ni supported on CeO$_2$ as selective bimetallic catalyst for dry reforming of methane." *International Journal of Hydrogen Energy* no. 37 (21):15992–15999.

Ma. De Jesus Soria Aguilar, Damaris Margarita Puente, Francisco R. Carrillo-Pedroza, L. A. Garcia-Cerda, and Jesus Velazquez Salazar. 2015. "Ozonation and sol-gel method to obtain Cu/CuO nanoparticles from cyanidation wastewater." *Revista Internacional de Contaminacion Ambiental* no. 31 (3):265–270.

Mackenzie, John D., and Eric P. Bescher. 2007. "Chemical routes in the synthesis of nanomaterials using the sol–gel process." *Accounts of Chemical Research* no. 40 (9):810–818.

Maeda, Kazuhiko, Kentaro Teramura, Daling Lu, Nobuo Saito, Yasunobu Inoue, and Kazunari Domen. 2006. "Noble-metal/Cr$_2$O$_3$ core/shell nanoparticles as a cocatalyst for photocatalytic overall water splitting." *Angewandte Chemie International Edition* no. 45 (46):7806–7809.

Maeda, Kazuhiko, Ryu Abe, and Kazunari Domen. 2011. "Role and function of ruthenium species as promoters with TaON-based photocatalysts for oxygen evolution in two-step water splitting under visible light." *The Journal of Physical Chemistry C* no. 115 (7):3057–3064.

Malka, Eyal, Ilana Perelshtein, Anat Lipovsky, Yakov Shalom, Livnat Naparstek, Nina Perkas, Tal Patick, Rachel Lubart, Yeshayahu Nitzan, and Ehud Banin. 2013. "Eradication of multi-drug resistant bacteria by a novel Zn-doped CuO nanocomposite." *Small* no. 9 (23):4069–4076.

Manna, Sujit, and Subodh Kumar De. 2010. "Room temperature ferromagnetism in Fe doped CuO nanorods." *Journal of Magnetism and Magnetic Materials* no. 322 (18):2749–2753.

Maragatha, Jothirajan, Somasundaram Rajendran, Takeshi Endo, and Subbian Karuppuchamy. 2017. "Microwave synthesis of metal doped TiO$_2$ for photocatalytic applications." *Journal of Materials Science: Materials in Electronics* no. 28 (7):5281–5287.

Marselin, M. Abila, and N. Victor Jaya. 2014. "Synthesis and characterization of pure and cobalt-doped NiO nanoparticles." *International Journal of ChemTech Research* no. 7 (6):2654–2659.

Mehran Vafaee Khanjani, and Morteza Sasani Ghamsari. 2007. "Preparation and characterization of ZnO nanoparticles by a novel sol–gel route." *Materials Letters* no. 61 (14):3265–3268.

Mirzaei, Ali, Janghorban, Kamal, Hashemi, Behrooz, Bonavita, Anna, Bonyani, Maryam, Leonardi, Salvatore Gianluca Leonardi, Neri, Giovanni. 2015. "Synthesis, characterization and gas sensing properties of Ag@α-Fe2O3 core–shell nanocomposites." Nanomaterials no. 5:737–749.

Mizukoshi, Yoshiteru, Yoji Makise, Tatsuya Shuto, Jinwei Hu, Aki Tominaga, Sayoko Shironita, and Shuji Tanabe. 2007. "Immobilization of noble metal nanoparticles on the surface of TiO$_2$ by the sonochemical method: Photocatalytic production of hydrogen from an aqueous solution of ethanol." *Ultrasonics Sonochemistry* no. 14 (3):387–392.

Mohebbi, Sajjad, Somayeh Molaei, and Azar Amir Reza Judy. 2013. "Preparation and study of Sn-doped CuO nanoparticles as semiconductor." *Journal of Applied Chemistry no.* 8 (27):27–30.

Moncada, Edwin, Raul Quijada Abarca, and Jaime Retuert. 2007. "Nanoparticles prepared by the sol–gel method and their use in the formation of nanocomposites with polypropylene." *Nanotechnology* no. 18 (33):335606.

Moon, Jooho, Hidenori Takagi, Yoshinobu Fujishiro, and Masanobu Awano. 2001. "Preparation and characterization of the Sb-doped TiO$_2$ photocatalysts." *Journal of Materials Science* no. 36 (4):949–955.

Mu, Wanjun, Xiang Xie, Xingliang Li, Rui Zhang, Qianhong Yu, Kai Lv, Hongyuan Wei, and Yuan Jian. 2014. "Characterizations of Nb-doped WO$_3$ nanomaterials and their enhanced photocatalytic performance." *RSC Advances* no. 4 (68):36064–36070.

Mukhtar, Mergoramadhayenty, Lusitra Munisa, and Rosari Saleh. 2012. "Co-precipitation synthesis and characterization of nanocrystalline zinc oxide particles doped with Cu^{2+} ions." *Materials Sciences and Applications* no. 3 (08):543.

Murata, Akiyo, Nobuto Oka, Shinichi Nakamura, and Yuzo Shigesato. 2012. "Visible-light active photocatalytic WO$_3$ films loaded with Pt nanoparticles deposited by sputtering." *Journal of Nanoscience and Nanotechnology* no. 12 (6):5082–5086.

Nagarale, Rajaram K., G.S. Gohil, Vinod K. Shahi, and Roopesh Rangarajan. 2004. "Organic–inorganic hybrid membrane: thermally stable cation-exchange membrane prepared by the sol–gel method." *Macromolecules* no. 37 (26):10023–10030.

Naz, Gul Jabeen, Usman Khan, Saira Riaz, and Shahzad Naseem. 2015. "Effect of transition bi-metal doping on ZnO thin films." *Materials Today: Proceedings* no. 2 (10):5160–5165.

Ng, Law Yong, Abdul Wahab Mohammad, Choe Peng Leo, and Nidal Hilal. 2013. "Polymeric membranes incorporated with metal/metal oxide nanoparticles: A comprehensive review." *Desalination* no. 308:15–33.

Parada, Carmen, and Emilio Morán. 2006. "Microwave-assisted synthesis and magnetic study of nanosized Ni/NiO materials." *Chemistry of Materials* no. 18 (11):2719–2725.

Palchik, Oleg, Sigalit Avivi, Dalia Pinkert, and Aharon Gedanken. 1999. "Preparation and characterization of Ni/NiO composite using microwave irradiation and sonication." *Nanostructured Materials* no. 11 (3):415–420.

Panda, Jagannath, I. Sasmal, and Tapan K. Nath. 2016. "Magnetic and optical properties of Mn-doped ZnO vertically aligned nanorods synthesized by hydrothermal technique." *AIP Advances* no. 6 (3):035118.

Pang, Cholsong, Ji Luo, Zhimeng Guo, Min Guo, and Ting Hou. 2010. "Inhibition of tungsten particle growth during reduction of V-doped WO_3 nanoparticles prepared by co-precipitation method." *International Journal of Refractory Metals and Hard Materials* no. 28 (3):343–348.

Pasha, Nayeem, Nakka Lingaiah, Putluru Siva Sankar Reddy, and Potharaju S Sai Prasad. 2009. "Direct decomposition of N_2O over cesium-doped CuO catalysts." *Catalysis Letters* no. 127 (1–2):101–106.

Peled, Aaron. 1995. Photodeposition from liquid phase. *ROMOPTO'94: Fourth Conference in Optics* no. 2461:13.

Perkas, Nina, Hadar Rotter, Leonid Vradman, Miron V Landau, and Aharon Gedanken. 2006. "Sonochemically prepared Pt/CeO_2 and its application as a catalyst in ethyl acetate combustion." *Langmuir* no. 22 (16):7072–7077.

Perkas, Nina, Ziyi Zhong, Luwei Chen, Michele Besson, and Aharon Gedanken. 2005. "Sonochemically prepared high dispersed Ru/TiO_2 mesoporous catalyst for partial oxidation of methane to syngas." *Catalysis Letters* no. 103 (1–2):9–14.

Piera, Eva, M. Isabel Tejedor-Tejedor, Michael E. Zorn, and Marc A. Anderson. 2003. "Relationship concerning the nature and concentration of Fe (III) species on the surface of TiO_2 particles and photocatalytic activity of the catalyst." *Applied Catalysis B: Environmental* no. 46 (4):671–685.

Rahman, Mohammed M, Aslam Jamal, Sher Bahadar Khan, and Mohd Faisal. 2011. "Fabrication of highly sensitive ethanol chemical sensor based on Sm-doped Co_3O_4 nanokernels by a hydrothermal method." *The Journal of Physical Chemistry C* no. 115 (19):9503–9510.

Ramakrishnan, Vivek, Hyun Kim, Jucheol Park, and Beelyong Yang. 2016. "Cobalt oxide nanoparticles on TiO_2 nanorod/FTO as a photoanode with enhanced visible light sensitization." *RSC Advances* no. 6 (12):9789–9795. doi:10.1039/C5RA23200G.

Ramakrishnan, Vivek, Chandraraj Alex, Aruna N. Nair, and Neena S. John. 2018. "Designing metallic MoO2 nanostructures on rigid substrates for electrochemical water activation." *Chemistry-A European Journal* no. 24:18003–18011.

Ranjit, Koodali T., and Balasubramanian Viswanathan. 1997. "Synthesis, characterization and photocatalytic properties of iron-doped TiO_2 catalysts." *Journal of Photochemistry and Photobiology A: Chemistry* no. 108 (1):79–84.

Rao, Kalya Jagannatha, Bala Vaidhyanathan, Munia Ganguli, and P.A. Ramakrishnan. 1999. "Synthesis of inorganic solids using microwaves." *Chemistry of Materials* no. 11 (4):882–895.

Raudonyte-Svirbutaviciene, Eva, Alexandra Neagu, Vida Vickackaite, Vitalija Jasulaitiene, Aleksej Zarkov, Cheuk-Wai Tai, and Arturas Katelnikovas. 2018. "Two-step photochemical inorganic approach to the synthesis of $Ag-CeO_2$ nanoheterostructures and their photocatalytic activity on tributyltin degradation." *Journal of Photochemistry and Photobiology A: Chemistry* no. 351:29–41.

Ray, Chaiti, and Tarasankar Pal. 2017. "Recent advances of metal–metal oxide nanocomposites and their tailored nanostructures in numerous catalytic applications." *Journal of Materials Chemistry A* no. 5 (20):9465–9487.

Rejitha, S.G., and C. Krishnan. 2013. "Synthesis of cadmium-doped copper oxide nanoparticles: Optical and structural characterizations." *Advances in Applied Science Research* no. 4 (2):103.

Ren, Shoutian, and Wenjun Liu. 2016. "One-step photochemical deposition of PdAu alloyed nanoparticles on TiO_2 nanowires for ultra-sensitive H_2 detection." *Journal of Materials Chemistry A* no. 4 (6):2236–2245.

Sakurai, Hiroaki, Atsushi Ueda, Tetsuhiko Kobayashi, and Masatake Haruta. 1997. "Low-temperature water–gas shift reaction over gold deposited on TiO_2." *Chemical Communications* 3:271–272.

Sanchez, Enrique Mora, Tessy Maria Lopez Goerne, Ricardo Gomez, Antonio Morales, and Octavio Augusto Novaro. 1996. "Synthesis and characterization of sol–gel Pt/TiO_2 catalyst." *Journal of Solid State Chemistry* no. 122 (2):309–314.

Schmidt, Helmut, Gerhard Jonschker, Stefan Goedicke, and Martin Mennig. 2000. "The sol-gel process as a basic technology for nanoparticle-dispersed inorganic-organic composites." *Journal of Sol-Gel Science and Technology* no. 19 (1):39–51.

Shaikh, Jasmin, Rajendra C. Pawar, Rupesh S. Devan, Yuan-Ron Ma, Prathmesh P. Salvi, Sanjay S. Kolekar, and Pramod S. Patil. 2011. "Synthesis and characterization of Ru doped CuO thin films for supercapacitor based on Bronsted acidic ionic liquid." *Electrochimica Acta* no. 56 (5):2127–2134.

Sharma, Neha, Anurag Gaur, and Ravinder K. Kotnala. 2015. "Signature of weak ferroelectricity and ferromagnetism in Mn doped CuO nanostructures." *Journal of Magnetism and Magnetic Materials* no. 377:183–189.

Singh, Kaushalendra K., Vivek Ramakrishnan, Ramya Prabhu, and Neena S John. 2017. "Rapid augmentation of vertically aligned MoO_3 nanorods via microwave irradiation." *CrystEngComm* no. 19 (44):6568–6572.

Sonia, Suganthiraja, I Jose Annsi, Palaniswamy Suresh Kumar, Devanesan Mangalaraj, Chinnuswamy Viswanathan, and Nagamony Ponpandian. 2015. "Hydrothermal synthesis of novel Zn doped CuO nanoflowers as an efficient photodegradation material for textile dyes." *Materials Letters* no. 144:127–130.

Sreeprasad, Theruvakkattil Sreenivasan, Shihabudheen M. Maliyekkal, Kinattukara Parambil Lisha, and Thalappil Pradeep. 2011. "Reduced graphene oxide–metal/metal oxide composites: facile synthesis and application in water purification." *Journal of Hazardous Materials* no. 186 (1):921–931.

Sudarsan, Vasanthakumaran, Sri Sivakumar, Frank CJM van Veggel, and Mati Raudsepp. 2005. "General and convenient method for making highly luminescent sol–gel derived silica and alumina films by using LaF3 nanoparticles doped with lanthanide ions (Er^{3+}, Nd^{3+}, and Ho^{3+})." *Chemistry of Materials* no. 17 (18):4736–4742.

Sun, Fazhe, Xueliang Qiao, Fatang Tan, Wei Wang, and Xiaolin Qiu. 2012. "One-step microwave synthesis of Ag/ZnO nanocomposites with enhanced photocatalytic performance." *Journal of Materials Science* no. 47 (20):7262–7268.

Sun, Lan, Jing Li, Chenglin Wang, Sifang Li, Yuekun Lai, Hongbo Chen, and Changjian Lin. 2009. "Ultrasound aided photochemical synthesis of Ag loaded TiO_2 nanotube arrays to enhance photocatalytic activity." *Journal of Hazardous Materials* no. 171 (1–3):1045–1050.

Sun, Ming, Fei Ye, Bang Lan, Lin Yu, Xiaoling Cheng, Shengnan Liu, and Xiaoqing Zhang. 2012. "One-step hydrothermal synthesis of Sn-doped OMS-2 and their electrochemical performance." *International Journal of Electrochemical Science* no. 7:9278–9289.

Sun, Ming, Ting Lin, Gao Cheng, Fei Ye, and Lin Yu. 2014. "Hydrothermal synthesis of boron-doped MnO_2 and its decolorization performance." *Journal of Nanomaterials* no. 2014:150.

Sun, Yu-Feng, Shao-Bo Liu, Fan-Li Meng, Jin-Yun Liu, Zhen Jin, Ling-Tao Kong, and Jin-Huai Liu. 2012. "Metal oxide nanostructures and their gas sensing properties: A review." *Sensors* no. 12 (3):2610–2631.

Tada, Hiroaki, Musashi Fujishima, and Hisayoshi Kobayashi. 2011. "Photodeposition of metal sulfide quantum dots on titanium (IV) dioxide and the applications to solar energy conversion." *Chemical Society Reviews* no. 40 (7):4232–4243.

Tan, Yong Nian, Chung Leng Wong, and Abdul Rahman Mohamed. 2011. "An overview on the photocatalytic activity of nano-doped-TiO_2 in the degradation of organic pollutants." *ISRN Materials Science* no. 2011: 1–18.

Thangeeswari, Tharmar, Jagadeesan Velmurugan, Priya Murugasen. 2014. Structural and luminescence properties of Ni, Pb, Cu-doped ZnO nano particles by co-precipitation method—Opto electronic devices. *Science Engineering and Management Research (ICSEMR), 2014 International Conference on* IEEE, Chennai, India.

Theiss, Frederick L., Godwin A. Ayoko, and Ray L. Frost. 2016. "Synthesis of layered double hydroxides containing Mg^{2+}, Zn^{2+}, Ca^{2+} and Al^{3+} layer cations by co-precipitation methods—A review." *Applied Surface Science* no. 383:200–213.

Thota, Subhash, and Jitendra Kumar. 2007. "Sol–gel synthesis and anomalous magnetic behaviour of NiO nanoparticles." *Journal of Physics and Chemistry of Solids* no. 68 (10):1951–1964.

Titirici, Maria-Magdalena, Markus Antonietti, and Arne Thomas. 2006. "A generalized synthesis of metal oxide hollow spheres using a hydrothermal approach." *Chemistry of Materials* no. 18 (16):3808–3812.

Upadhyay, Shipra B., Rajneesh K. Mishra, and Pradosh P. Sahay. 2014. "Structural and alcohol response characteristics of Sn-doped WO_3 nanosheets." *Sensors and Actuators B: Chemical* no. 193:19–27.

Vasylkiv, Oleg, Yoshio Sakka, Yasuaki Maeda, and Valeriy V Skorokhod. 2004. "Nano-engineering of zirconia–noble metals composites." *Journal of the European Ceramic Society* no. 24 (2):469–473.

Vives, S, Christian Guizard, Louis Cot, and C. Oberlin. 1999. "Sol-gel/co-precipitation method for the preparation and characterization of zirconia-tungsten composite powders." *Journal of Materials Science* no. 34 (13):3127–3135.

Wang, Donghai, Daiwon Choi, Juan Li, Zhenguo Yang, Zimin Nie, Rong Kou, Dehong Hu, Chongmin Wang, Laxmikant V Saraf, and Jiguang Zhang. 2009. "Self-assembled TiO_2–graphene hybrid nanostructures for enhanced Li-ion insertion." *ACS Nano* no. 3 (4):907–914.

Wang, Jixin, Rusheng Yuan, Liyan Xie, Qinfen Tian, Shuying Zhu, Yanhua Hu, Ping Liu, Xicheng Shi, and Donghui Wang. 2012. "Photochemical treatment of As (III) with α-Fe_2O_3 synthesized from Jarosite waste." *RSC Advances* no. 2 (3):1112–1118.

Wang, Yanqin Q., Lunxiang X. Yin, Oleg Palchik, Yaron Rosenfeld Hacohen, Yuri Koltypin, Aharon Gedanken. 2001b. "Rapid synthesis of mesoporous yttrium–zirconium oxides with ultrasound irradiation." *Langmuir* no. 17 (13):4131–4133.

Wang, Yanqin, Lunxiang Yin, Oleg Palchik, Yaron Rosenfeld Hacohen, Yuri Koltypin, and Aharon Gedanken. 2001a. "Sonochemical synthesis of layered and hexagonal yttrium–zirconium oxides." *Chemistry of Materials* no. 13 (4):1248–1251.

Wang, Zhiyu, and Liang Zhou. 2012. "Metal oxide hollow nanostructures for lithium-ion batteries." *Advanced Materials* no. 24 (14):1903–1911.

Wen, Jianye, and Garth L. Wilkes. 1996. "Organic/inorganic hybrid network materials by the sol–gel approach." *Chemistry of Materials* no. 8 (8):1667–1681.

Wenderich, Kasper, and Guido Mul. 2016. "Methods, mechanism, and applications of photodeposition in photocatalysis: A review." *Chemical Reviews* no. 116 (23):14587–14619.

Whittingham, M. Stanley. 1996. "Hydrothermal synthesis of transition metal oxides under mild conditions." *Current Opinion in Solid State and Materials Science* no. 1 (2):227–232.

Wilson Robert J., and Frances A. Houle. 1985. "Composition, structure, and electric field variations in photodeposition." *Physical Review Letters* no. 55 (20):2184.

Wong, Wai Leong Eugene, Subburathinam Karthik, and Manoj Gupta. 2005b. "Development of hybrid Mg/Al$_2$O$_3$ composites with improved properties using microwave assisted rapid sintering route." *Journal of Materials Science* no. 40 (13):3395–3402.

Wong, Wai Leong Eugene, S. Karthik, and Manoj Gupta. 2005a. "Development of high performance Mg–Al$_2$O$_3$ composites containing Al$_2$O$_3$ in submicron length scale using microwave assisted rapid sintering." *Materials Science and Technology* no. 21 (9):1063–1070.

Woo, Kyoungja, Ho Jin Lee, J-P Ahn, and Yong Sung Park. 2003. "Sol–gel mediated synthesis of Fe$_2$O$_3$ nanorods." *Advanced Materials* no. 15 (20):1761–1764.

Wu, Chun-Guey, Chia-Cheng Chao, and Fang-Ting Kuo. 2004. "Enhancement of the photo catalytic performance of TiO$_2$ catalysts via transition metal modification." *Catalysis Today* no. 97 (2–3):103–112.

Wu, Mi, Wei-Jian Chen, Yu-Hua Shen, Fang-Zhi Huang, Chuan-Hao Li, and Shi-Kuo Li. 2014. "In situ growth of matchlike ZnO/Au plasmonic heterostructure for enhanced photoelectrochemical water splitting." *ACS Applied Materials & Interfaces* no. 6 (17):15052–15060.

Xiang, Quanjun, Jiaguo Yu, Bei Cheng, and HC Ong. 2010. "Microwave-hydrothermal preparation and visible-light photoactivity of plasmonic photocatalyst Ag-TiO$_2$ nanocomposite hollow spheres." *Chemistry–An Asian Journal* no. 5 (6):1466–1474.

Xiong, Huan-Ming, Dmitry G Shchukin, Helmuth Möhwald, Yang Xu, and Yong-Yao Xia. 2009. "Sonochemical synthesis of highly luminescent zinc oxide nanoparticles doped with magnesium (II)." *Angewandte Chemie International Edition* no. 48 (15):2727–2731.

Xu, Haitao, Qidong Zhao, Hua Yang, and Yan Chen. 2009. "Study of magnetic properties of ZnO nanoparticles codoped with Co and Cu." *Journal of Nanoparticle Research* no. 11 (3):615–621.

Xu, Hui, Min Zeng, Jing Li, and Xiaoling Tong. 2015. "Facile hydrothermal synthesis of flower-like Co-doped NiO hierarchical nanosheets as anode materials for lithium-ion batteries." *RSC Advances* no. 5 (111):91493–91499.

Xu, Shiping, and Darren Delai Sun. 2009. "Significant improvement of photocatalytic hydrogen generation rate over TiO$_2$ with deposited CuO." *International Journal of Hydrogen Energy* no. 34 (15):6096–6104.

Yang, Chunzhen, Nicole K van der Laak, Kwong-Yu Chan, and Xin Zhang. 2012. "Microwave-assisted microemulsion synthesis of carbon supported Pt-WO$_3$ nanoparticles as an electrocatalyst for methanol oxidation." *Electrochimica Acta* no. 75:262–272.

Yang, Daniel, Sang-Eun Park, Joong-Kee Lee, and Sang-Wha Lee. 2009. "Sonochemical deposition of nanosized Au on titanium oxides with different surface coverage and their photocatalytic activity." *Journal of Crystal Growth* no. 311 (3):508–511.

Yang, Heqing, Daming Huang, Xingjun Wang, Xiaoxiao Gu, Fujian Wang, Songhai Xie, and Xi Yao. 2005. "Sol–Gel synthesis of luminescent InP nanocrystals embedded in silica glasses." *Journal of Nanoscience and Nanotechnology* no. 5 (10):1737–1740.

Yang, Yanan, and Peng Wang. 2006. "Preparation and characterizations of a new PS/TiO$_2$ hybrid membrane by sol–gel process." *Polymer* no. 47 (8):2683–2688.

Yao, Yao, Fangxu Ji, Mingli Yin, Xianpei Ren, Qiang Ma, Junqing Yan, and Shengzhong Frank Liu. 2016. "Ag nanoparticle-sensitized WO$_3$ hollow nanosphere for localized surface plasmon enhanced gas sensors." *ACS Applied Materials & Interfaces* no. 8 (28):18165–18172.

Yayapao, Oranuch, Somchai Thongtem, Anukorn Phuruangrat, and Titipun Thongtem. 2013a. "Sonochemical synthesis, photocatalysis and photonic properties of 3% Ce-doped ZnO nanoneedles." *Ceramics International* no. 39:S563–S568.

Yayapao, Oranuch, Titipun Thongtem, Anukorn Phuruangrat, and Somchai Thongtem. 2013b. "Sonochemical synthesis of Dy-doped ZnO nanostructures and their photocatalytic properties." *Journal of Alloys and Compounds* no. 576:72–79.

Ye, Jilei, Jianguo Liu, Zhigang Zou, Jun Gu, and Tao Yu. 2010. "Preparation of Pt supported on WO_3–C with enhanced catalytic activity by microwave-pyrolysis method." *Journal of Power Sources* no. 195 (9):2633–2637.

Yildiz, Abdullah, Şeyda Horzum, Necmi Serin, and Tulay Serin. 2014. "Hopping conduction in In-doped CuO thin films." *Applied Surface Science* no. 318:105–107.

Yu, Jiaguo, and Xiaoxiao Yu. 2008. "Hydrothermal synthesis and photocatalytic activity of zinc oxide hollow spheres." *Environmental Science & Technology* no. 42 (13):4902–4907.

Yu, Yeon-Tae, and Prabir Dutta. 2011. "Examination of Au/SnO_2 core-shell architecture nanoparticle for low temperature gas sensing applications." *Sensors and Actuators B: Chemical* no. 157 (2):444–449.

Yuan, Changzhou, Hao Bin Wu, Yi Xie, and Xiong Wen David Lou. 2014. "Mixed transition-metal oxides: Design, synthesis, and energy-related applications." *Angewandte Chemie International Edition* no. 53 (6):1488–1504.

Yuasa, Masayoshi, Tetsuya Kida, and Kengo Shimanoe. 2012. "Preparation of a stable sol suspension of Pd-loaded SnO_2 nanocrystals by a photochemical deposition method for highly sensitive semiconductor gas sensors." *ACS applied Materials & Interfaces* no. 4 (8):4231–4236.

Zhang, Fanli, Zhiqiang Cheng, Liying Cui, Tingting Duan, Ahmed Anan, Chunfeng Zhang, and Lijuan Kang. 2016. "Controllable synthesis of Ag@ TiO_2 heterostructures with enhanced photocatalytic activities under UV and visible excitation." *RSC Advances* no. 6 (3):1844–1850.

Zhang, Shicheng, Zhijian Zheng, Jinhe Wang, and Jianmin Chen. 2006. "Heterogeneous photocatalytic decomposition of benzene on lanthanum-doped TiO_2 film at ambient temperature." *Chemosphere* no. 65 (11):2282–2288.

Zhi, Mingjia, Chengcheng Xiang, Jiangtian Li, Ming Li, and Nianqiang Wu. 2013. "Nanostructured carbon–metal oxide composite electrodes for supercapacitors: A review." *Nanoscale* no. 5 (1):72–88.

Zhong, M, S Wang, Y Li, W Li, Y Hu, M Zhu, and H Jin. 2015. Transition metal-doped ZnO diluted magnetic semiconductors tuned by high pulsed magnetic field. *Magnetics Conference (INTERMAG)*, 2015 IEEE.

Zhu, Ying-Jie, and Feng Chen. 2014. "Microwave-assisted preparation of inorganic nanostructures in liquid phase." *Chemical Reviews* no. 114 (12):6462–6555.

9 Inorganic One-Dimensional Nanomaterials for Supercapacitor Electrode Applications

K. K. Purushothaman, B. Saravanakumar, and B. Sethuraman

CONTENTS

9.1 INTRODUCTION

Over the past decades, energy demand has increased rapidly due to increase in population, vehicles, and electrical energy-based appliances. In general, the energy production relies mostly on the combustion of fossil fuels and their lesser availability. However, the environmental impact of using fossil fuels is driving the researchers to develop renewable and highly efficient methods for energy conversion/storage without any impact on the environment. Conversion of energy from renewable resources depends on the availability of resources during different seasons. This issue forced the researchers to think about the advanced energy storage devices. Among the various energy storage devices, electrochemical energy storage (EES) devices including batteries and supercapacitors have attracted the researchers due to their high efficiency and versatility [1].

Electrochemical capacitors are also called supercapacitors (SCs), ultracapacitors, power capacitors, gold capacitors, and power caches. Higher power density of SCs made them available for the applications reserved for batteries. Environmental-friendly nature, safety, and light weights are other advantages of SCs. The most significant feature of SCs is the ability to charge and discharge in seconds continuously without degradation [1,2]. Recently, hybrid energy storage systems, i.e., SCs coupled with other energy devices such as batteries and fuel cells, are focused in order to increase the overall energy efficiency of a system. SCs in a hybrid system can handle the peak power demands and recover the energy during braking. These interesting features have created a great interest toward the applications of SCs in the consumer electronics, heavy electric vehicles, and industrial power management [3].

The overall performance of an energy storage device depends on the energy density and power density. The electrochemical characteristics of batteries, SCs, and conventional capacitors are presented in Table 9.1. SCs stand between batteries and conventional solid-state/electrolytic capacitors [4].

Notably, SCs are considered as alternative candidates for batteries due to superior operating lifetime, ultrafast charging–discharging rates and high power densities, very less maintenance, and so on [5–7], but SCs suffer with lower energy density due to the storage of charges at the electrode surface and not in an entire electrode [8].

SCs are classified into two major classes based on their charge storage mechanism. They are electric double-layer capacitors (EDLCs) and pseudocapacitors. The storage of energy in EDLCs is based on the separation of charge between electrode and electrolyte, i.e., reversible adsorption of ions from an electrolyte onto electrodes. The EDLCs use high-surface area carbon-based materials such as carbon nanotubes, activated carbon, graphene, and porous carbon as electrodes [1,9]. EDLC exhibits excellent cyclic stability and long service lifetime because there is no chemical change in the electrode material during charge/discharge processes. Lifetime is measured in hundreds of thousands to millions of cycles, whereas batteries have hundreds or thousands of cycles. Electrode materials of EDLC are inexpensive and have good resistance to corrosion. EDLC displays higher power density and lower capacitance. Pseudocapacitors/SCs are different from EDLCs, as pseudocapacitance arises due to the reversible surface redox reactions [1,10], i.e., valance electrons of

TABLE 9.1

Comparison of the Characteristics of Batteries, Supercapacitors, and Conventional Capacitors

Electrochemical Parameters	Batteries	Supercapacitors	Capacitors
Specific energy (Wh kg^{-1})	10–100	1–10	<0.1
Charging time	$1 < t < 5$ hr	1–30 s	$10^{-3} < t < 10^{-6}$ s
Discharging time	$t > 0.3$ h	1–30 s	$10^{-3} < t < 10^{-6}$ s
Lifetime (cycles)	1000	10^6	10^6
Specific power (W kg^{-1})	<1000	10,000	10^6
Charge/discharge efficiency	0.7–0.85	0.85–0.98	>0.95

electroactive materials are transformed across the electrode/electrolyte interface. The pseudocapacitors employ transition metal oxides (RuO_2, Co_3O_4, MnO_2, Ni, etc.) and conducting polymers (polyaniline (PANI), polypyrrole (PPy), Poly(3,4-ethylenedioxythiophene) (PEDOT), etc.) as the electrode materials.

Pseudocapacitors exhibit higher capacitance and energy density than EDLCs. The Faradic supercapacitor (FS) exhibits higher capacitance than the electrostatic capacitance but lower power density and lack of stability as surface redox reaction during cyclic process hinders its commercial applications. Supercapacitor possesses two electrodes sandwiched between a porous separator; both are impregnated in the electrolyte (Figure 9.1).

The electrochemical capacitors mainly depend on the nature of the electrode material. Hence, there is an urgent need to fully understand the properties of electrode materials. Development of new cost-effective electrode materials, with increased efficiency and durability, will make a significant impact on supercapacitor research. In this regard, nanostructured materials are put forward due to their great advantages [7,8,11]. Reduced dimension of the electrode materials can increase the electrode/electrolyte contact area, resulting in more ion adsorption sites and better charge transfer reactions. Innovative morphologies and porosity of the nanostructured materials play a vital role in deciding the diffusion time of ions and charge/discharge rate capability. Porous electrode material maximizes ion-accessible area, and hence, the kinetics of electrode/electrolyte reactions is sufficiently fast over the electrode surface. The confinement of the material displays the tolerance to strain and structural distortion when the material undergoes many cycles at high current rates [12–15].

Among other kinds of electrode materials, transition metal oxides (TMOs) with nanostructure attracted the attention of the researchers. TMOs are classified as either noble or base metal oxides. Noble metal oxides, such as RuO_2, IrO_2, and so on, exhibit good conductivity and excellent power densities; however, their high cost and harmful effect on the environment have limited their industrial applications in SCs.

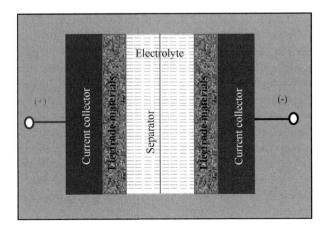

FIGURE 9.1 Schematic view of a supercapacitor device.

Base metal oxides, such as MnO_2, Co_3O_4, NiO, CuO, and so on, are attractive materials for SC applications due to high theoretical capacitance, availability, environmental compatibility, and low cost [16].

Various nanostructures from 0 to 3D have been synthesized and analyzed for supercapacitor electrode application. Among other dimensions, 1D materials have attracted great attention due to their excellent physical–chemical properties and great potential in various applications. In energy storage devices, 1D nanostructures may provide direct current pathways, lower the charge/discharge time, shorten the ion diffusion distance, increase the contact area of electrode and electrolyte, accommodate volume expansion, and limit the mechanical degradation. Typical 1D nanostructure includes nanorods, nanowires, nanotubes, and nanobelts [17–20]. Furthermore, 1D heterostructures consist of multiple components with core–shell structure, branched nanowires, array structure, and hollow nanostructure [21–23]. In this chapter, supercapacitor features like specific capacitance, energy density, power density, rate capability, cyclic stability, and electrochemical impedance of various inorganic 1D transition metal oxide nanostructures are discussed.

9.2 INORGANIC ONE-DIMENSIONAL METAL OXIDES

9.2.1 NICKEL OXIDE-BASED SC ELECTRODES

Nickel oxide (NiO) is used for various applications such as electrochromics, catalysis, chemical sensors, and electrochemical energy storage systems [24–26]. The outstanding theoretical capacitance (3750 Fg^{-1} within 0.5 V) with fast reversible oxidation/reduction reaction between Ni^{2+} and Ni^{3+} made nickel oxide as an interesting candidate for supercapacitor electrode application. However, poor electrical conductivity and low accessible area result in lower capacitance [27–29]. For instance, Cao et al. [30] reported the preparation of porous NiO nanotube arrays on nickel foam using ZnO nanorod as template. The NiO nanotube arrays exhibited a maximum specific capacitance of 675 Fg^{-1} at a constant current density of 2 Ag^{-1} with good cyclic stability of 93.2% after 10,000 cycles. The NiO nanorods were successfully grown by Kannan et al. [31] using oblique angle deposition technique in an electron beam evaporator method. The NiO rods prepared at the angle of 75° showed highest specific capacitance of 344 Fg^{-1} with the energy and power density of 8.78 Wh kg^{-1} and 2.5 kW kg^{-1}, respectively, in the 2 M KOH electrolyte within the potential range of 0–0.5 V. The electrode revealed very good electrochemical stability without any degradation even after 5000 cycles. The NiO nanofibers were prepared through sol–gel-based electrospinning method by Kolathodi et al. [32]. The nanofibers exhibited the specific capacitance of 248 Fg^{-1} at 1 Ag^{-1} in the 6 M KOH electrolyte. They fabricated asymmetric supercapacitors and reported the gravimetric capacitance of 141 Fg^{-1}, energy density of 43.75 Wh kg^{-1}, and power density of 7.5 kW kg^{-1} in the potential range of 0–1.5 V. This device retained 93% of initial capacitance after 5000 cycles.

Porous NiO nanofibers (average diameter of 280–400 nm) were synthesized by Kundu et al. using electrospinning method on Ni foam current collector [33].

The binder-free electrode exhibited the specific capacitance of 737, 731, 679, 626, 591, and 570 Fg^{-1} at the current densities of 2, 5, 10, 20, 30, and 40 Ag^{-1}, respectively. The NiO nanofibers exhibited the charge transfer resistance of 0.9 Ω with the energy density of 13.8 Wh Kg^{-1} and the power density of 7.6 kW kg^{-1}. The formation of NiO nanotubes due to the self-assembly of nanosheets via template-free solvothermal method was reported by Liu et al. [34]. As-synthesized nanotubes had the length of 2–4 μm with an internal tubular diameter of 350 nm. These NiO nanotubes showed the specific capacitance of 919, 891, 813, 761, and 658 Fg^{-1} at 1, 2, 3, 5, and 10 only Ag^{-1} respectively. The material displayed 5.3% of capacitance loss after 2000 cycles. NiO nanorods were prepared on indium tin oxide thin films by hot filament metal oxide vapor deposition technique and tested for supercapacitor electrode application by Patil et al. [35]. The nanorods had the width of 100 nm and length of 500 nm, which exhibited the specific capacitance of 230 Fg^{-1} at 3 Ag^{-1}.

One-dimensional core–shell Ni/NiO nanoarchitecture electrode exhibited the specific capacitance of 717 Fg^{-1} and 1635 Fg^{-1} after hydrogenation [36]. The presence of hydroxyl groups on the NiO surface enhanced the specific capacitance, and the metallic Ni nanowire core improved the electron conduction to the current collector. The electrode exhibited the energy density of 49.35 Wh kg^{-1} and power density of 7.9 kW kg^{-1} at a current density of 15.1 Ag^{-1}. The charge transfer resistance reduced from 1.004 to 0.444 Ω after hydrogenation. There was 5.2% of capacitance loss after 1200 cycles. Ribbon-like NiO nanostructure was synthesized using mesoporous carbon as hard template by Yao et al. [37], which exhibited the maximum specific capacitance of 1260 Fg^{-1} at a current density of 1 Ag^{-1} and 748 Fg^{-1} even at a higher current density of 20 Ag^{-1}. This electrode retained 95% of the capacitance after 5000 cycles.

$Ni(OH)_2$ nanobelts were synthesized via hydrothermal method, and in the second step, gold nanoparticles were decorated into NiO nanobelts using ultrasonification followed by heat treatment at 500°C [38]. The effect of amount of gold nanoparticles (0–17.38 wt%) on the supercapacitor behavior of NiO nanobelts has been studied. The maximum specific capacitance of 597 Fg^{-1} was obtained for the NiO nanobelt with 1.54 wt% of Au content. Interestingly, this nanobelt-based asymmetric device exhibited the energy density of 18 Wh kg^{-1} with the capacitance retention of 84% after 22,000 cycles. To enhance the electrical conductivity, one-dimensional coaxial architecture composed of silver nanowire as core and $Ni(OH)_2$ as shell has been synthesized [39]. The core–shell structure showed the specific capacitance of 1165.2 Fg^{-1} at a current density of 3 Ag^{-1} with the capacity retention of 93% after 3000 cycles.

NiO hollow nanofibers were synthesized using electrospun poly (amic acid) nanofiber templates through ion exchange process and subsequent thermal annealing by Zhang et al. [40]. Nanofibers possessed the BET surface area of 117.1 m^2g^{-1} with a pore volume of 0.66 m^3g^{-1}. The nanofibers exhibited the specific capacitance of 700 Fg^{-1} at a discharge current of 2 Ag^{-1} with 96% capacity retention after 5000 cycles at a current density of 5 Ag^{-1} and 80% capacitance was retained when current density was increased from 1 to 5 Ag^{-1} in the 0–0.4 V window. The electrospinning process has been used to synthesize citric acid-modified NiO nanofibers [41]. The BET surface area increased from 115.8 to 212.1 m^2g^{-1} after citric acid modification. The calculated specific capacitance was 336, 258, and 181 Fg^{-1} at current densities of 5, 10, and 20 mA cm^{-2}, respectively, in the potential window of 0–0.45 V in 6 M KOH electrolyte. This electrode showed the charge transfer resistance of 0.23 Ω. The initial capacitance decreased by 13% after 1000 cycles. The NiO nanowires with the diameter of 50–70 nm containing densely packed cuboidal grains (10–20 nm) were fabricated using electrospinning method by Vidhyadharan et al. [42]. The CV analysis showed the specific capacitance of 746 Fg^{-1} at the scan rate of 1 mV s^{-1} in the 6 M KOH electrolyte in potential range of 0–0.5 V. The charge/discharge measurement revealed the specific capacitance of 670 Fg^{-1} with 100% retention of capacitance after 1000 cycles and 98% columbic efficiency. The impedance analysis showed the charge transfer resistance of 0.45 Ω and relaxation time of 43 ms.

Mesoporous NiO nanobelts prepared by Wang et al. [43] exhibited the specific capacitance of 735, 675, 576, 478, 348, and 183 Fg^{-1} at the scan rate of 1, 2, 5, 10, 20, and 50 m Vs^{-1}, respectively. This electrode exhibited 600 Fg^{-1} even at 5 Ag^{-1} with the capacitance retention of 95% after 2000 cycles. Xiong et al. [44] synthesized different NiO hierarchical nanostructures using hydrothermal method and subsequent thermal decomposition. The NiO nanotubes delivered a specific capacitance of 405 Fg^{-1} at the current density of 0.5 Ag^{-1} and retained 91% of initial capacitance after 1500 charge/discharge cycles.

The NiO nanofibers were synthesized by Zang et al. [45]. The material displayed the specific surface area of 118 and 52 m^{2}g^{-1} at 300°C and 500°C of calcination temperature, respectively. The NiO nanofibers showed the specific capacitance of 884 Fg^{-1} and 452 Fg^{-1} at a current density of 0.5 Ag^{-1} and 20 Ag^{-1}, respectively. The authors observed 13% of loss after 1000 cycles [45]. The NiCo$_2$O$_4$ nanorod arrays were synthesized using low-temperature solvothermal approach followed by post-calcination treatment [46]. The electrode showed the specific capacitance of 440 Fg^{-1} at a scan rate of 5 mV s^{-1}. This electrode retained 94% of initial capacitance after 2000 cycles at a current density of 8 Ag^{-1}. A symmetrical device using this material exhibited the energy and power density of 12.6 Wh kg^{-1} and 4003 W kg^{-1}, respectively. NiCo$_2$O$_4$ nanorods were synthesized via solid-state reactions at room temperature followed by annealing at 300°C for 2 hrs by Zhu et al. [47]. The BET surface area of the nanorods was 71.646 m^{2}g^{-1} with the pore size of 19 nm. The NiCo$_2$O$_4$ nanorods exhibited the specific capacitance of 565 Fg^{-1} at a current density of 1 Ag^{-1}.

9.2.2 COBALT OXIDE-BASED SC ELECTRODES

Cobalt oxides are attractive pseudocapacitive electrode materials because of their high theoretical capacitance (3560 Fg^{-1}) and superior electrochemical properties [48–50]. Electron and ion transportation efficiency for charge storage in cobalt oxide mainly depends on surface area and morphology of electrode materials. Hence, the research is focused toward the synthesis of various nanostructures of cobalt oxide to enhance the supercapacitor performance.

Nanorod-assembled, multi-shelled Co$_3$O$_4$ hollow microspheres with complex interiors have been synthesized with the assistance of carbon spheres as hard templates [51]. Thermogravimetric analysis confirmed the complete removal of carbon spheres and the conversion of precursors into Co$_3$O$_4$ after 400°C. The electrochemical performance of as-prepared multi-shelled cobalt oxide has been evaluated by cyclic voltammetry (CV) in 2 M KOH aqueous solution between the voltage window of 0 and 0.5 V at different scan rates. Multi-shelled Co_3O_4 hollow microspheres exhibited the specific capacitance of 394.4 Fg^{-1} at the current density of 2 Ag^{-1}.

Porous $Ni_xCo_{3-x}O_4$ nanowires synthesized by Wang et al. [52] displayed the specific capacitance of 1479 Fg^{-1} at the current density of 1 Ag^{-1}. An asymmetric device was fabricated using these materials, and the electrochemical performance was evaluated in the potential range of 0–1.6 V. Asymmetric device displayed the specific capacitance of 105 Fg^{-1} at a current density of 3.6 mA cm^{-2} with an energy density of 37.4 Wh kg^{-1} and a power density of 163 W kg^{-1}. It retained 82% of initial capacitance over 3000 cycles. Wang et al. [53] synthesized a large quantity of nonporous Co_3O_4 nanorods via hydrothermal method, and the nanorods revealed the specific capacitance of 281 Fg^{-1} at a scan rate of 5 mV s^{-1}.

Co_3O_4 nanofibers were prepared by Kumar et al. [54]. The N_2 adsorption–desorption isotherm of the Co_3O_4 nanofibers showed type II hysteresis loop with a BET surface area of 67.0 m g^{-2}. The nanofibers displayed a specific capacitance of 407 Fg^{-1} at a scan rate of 5 mV s^{-1} in 6 M KOH electrolyte. The material showed the capacity retention of 94% after 1000 continuous charge/discharge cycles at a constant current density of 1 Ag^{-1}. Deng et al. [55] prepared $Co(OH)_2$ nanotube arrays and nanoporous films on Ni foam by electrochemical method. The specific capacitance value of the nanotubes and nanoporous films was 2500 and 2900 Fg^{-1}, respectively, in the potential range of −0.2 to +0.45 V in 1 M KOH electrolyte at 27°C. These electrodes retained 95% and 92% of the initial capacitance, respectively, even after 10,000 cycles.

Hierarchical Co_3O_4@$CoMoO_4$ core–shell nanowire arrays on nickel foam were fabricated via ion exchange hydrothermal route by Gu et al. [56]. This Co_3O_4@$CoMoO_4$ electrode exhibited the specific capacitance of 1040 Fg^{-1} at 1 Ag^{-1}. The aqueous symmetric supercapacitor based on Co_3O_4@$CoMoO_4$ hybrid electrode delivered an energy density of 92.44 Wh kg^{-1} at a power density of 6550 W kg^{-1} and retained 91.22% of its initial specific capacitance after 5000 cycles. One-step hydrothermal method was used by Li et al. to prepare nanoflakes/nanorods-assembled 3D flower-like $Co(OH)_2$ microspheres on nickel foam [57]. The electrode displayed a specific capacitance of 1350 Fg^{-1} at 5 mV s^{-1}. The aerial capacitance calculated from the discharge curve was 14.7, 13.7, 11.3, 10.4, 9.8, and 9.1 Fcm^{-2} at different current densities of 5, 10, 20, 30, 40, and 50 mA cm^{-2}, respectively. It retained 84.6% of initial capacitance after 2000 cycles.

He et al. [58] fabricated one-dimensional porous NiO-decorated Co_3O_4 composites. This NiO-decorated Co_3O_4 composite yielded a specific capacitance of 1112 Fg^{-1} at a current density of 2 Ag^{-1} with the retention of 97% after 1000 cycles. The nanorods of cobalt (Co)-composites were produced under the existence of H_3BTC and PHEN, and they were calcinated to get Co_3O_4 nanorods by Qian et al. [59]. SEM images displayed nanorods of Co-composites with the diameter of 50 nm and a length of 1 μm. The calculated specific capacitance was 262 Fg^{-1} at 5 mV s^{-1} in the potential range of 0.2–0.7 V, and nearly no degradation was observed after 500 cycles at 50 mV s^{-1}. The asymmetrical Co_3O_4//AC button cell exhibited the energy density value of 42.81 Wh kg^{-1} at the discharge current of 1.5 Ag^{-1} in the 1.5 V window.

Interestingly, Li et al. [60] prepared 0D, 1D, and 2D Co_3O_4 via hydrothermal method using chicken eggshell membrane as template and dopant. Co_3O_4 nanowires exhibited better electrochemical behavior while compared with Co_3O_4 nanoparticles and Co_3O_4 nanosheets. Co_3O_4 nanowires showed a specific capacitance of 1498 Fg^{-1} at a current density of 0.5 Ag^{-1} in 6 M KOH solution with a rate capability of 795 Fg^{-1} at a high current density of 10 Ag^{-1}. An asymmetric supercapacitor using this electrode exhibited the energy density of 29.5 Wh Kg^{-1} with 91.4% of capacitance retention over 1000 cycles. Nanocomposite nanowires consisting of two ceramic components such as CuO and Co_3O_4 have been synthesized via electrospinning technique by Harilal et al. [61]. The specific capacitance values at the scan rate of 2 mV s^{-1} were 712, 1104, and 1242 Fg^{-1} for CuO, Co_3O_4, and CuO-Co_3O_4 electrodes, respectively. Equivalent electrical resistance of the electrodes CuO, Co_3O_4, and CuO-Co_3O_4 was calculated from charge and discharge curves, and the values were 1.8, 2.3, and 0.81 Ω, respectively. Asymmetric supercapacitors fabricated using CuO-Co_3O_4 electrode delivered 158 Fg^{-1} with the specific energy density of 52.8 Wh kg^{-1} at the specific power density of 1620 W kg^{-1}.

Co_3O_4 nanowire prepared on a flexible carbon fabric by hydrothermal route was reported by Howli et al. [62]. Co_3O_4 nanowire delivered a specific capacitance of 3290 Fg^{-1} at a scan rate of 5 mV s^{-1}. A solid-state symmetric supercapacitor based on Co_3O_4 nanowire delivered a maximum energy and power density of 6.7 Wh kg^{-1} and 5000 W kg^{-1}. The capacity retention was 95.3% after 5000 cycles. Chen et al. [63] fabricated the Co_3O_4 nanorod arrays with the diameter of 450 nm as electrode material for supercapacitor applications. Electrochemical impedance analysis showed the solution resistance of 1.12 Ω and the charge transfer resistance of 1.8 Ω for Co_3O_4 nanorods. These nanorods exhibited the specific capacitance of 387.25 Fg^{-1} at 1 Ag^{-1} with 88% of stability after 1000 repeated charge/discharge cycles.

One-dimensional porous ZnO/Co_3O_4 heterojunction nanorods prepared at 450°C by Gao et al. [64] displayed the specific capacitance of 1135 Fg^{-1} at 1 Ag^{-1}, which was 1.4 times higher than Co_3O_4 (814 Fg^{-1}), and better rate performance, i.e., 4.9 times higher than Co_3O_4. The material retained 83% of the initial capacitance after 5000 cycles at 10 Ag^{-1}. The ZnO/Co_3O_4-based asymmetric supercapacitor delivered an energy density of 47.7 Wh kg^{-1} and a power density of 7500 W kg^{-1}. Cobalt oxide (Co_3O_4) nanofilm was grown by an oblique angle deposition technique in an e-beam evaporator by Kannana et al. [65]. The fabricated electrodes at 80°C exhibited a specific capacitance of 2875 Fg^{-1} with the capacitance retention of 62% after 2000 cycles. These electrodes also showed the energy density of 57.7 Wh kg^{-1} and power density of 9.5 kW kg^{-1}. Calcination-assisted hydrothermal method was used to synthesize Co_3O_4 nanorods with porous structure by Pan et al. [66]. Co_3O_4 nanorods exhibited a specific capacity of 1272 Cg^{-1} at a current density of 1.89 Ag^{-1} and showed 90.3% of capacity retention after 18,000 cycles at 23.58 Ag^{-1}. The Co_3O_4/active carbon (AC) device showed the energy density and power density of 55.4 Wh kg^{-1} and 4490 W kg^{-1}, respectively. This device retained 90.4% of initial capacity after 36,000 cycles.

Two-step solution-based method was used to synthesize 1D porous NiO-decorated Co_3O_4 composites. N_2 adsorption/desorption isotherms exhibited H3-type hysteresis loop which confirmed the presence of mesopores in the prepared sample [67]. The specific surface area of the composite was 70.0 m^2g^{-1}, and the pure Co_3O_4 was

53.2 m^2g^{-1}. The specific capacitance values were 304 and 888 Fg^{-1} at a scan rate of 5 mV s^{-1} for pure and NiO-decorated Co_3O_4, respectively, in the 0–0.6 V range. Galvanostatic charge/discharge study exhibited the specific capacitance of 1112 Fg^{-1} at a current density of 2 Ag^{-1}, with the retention of 97% after 1000 cycles for NiO-decorated Co_3O_4.

Porous one-dimensional CoO nanobelts and two-dimensional CoO nanoplates were fabricated by Xiao et al. [68]. Porous CoO nanobelts and porous CoO nanoplates exhibited the specific capacitances of 1178 and 640 Fg^{-1}, and 978 and 480 Fg^{-1} at 1 and 40 Ag^{-1}, respectively. Asymmetric supercapacitor based on CoO nanobelts in the aqueous electrolyte displayed the energy density of 68.5 Wh kg^{-1} at a power density of 850.0 W kg^{-1}. Co_3O_4 nanowires@NiO nanosheet arrays have been synthesized on Ni foam by a sequential hydrothermal process. Co_3O_4 nanowires@NiO nanosheet arrays revealed an aerial capacitance of 2018 mF cm^{-2} at a current density of 2 mA cm^{-2}. A flexible asymmetric supercapacitor based on Co_3O_4 nanowire@NiO showed an aerial capacitance of 134.6 mF cm^{-2} at a current density of 2 mA cm^{-2} [69].

9.2.3 Manganese Oxide-Based SC Electrodes

Manganese oxide is a versatile compound for energy storage applications due to its advantageous physiochemical features. Manganese can exist as a variety of stable oxides (MnO, Mn_3O_4, Mn_2O_3, and MnO_2) with various crystal structures. MnO_x may be synthesized in different ways with desirable structures for various applications [70–72]. First, MnO_x were used to develop Leclanche alkaline battery cell, and recently, they are being used as cathode materials for Li-ion batteries and fast-charging material for SCs due to their multiple oxidation states such as +2, +3, +4, +6, and +7.

Bai et al. [73] synthesized hierarchical multidimensional MnO_2 nanostructures via hydrothermal method by varying the hydrothermal temperature and dwell time. Needle-like α-MnO_2 nanorods showed higher specific surface area (114 m^2g^{-1}) while compared with δ-MnO_2. The nanorods delivered the specific capacitance of 311.52 Fg^{-1} at a current density of 0.3 Ag^{-1}. The different crystallographic forms of MnO_2 with various morphologies like needles, rods, and spindles were fabricated by Chen et al. [74] using a quick precipitation method at low temperature without any template and surfactant. The needle-like structure exhibited the specific capacitance of 233.5 and 83.1 Fg^{-1}, and spindle-like structure delivered 95.5 and 29.3 Fg^{-1} at the scan rate of 5 and 100 mV s^{-1}, respectively, in the 1 M Na_2SO_4 electrolyte.

MnO_2 nanoneedle with 2 nm thickness has been prepared by Davaglio et al. [75]. They have studied the capacitive behavior at various voltage ranges from 0 to 0.6, 0.7, 0.8, 0.9, and 1 V in 1 M Na_2SO_4 electrolyte and reported that 0–1 V range provides the maximum faradic charges. This electrode exhibited the specific capacitance of 287 Fg^{-1} at 10 mV s^{-1}, and 40% loss was observed while scan rate was increased to 200 mV s^{-1} (133 Fg^{-1}). The Galvanostatic Charge/Discharge (GCD) analysis showed the specific capacitance of 289 Fg^{-1} at a current density of 0.5 Ag^{-1}. This material retained 88% of the initial capacitance after 10,000 cycles. Electrodeposition method has been utilized to prepare MnO_2 nanofibril/nanowire by Duay et al. [76]. They have tested the charge storage behavior in aqueous and organic electrolyte in the voltage range of 0–1 V. The addition of nanofibrils increased the specific capacitance of nanowires from 289 to 345 Fg^{-1} in the aqueous electrolyte and 324 to 387 Fg^{-1} in the organic electrolyte at a scan rate of 0.5 mV s^{-1}. The material retained 85.2% of the initial capacitance after 1000 cycles. MnO_2 nanorods with the diameter of less than 20 nm and the average length of 100 nm were synthesized using ZnO nanorod as template by Gong et al. [77]. The nanorods exhibited the specific capacitance of 167 Fg^{-1} at 0.4 Ag^{-1} with the cyclic stability of 80% after 1000 cycles.

Hydrothermal method has been used to synthesize MnO_2 nanorods by Kumar et al. [78]. The electrode exhibited the specific capacitance of 643.5 Fg^{-1} at 15 Ag^{-1}, 270 Fg^{-1} at 20 Ag^{-1}, and 185 Fg^{-1} at 25 Ag^{-1}. Further, it retained 90.5% of initial capacitance after 4000 cycles at 25 Ag^{-1} in the −0.4–0.6 V range. One-dimensional tubular Ag/MnO_x nanocomposites were fabricated via solvothermal method using Kirkendall effect between $KMnO_4$ and Ag nanowire by Li et al. [79]. The pH level has been varied to alter the morphology and electrochemical properties. In the neutral environment, MnO_2 nanotubes with Ag nanoparticles were formed. This electrode exhibited the charge transfer resistance of 11.5 Ω and showed the specific capacitance of 133.5 Fg^{-1} at 0.1 Ag^{-1} with 90% of stability at 1000 cycles. Polycarbonate membrane has been used in the hydrothermal method to synthesize one-dimensional tubular MnO_2 array-assembled ultrathin sheet [80]. The maximum specific capacitance of 411.9 Fg^{-1} at a current density of 0.25 Ag^{-1} was reported. MnO_2 nanotubes-based asymmetric device delivered an energy density of 22.6 Wh kg^{-1} with a power density of 225.3 W kg^{-1} between 0 and 1.8 V potential window. Li et al. [81] synthesized MnO_2 with different morphologies like nanorods, hollow urchin, and smooth ball via hydrothermal method. The nanorods exhibited the higher specific capacitance (317 Fg^{-1}) at 5 mV s^{-1}, while compared with hollow urchin (204 Fg^{-1}) and smooth ball (276 Fg^{-1}). The nanorods-based electrode retained 70% of the initial capacitance even after 2000 continuous charge/discharge cycles.

Successive ionic layer deposition method has been used to coat manganite γ-MnOOH on a substrate [82]. The coated layers were formed by aggregates of nanorods with 80–100 nm of length and 8–10 nm of diameter. The nanorods on the Ni foam delivered the specific capacitance of 1120 Fg^{-1}, and the observed degradation after 1000 cycles was less than 3%. Ma et al. [83] constructed core–shell structure of α-MnO_2 on the δ-MnO_2 nanosheets by solution-phase technique. The core–shell structure (91.5 m^2g^{-1}) was having higher BET surface area than nanowires (30.3 m^2g^{-1}), and the same showed the specific capacitance of 153.8 Fg^{-1} at the current density of 20 Ag^{-1}, with a cyclic stability of 98.1% after 10,000 cycles. The core–shell structure-based asymmetric device exhibited the energy density of 78 Wh kg^{-1} and power density of 21.7 k W kg^{-1}. Interconnected network of MnO_2 nanowires with a cocoon-like morphology exhibited the BET specific surface area of 80.3 m^2g^{-1} with the average pore size of 4 nm [84]. The nanowire network-based device delivered the maximum specific capacitance of 775 Fg^{-1} at a scan rate of 2 mV^{-1} in the 3 M KOH electrolyte for the potential window of −1 to 1 V. They showed seven times increase in the energy density due to the addition of K_4Fe $(CN)_6$ as a redox active additive to KOH. Purushothaman et al. [85] prepared α-$MnMoO_4$ nanorods via sol–gel spin-coating method and studied the supercapacitor behavior in the three different acidic electrolytes, namely para-toluenesulfonic acid (p-TSA), sulfuric acid

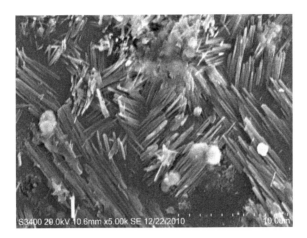

FIGURE 9.2 SEM image of α-MnMoO$_4$ nanorods.

(H_2SO_4), and hydrochloric acid. The nanorods exhibited the specific capacitance of 998 Fg^{-1}, 784 Fg^{-1}, and 530 Fg^{-1} at a scan rate of 5 mV s^{-1} in the H_2SO_4, p-TSA, and HCl electrolytes, respectively (Figure 9.2).

MnO$_2$ nanorods were prepared by Qu et al. [86] using precipitation method, and their electrochemical performance was analyzed in 0.5 molL^{-1} Li$_2$SO$_4$, Na$_2$SO$_4$, and K$_2$SO$_4$ aqueous electrolytes. At the slower scan rates, Li$_2$SO$_4$ electrolyte performed better, while at higher scan rates K$_2$SO$_4$ electrolyte was good. An asymmetric device based on K$_2$SO$_4$ electrolyte delivered the energy density of 17 Wh kg^{-1} at the power density of 2 kW kg^{-1} in the potential range of 0–1.8 V with the capacitance loss of 6% after 23,000 cycles.

CoMn-layered double-hydroxide nanoneedles grown on Ni foam exhibited the maximum specific capacitance of 2422 Fg^{-1} at a current density of 1 Ag^{-1} at 941st cycle and showed 2096 Fg^{-1} even after 3000 cycles [87]. Sung et al. [88] prepared 2D nanoplates and 1D nanowires/nanorods of manganese oxide at room temperature using one-pot oxidation reaction. The nanorods and nanowires had the surface area of 84 m^{2}g^{-1} and 53 m^{2}g^{-1}, respectively. The estimated specific capacitance of nanowires and nanorods was 110 Fg^{-1} and 94 Fg^{-1}, and 120 Fg^{-1} and 127 Fg^{-1} in the 1 M Li$_2$SO$_4$ and 1 M Na$_2$SO$_4$ electrolytes, respectively.

MnO$_2$ nanowires- and Fe$_2$O$_3$ nanotubes-based solid-state flexible asymmetric supercapacitor has been fabricated [89]. This device delivered the maximum specific capacitance of 91.3 Fg^{-1} and aerial capacitance of 1.5 Fcm^{-3} at the current density of 2 mA cm^{-2}. The maximum power density of 139.1 mW cm^{-3} was obtained at the energy density of 0.32 mWh cm^{-3} at a constant current of 10 mA cm^{-2}.

9.2.4 TITANIUM-BASED SC ELECTRODES

TiO$_2$ has been used in various fields like photochromic devices, self-cleaning, sensors, pigments, lithium-ion batteries, supercapacitors, and solar cells. Nanoscaled TiO$_2$ has high surface area, good physical–chemical stability, and good rate capability [90].

Ordered V_2O_5–TiO_2 nanotubes have been prepared by self-organizing anodization of Ti–V alloys. V_2O_5–TiO_2 nanotubes delivered a specific capacitance of 220 Fg^{-1} and energy density of 19.56 Wh kg^{-1} and were stable up to 500 cycles [91]. Hydrothermal method has been used to fabricate the vertically aligned TiO_2 nanorod arrays on the fluorine-doped tin oxide substrates by Ramadoss et al. [92]. TiO_2 nanorod arrays exhibited the specific capacitance of 8.5 Fg^{-1}at the scan rate of 5 mV s^{-1} in 1 M Na_2SO_4 in the potential window of 0–0.8 V. The energy density, power density, and the columbic efficiency of TiO_2 nanorods were 0.22 Wh kg^{-1}, 20 W kg^{-1}, and 87%, respectively.

Zhou et al. synthesized self-doped TiO_2 nanotube arrays and MnO_2/TiO_2 composite [93]. Self-doped TiO_2 nanotubes exhibited the aerial capacitance of 1.84 mF cm^{-2} at a sweep rate of 5 mV s^{-1} with capacitance retention of 93.1% after 2000 cycles. The MnO_2/TiO_2 composite achieved the specific capacitance of 1232 Fg^{-1} at the sweep rate of 5 mV s^{-1}.

TiO_2 with various morphologies were synthesized via hydrothermal method by Choi et al. [94]. The capacitances of the anatase nanoparticle, nanorod, flower-like, and urchin-like TiO_2 were 50.2, 55.4, 58.6, and 61.1 Fg^{-1}, respectively, at a current density of 0.5 Ag^{-1}. Urchin-like TiO_2-based hybrid supercapacitor delivered a maximum power density of 12224.356 W kg^{-1} and maximum energy density of 50.648 Wh kg^{-1}. The urchin-like TiO_2 showed the minimum charge transfer resistance (0.028 Ω) and higher cyclic stability (93% at 3.0 Ag^{-1} after 2000 cycles), while compared with other structures. He et al. used electrospinning method to prepare TiO_2 nanofibers and a posttreatment by KOH. KOH treatment enhanced the conductivity of TiO_2 nanofibers. The KOH treatment enhanced the specific capacitance from 0.04 Fg^{-1} to 65.84 Fg^{-1} at 1 mV s^{-1} with 90% of retention after 10,000 cycles [90]. Electrochemically doped TiO_2 nanotube arrays were fabricated through cyclic voltammetry method, and then Cu_2O nanoparticles were deposited onto TiO_2 nanotubes [95]. The TiO_2 nanotubes exhibited an aerial capacitance of 5.42 mF cm^{-2} at a scan rate of 10 mV s^{-1}, and Cu_2O/TiO_2 nanotubes electrode exhibited a specific capacitance of 198.7 Fg^{-1} at the current density of 0.2 Ag^{-1}. Approximately 88.7% of the initial capacitance was retained after 5000 cycles in 0.5 M Na_2SO_4 solution.

The anodization method has been employed to synthesize TiO_2 nanotube arrays by Cui et al. [96]. They have studied the effect of concentration of $(NH_4)_2TiF_6$ dilute solutions on the TiO_2 nanotubes under hydrothermal condition. Samples treated in 0.01 M $(NH_4)_2TiF_6$ solution displayed the aerial capacitance of 31.12 mF cm^{-2}. The cyclic stability test showed nearly 20% loss in specific capacitance after 500 cycles, and finally 74% was retained after 2000 cycles.

K-doped mixed-phase (anatase and rutile) TiO_2 nanofibers were grown on Ti metal foil via KOH-assisted hydrothermal method by Barai et al. [97]. The aerial capacitance of K-doped TiO_2 nanofibers prepared using 4 M, 5 M, and 6 M KOH was 81.36, 95.36, and 102.12 mF cm^{-2} at a scan rate of 5 mV s^{-1}, respectively, in the potential range of −0.8 to +0.4 V. The corresponding energy densities were 3.0, 3.70, and 4.30 mWh cm^{-2}, and power densities were 1680.70, 1640.60, and 1558.56 mW cm^{-2}. Further, the initial capacitance loss of 9.14, 11.70, and 8.60% was observed after 2500 cycles for TiO_2 nanofibers prepared using 4 M, 5 M, and 6 M KOH, respectively. Kim et al. [98] prepared $NiO-TiO_2$ nanotube arrays by the electrochemical anodization method. $NiO-TiO_2$ nanotubes annealed at 600°C exhibited good rate capacity and high stability during long-term cycling compared with film.

9.2.5 VANADIUM-BASED SC ELECTRODES

Vanadium oxides are favorable candidate for both lithium-ion batteries and supercapacitors, since they have most accessible layered structure with various oxidation states (V^{2+} to V^{5+}) which leads to high specific capacity, high energy density, and wide potential window [99–101]. However, vanadium oxides have relatively low electronic conductivity compared to RuO_2, which decreases the charge transfer rate during the charging/discharging process and limits their commercial use. Balamuralitharan et al. [102] synthesized V_2O_5 nanorods through hydrothermal approach. The structural investigation was made on the prepared nanorods, and it confirmed the orthorhombic structure of V_2O_5 with lattice constants a = 11.51 A°, b = 3.56 A°, and c = 4.37 A°. Furthermore, the binding energy gap (V $2p_{3/2}$ and V $2p_{1/2}$) was found to be 12.8 eV, and it was confirmed by X-ray photoelectron spectroscopy (XPS) analysis. V_2O_5 nanorods exhibited the specific capacitance of 417 mF cm^{-2} at a scan rate of 5 mV s^{-1} in 0.5 M Na_2SO_4 neutral electrolyte with 80% capacitance retention even after 5000 continuous charge/discharge cycles.

Khoo et al. studied the supercapacitive characteristics (in $LiClO_4$/PC electrolyte) of sodium-doped V_2O_5 nanobelts prepared via hydrothermal route [103]. The structural change in the prepared system was expected to enhance the ion intercalation. The morphological investigation revealed the formation of interconnected nanobelts. A maximum specific capacitance of 320 Fg^{-1} at a scan rate of 5 mV s^{-1} has been obtained for V_2O_5 nanobelts. In addition to that, the charge storage mechanism was explained by Dunn's method. A capacitance degradation of 34% has been observed after 4000 charge/discharge cycles at a current density of 10 Ag^{-1}. Mu et al. reported electrochemical performance of V_2O_5 nanomaterials with different dimensions [104]. Among the different dimensions, one-dimensional nanorods prepared in acid medium portrayed excellent electrochemical performance and cyclic life. V_2O_5 nanorods exhibited orthorhombic structure with (001) as a predominant plane. Quasi-rectangular CV curves suggested the ideal-capacitive behavior of prepared nanorods which delivered a specific capacitance of 235 Fg^{-1} at a constant current of 1 Ag^{-1} in 1 M Na_2SO_4 electrolyte.

Rudra et al. [105] synthesized $Au-V_2O_5$ nanowires via a simple and facile hydrothermal procedure by utilizing vanadium metal (III) complex and gold chloride as precursors. The electrochemical analysis stated that the prepared $Au-V_2O_5$ nanowires were capable of delivering a specific capacitance of 419 Fg^{-1} at a constant current density of 1 Ag^{-1}. Further, 88% capacitance retention has been observed after 5000 charge/discharge cycles at 10 Ag^{-1}. $Au-V_2O_5$ nanowires exhibited an energy density of 53.33 Wh kg^{-1} with a power density of 3.85 kW kg^{-1}.

V_2O_5 nanofibers were synthesized via electrospinning method by Wee et al. [106]. They studied the effect of annealing temperature on the microstructure and morphology of V_2O_5 nanofibers. The maximum specific capacitance (250 Fg^{-1}) was achieved for V_2O_5 nanofibers annealed at 400°C in organic electrolyte (1 M $LiClO_4$

in PC) with an energy density of 78 Wh kg^{-1}. Xu et al. [107] proposed a one-step hydrothermal method to construct ordered V_2O_5 nanobelt arrays on Ni foam without any additive. V_2O_5 nanobelt array as binder-free electrode material showed the specific capacitance of 498 Fg^{-1} with a cycling stability of 88.8% after 5000 cycles. Further, it exhibited the charge transfer resistance of 14.2 Ω. Zhang et al. synthesized V_2O_5 nanobelts by hydrothermal method combined with calcination [108]. The morphology of the V_2O_5 depended on the quantity of oxalic acid. V_2O_5 nanobelts were obtained when 0.63 g of oxalic acid was used in the synthesis. V_2O_5 nanobelts delivered a specific capacitance of 140 Fg^{-1} in 1 M LiNO$_3$ electrolyte.

Saravanakumar et al. [109] prepared carbon/V_2O_5 nanorods by using sol–gel-assisted hydrothermal procedure. The SEM images of the sample are presented in Figure 9.3. This material showed a specific capacitance of 417 Fg^{-1} at a current density of 0.5 Ag^{-1}. An asymmetric supercapacitor device using this material delivered an energy density of 9.4 Wh kg^{-1} at a power density of 170 W kg^{-1}.

9.2.6 Other Inorganic Materials for SC Electrode Applications

NiMoO$_4$ nanorods with 80 nm diameter and about 300 nm to 1 μm length were prepared by Cai et al. [110]. The maximum specific capacitance of 945 Fg^{-1} at a current density of 1 Ag^{-1} was reported for NiMoO$_4$ nanorods. After 2000 cycles, the NiMoO$_4$ nanorods displayed 52.4% of capacitance retention at a current density of 5 Ag^{-1}. CoMoO$_4$ nanoneedle prepared by Fang et al. [111] exhibited a maximum specific capacitance of 1628.1 Cg^{-1} at a current density of 2 mA cm^{-2} with a rate capability of 874.8 Cg^{-1} at 50 mA cm^{-2}, and 91% capacity retention after 5000 cycles at 10 mA cm^{-2}.

Lu et al. [112] synthesized one-dimensional NiMoO$_4$ nanorods with the specific surface area of 35 m^{2}g^{-1} via solid-state chemical route. The NiMoO$_4$ nanorods delivered specific capacitance of 1415 Fg^{-1} and retained 80.2% of the initial capacitance after 1000 charge/discharge cycles. The NiMoO$_4$/rGO asymmetric supercapacitor

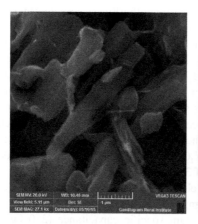
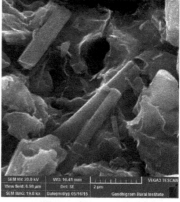

FIGURE 9.3   SEM images of carbon/V_2O_5 nanorods.

exhibited the specific energy density of 29.3 Wh kg^{-1} at a power density of 187 W kg^{-1}. The NiMoO$_4$/CoMoO$_4$ nanorods synthesized by Nti et al. [113] exhibited a specific capacitance of 1445 Fg^{-1} at 1 Ag^{-1} with the retention of 79% of its initial capacitance after 3000 cycles at 10 Ag^{-1}. It also exhibited an internal resistance of 1 Ω. MoO$_3$ nanobelts (surface area of 59.285 m^{2}g^{-1})-based supercapacitor device showed a specific capacitance of 257 Fg^{-1} at a scan rate of 5 mV s^{-1} in 0.5 M Na$_2$SO$_4$ aqueous electrolyte solution. This electrode exhibited an equivalent series resistance of 1.2 Ω and charge transfer resistance of 3.9 Ω [114]. In addition, MoO$_3$ supercapacitor electrode could withstand over 200 cycles with no obvious change in the CV curves.

Pham et al. [115] synthesized 1D molybdenum oxide nanorods on indium tin oxide (ITO)/glass substrates using the Hot-Filament Metal Oxide Vapor Deposition (HFMOVD) technique. The MoO$_2$ nanorods synthesized at 1200°C provided a discharge time of ~158 s (at 4 Ag^{-1}). Moreover, MoO$_2$ nanorod-based pseudocapacitors also possess excellent capacitance retention. Further, MoO$_2$ electrode delivered the power and energy densities of ~1800 W Kg^{-1} and ~7 Wh Kg^{-1} at a current density of 4 Ag^{-1}.

Rajeswari et al. [116] synthesized one-dimensional (1D) MoO$_2$ nanorods via simple thermal decomposition method. This material showed a specific capacitance of 140 Fg^{-1} in 1 M H$_2$SO$_4$ electrolyte. Xu et al. [117] synthesized MoO$_3$ nanorods on Ni nanowire array, which delivered a maximum aerial capacity of 477 mF cm^{-2} and 5% capacitance loss after 20000 cycles in 1 M LiSO$_4$ solution. Furthermore, an asymmetric supercapacitor has been fabricated using crystalline VO$_2$ (cathode) and MoO$_3$ (anode). This device can deliver an open circuit voltage of 1.6 V in an aqueous electrolyte and an energy density of 2.19 mW hcm^{-3} at the power density of 8.2 mW cm^{-3}. V$_2$O$_5$-doped Fe$_2$O$_3$ nanotubes were prepared via electrospinning method by Nie et al. [118]. The pure Fe$_2$O$_3$ nanotubes delivered 100.5 Fg^{-1} and 1% V$_2$O$_5$-doped nanotubes exhibit 183 Fg^{-1} at the current density of 1 Ag^{-1}. Doped nanotubes retained 81.5% of the initial capacitance after 2000 cycles. Rudra et al. [119] synthesized Au-Fe$_2$O$_3$ nanorods via modified hydrothermal method. CV analysis showed the maximum specific capacitance value of 785 Fg^{-1} at a scan rate of 5 mV s^{-1}. This electrode exhibited the specific capacitance of 570 Fg^{-1} at a current density of 1 Ag^{-1} in the 0.5 M H$_2$SO$_4$ electrolyte. The solid-state asymmetric device based on Au-Fe$_2$O$_3$ nanorods revealed the maximum energy density of 34.2 Wh kg^{-1} and power density of 2.73 kW kg^{-1} at 1 Ag^{-1} and 10 Ag^{-1}, respectively. In the acidic medium, 91% of the capacitance was retained after 5000 cycles. α-Fe$_2$O$_3$ nanorods were synthesized via hydrothermal method by Zhang et al. [120]. The nanorods exhibited the aerial capacitance of 500, 412, 300, 137, and 75 mF cm^{-2} at 4, 6, 8, 10, and 12 mA cm^{-2}, respectively, with charge transfer resistance of 2.88 Ω and cyclic stability of 92.1% after 3000 charge/discharge cycles.

CuO nanorods prepared by SILAR method exhibited the specific capacitance of 695 Fg^{-1} at 5 mV s^{-1}, and the ESR value was 3 Ω [121]. Core–shell nanowire arrays of CuO/Cu$_2$O@CoO prepared via chemical deposition method delivered an aerial capacitance of 280 mF cm^{-2} at a current density of 1 mA cm^{-2}. The material retained 90.7% of initial capacitance after 3000 cycles [122]. Ultrasonification-assisted chemical reduction method has been used to synthesize Ag-anchored ZnSb$_2$O$_6$ nanorods [123], which exhibited the specific capacitance of 165.9 Fg^{-1} at a current density of 1 Ag^{-1}

with 98% of capacitance retention after 2000 cycles. Urchin-like $ZnO@MnO_2$ core–shell structure showed the specific capacitance of 262.4 Fg^{-1} at 200 mAg^{-1} with 111.3% of initial capacitance retained after 5000 cycles [124]. Al-doped $ZnO@NiO$ composite exhibited the aerial capacitance of 49 mF cm^{-2} at a constant current of 1 mA cm^{-2} [125]. WO_3 nanorods prepared on carbon cloth exhibited the specific capacitance of 694 Fg^{-1} with the energy density of 25 Wh kg^{-1} and 87% capacitance retention after 2000 cycles [126]. Hexagonal WO_3 nanorods delivered a specific capacitance of 538 Fg^{-1} at 5 mV s^{-1} with the energy and power density of 48 Wh kg^{-1} and 1385 W kg^{-1}, respectively, and cycling stability of 85% over 2000 cycles [127].

9.3　CONCLUSIONS AND FUTURE PERSPECTIVE

Supercapacitors are evolving as promising energy storage device due to their unique electrochemical features. An intensive research is going on to promote the supercapacitors for future energy needs. In this view, pseudocapacitive inorganic metal oxides with one-dimensional nanostructures like nanorods, nanotubes, nanobelts, and nanowires have been tested extensively. To improve the supercapacitive features like specific capacitance, cycling stability, energy–power density, and rate capability, the researchers have made binary and ternary metal oxides with hierarchical one-dimensional nanostructures. The large surface area, easy accessibility, and short diffusion path for electrolyte ions make these one-dimensional nanostructured materials as promising candidate for high-performance supercapacitors. However, high energy density along with high power density is not achievable till now; therefore, there are numerous scopes for further research to enhance the electrochemical features of one-dimensional metal oxide nanostructures.

REFERENCES

1. Conway, B.E. 1999. *Electrochemical Supercapacitors: Scientific Fundamentals and Technological Applications*, Kluwer Academic Publishers/Plenum Press, New York.
2. Balducci, A., R. Dugas, P. Taberna, P. Simon, D. Plee, M. Mastragostino, and S. Passerini. 2007. High Temperature Carbon-Carbon Supercapacitor Using Ionic Liquid as Electrolyte. *J. Power Sources*. 165(2): 922–927.
3. Nishino, A. 1996. Capacitors: Operating Principles, Current Market and Technical Trends. *J. Power Sources*. 60(2): 137–140.
4. Zhang, Y., H. Feng, X. Wu, L. Wang, A. Zhang, T. Xia, H. Dong, X. Li, and L. Zhang. 2009. Progress of Electrochemical Capacitor Electrode Materials: A Review. *Int. J. Hydrogen Energ*. 34(11): 4889–4899.
5. Simon, P., and Y. Gogotsi. 2008. Materials for Electrochemical Capacitors. *Nature Mater*. 7(11): 845–854.
6. Zhi, M., C. Xiang, J. Li, M. Li, and N.Wu. 2013. Nanostructured Carbon-Metal Oxide Composite Electrodes for Supercapacitors. *Nanoscale*. 5(1): 72–88.
7. Rolison, D.R. 2003. Catalytic Nanoarchitectures—The Importance of Nothing and the Unimportance of Periodicity. *Science*. 299(5613): 1698–1701.
8. Duan, X., C.M. Niu, V. Sahi, J. Chen, J.W. Parce. S. Empedocles, and J.L. Goldman. 2003. High Performance Thin-Film Transistors Using Semiconductor Nanowires and Nanoribbons. *Nature*. 425: 274–278.

9. Pan, H., J. Li, and Y.P. Feng. 2010. Carbon Nanotubes for Supercapacitor. *Nanoscale Res. Lett.* 5(3): 654–668.

10. Wu, M.S., and C.J. Chiang. 2004. Fabrication of Nanostructured Manganese Oxide Electrodes for Electrochemical Capacitors. *Electrochem. Solid St. Lett.* 7(6): A123–A126.

11. Bruce, P.G., B. Scrosati, and J.M. Tarascon. 2008. Nanomaterials for Rechargable Lithium Batteries. *Angew. Chem. Int. Edit.* 47(16): 2930–2946.

12. Tian, N., Z.Y. Zhou, S.G. Sun, Y. Ding, and Z.L. Wang. 2007. Synthesis of Tetrahedral Planium Nanocrystals with High-Index Facets and High Electro-Oxidation Activity. *Science.* 316(5825): 732–735.

13. Appapillai, A.T., A.N. Mansour, J. Cho, and H. Shao. 2007. Microstructure of $LiCoO_2$ with and without $AlPO_4$ Nanoparticle Coating: Combined STEM and XPS Studies. *Chem. Mater.* 19(23): 5748–5757.

14. Long, J.W., R.M. Stroud, K.E. Lyons, and D.R. Rolison. 2000. How to Make Electrocatalysts More Active for Direct Methanol PtRu Bimetallic Alloys Oxidization-Avoid. *J. Phy. Chem. B.* 104(42): 9772–9776.

15. Rolison, D.R., and A.B. Dunn. 2001. Electrically Conductive Oxide Aerogels: New Materials in Electrochemistry. *J. Mater. Chem.* 11(4): 963–980.

16. Deng, W., X. Ji, Q. Chen, and C.E. Banks. 2011. Electrochemical Capacitors Utilizing Transition Metal Oxides: An Update of Recent Developments. *RSC Adv.* 1(7): 1171–1178.

17. Dong, Y., S. Li, K. Zhao, C. Han, W. Chen, B. Wang, L. Wang B. Xu, Q. Wei, and L. Zhang. 2015. Hierarchical Zigzag $Na_{1.25}V_3O_8$ Nanowires with Topotactically Encoded Superior Performance for Sodium-ion Battery Cathodes. *Energy Environ. Sci.* 8(4): 1267–1275.

18. Li, R., Y. Wang, C. Zhou, C. Wang, X. Ba, Y. Li, X. Huang, and J. Liu, 2015. Carbon-Stabilized High-Capacity Ferroferric Oxide Nanorod Array for Flexible Solid-State Alkaline Battery–Supercapacitor Hybrid Device with High Environmental Suitability. *Adv. Funct. Mater.* 25(33): 5384–5394.

19. Niu, C., J. Meng, X. Wang, C. Han, M. Yan, K. Zhao, X. Xu, W. Ren, Y. Zhao, and L. Xu, 2015. General Synthesis of Complex Nanotubes by Gradient Electrospinning and Controlled Pyrolysis. *Nat. Commun.* 6: 1–9.

20. Cheng, C., and H.J. Fan, 2012. Branched Nanowires: Synthesis and Energy Applications. *Nano Today.* 7(4): 327–343.

21. Lu, X.U., X.Y. Chen, W. Zhou, Y.X. Tong, and G.R. Li. 2015. α-Fe_2O_3@PANI Core–Shell Nanowire Arrays as Negative Electrodes for Asymmetric Supercapacitors. *ACS Appl. Mater. Interfaces.* 7(27): 14843–14850.

22. Huang, J., H. Li, Y. Zhu, Q. Cheng, X. Yang, and C. Li. 2105. Sculpturing Metal Foams Toward Bifunctional 3D Copper Oxide Nanowire Arrays for Pseudo-Capacitance and Enzyme-Free Hydrogen Peroxide Detection. *J. Mater. Chem. A.* 3(16): 8734–8741.

23. Vu, A., Y. Qian, and A. Stein. 2012. Porous Electrode Materials for Lithium-Ion Batteries—How to Prepare Them and What Makes Them Special. *Adv. Energy Mater.* 2(9): 1056–1085.

24. Hammouche, A., E. Karben, and R.W. Doncker. 2004. Monitoring State-of-Charge of Ni-MH and Ni-Cd Batteries Using Impedance Spectroscopy. *J. Power Sources.* 127: 105–111.

25. Shukla, A.K., P. Ercius, A.R. Gautam, J. Cabana, and U. Dahmen. 2014. Electron Tomography Analysis of Reaction Path During Formation of Nanoporous NiO by Solid State Decomposition. *Cryst. Growth Des.* 14(5): 2453–2459.

26. Cao, M., X. He, J. Che, and C. Hu. 2007. Self Assembled Nickel Hydroxide Three Dimensional Nanostructures: A Nanomaterial for Alkaline Batteries. *Cryst. Growth Des.* 7(1): 170–174.

27. Yuan, C., X. Zhang, L. Su, B. Gao, and L. Shen. 2009. Facile Synthesis and Self Assembly of Hierarchical Porous NiO Nano/Micro Spherical superstructures for High Performance Supercapacitors. *J. Mater. Chem.* 19(32): 5772–5777.

28. Gund, G.S., B.P. Dupal, S.S. Shinde, and C.D. Lokhande. 2014. Architectured Morphologies of Chemically Prepared NiO/MWCNTs Nano Hybrid Thinfilms for High Performance Supercapacitors. *ACS Appl. Mater. Interfaces.* 6(5): 3176–3188.

29. Bi, R.R., X.L. Wu, F.F. Gao, L.Y. Jiang, Y.G. Guo, and L.J. Wan. 2010. Highly Dispersed RuO_2 Nanoparticles on Carbon Nanotubes: Facile Synthesis and Enhanced Supercapacitance Performance. *J. Phys. Chem.* 114(6): 2448–2451.

30. Cao, F., G.X. Pan, X.H. Xia, P.S. Tang and H.F. 2014. Chen Synthesis of Hierarchical Porous NiO Nanotube Arrays for Supercapacitor Application. *J. Power Sources.* 264: 161–167.

31. Kannan, V., A.I. Inamdar, S.M. Pawar, H.S. Kim, H.C. Park, H.S. Kim, H. Im, and Y.S. Chae. 2106. Facile Route to NiO Nanostructured Electrode Grown by Oblique Angle Deposition Technique for Supercapacitors. *ACS Appl. Mater. Interfaces.* 8(27): 17220–17225.

32. Kolathodi, M.S., M. Palei, and T.S. Natarajan. 2015. Electrospun NiO Nanofibers as Cathode Materials for High Performance Asymmetric Supercapacitors. *J. Mater. Chem. A.* 3(14): 7513–7522.

33. Kundu, M., and L. Liu. 2015. Binder-free Electrodes Consisting of Porous NiO Nanofibers Directly Electrospun on Nickel foam for High-Rate Supercapacitors. *Mater. Lett.* 144: 114–118.

34. Liu, A., H. Chen, Y. Mao, Y. Wang, J. Mu, C. Wu, Y. Bai, X. Zhang, and G. Wang. 2016. Template-Free Synthesis of One-Dimensional Hierarchical NiO Nanotubes Self-Assembled by Nanosheets for High-Performance Supercapacitors. *Ceram. Int.* 42: 11435–11441.

35. Patil, R.A., C.P. Chang, R.S. Devan, Y. Liou, and Y.R. Ma. 2016.The Impact of Nanosize on Supercapacitance: Study of 1D Nanorods and 2D Thin-Film of Nickel Oxide. *ACS Appl. Mater. Interfaces.* 8(15): 9872–9880.

36. Singh, A.K., D. Sarkar, G.G. Khan, and K. Mandal. 2013. Unique Hydrogenated Ni/NiO Core/Shell 1D Nanoheterostructures with Superior Electrochemical Performance as Supercapacitors. *J. Mater. Chem. A.* 1: 12759–12767.

37. Yao, M., Z. Hu, Z. Xu, Y. Liu, P. Liu, and Q. Zhang. 2015. Template Synthesis and Characterization of Nanostructured Hierarchical Mesoporous Ribbon-like NiO as High Performance Electrode Material for Supercapacitor. *Electrochim. Acta.* 158: 96–104.

38. Tan, Y., Y. Li, L. Kong, L. Kang, and F. Ran. 2018. Synthesis of Ultra-Small Gold Nanoparticles Decorated into NiO Nanobelts and Its High Electrochemical Performance. *Dalton Trans.* 47: 8078–8086.

39. Yuksel, R., S. Coskun, Y.E. Kalay, H.E. Unalan. 2016. Flexible, Silver Nanowire Network Nickel Hydroxide Core-Shell Electrodes for Supercapacitors. *J. Power Sources.* 328: 167–173.

40. Zhang, M., Q. Li, D. Fang, I.A. Ayhan, Y. Zhou, L. Dong, C. Xiong, and Q. Wang. 2015. NiO Hierarchical Hollow Nanofibers as High performance Supercapacitor Electrodes. *RSC Adv.* 5: 96205–96212.

41. Ren, B., M. Fan, Q. Liu, J. Wang, D. Song, and X. Bai. 2013. Hollow NiO Nanofibers Modified by Citric Acid and the Performances as Supercapacitor Electrode. *Electrochim. Acta.* 92: 197–204.

42. Vidhyadharan, B., N.K.M. Zain, I.I. Misnon, R.A. Aziz, J. Ismail, M.M. Yusoff, and R. Jose. 2014. High Performance Supercapacitor Electrodes from Electrospun Nickel Oxide Nanowires. *J. Alloy. compd.* 160: 143–150.

43. Wang, B., J.S. Chen, Z. Wang, S. Madhavi, and X.W. Lou. 2012. Green Synthesis of NiO Nanobelts with Exceptional Pseudo-Capacitive Properties. *Adv. Energy Mater.* 2(10): 1188–1192.

44. Xiong, S., C. Yuan, X. Zhang, and Y. Qian. 2011. Mesoporous NiO with Various Hierarchical Nanostructures by Quasi-nanotubes/Nanowires/Nanorods Self-assembly: Controllable Preparation and Application in Supercapacitors. *Cryst. Eng. Comm.* 13: 626–632.

45. Zang, L., J. Zhu, and Y. Xia. 2014. Facile Synthesis of Porous NiO for High Performance Supercapacitors. *JMEPEG.* 23: 679–683.

46. Sethi, M., and D.K. Bhat. 2019. Facile Solvothermal Synthesis and High Supercapacitor Performance of NiCo$_2$O$_4$ Nanorods. *J. Alloy. Compd.* 781: 1013–1020.

47. Zhu, Y., X. Pu, W. Song, Z. Wu, Z. Zhou, X. He, F. Lu, M. Jing, B. Tang, and X. Ji. 2014. High Capacity NiCo$_2$O$_4$ Nanorods as Electrode Materials for Supercapacitor. *J. Alloy. Compd.* 617: 988–993.

48. Lang, J.W., X.B. Yan, and Q.J. Xue. 2011. Facile Preparation and Electrochemical Characterization of Cobalt Oxide/Multi-Walled Carbon Nanotube Composites for Supercapacitors. *J. Power Sources.* 196(18): 7841–7846.

49. Liu, X., Q. Long, C. Jiang, B. Zhan, C. Li, S. Liu, Q. Zhao, W. Huang, and X. Dong. 2013. Facile and Green Synthesis of Mesoporous Co$_3$O$_4$ Nanocubes and Their Applications for Supercapacitors. *Nanoscale.* 5(14): 6525–6529.

50. Cao, L., F. Xu, Y.Y. Liang, and H.L. Li. 2004. Preparation of the Novel Nanocomposite Co(OH)$_2$/Ultra-Stable Y Zeolite and its Application as a Supercapacitor with High Energy Density. *Adv. Mater.* 16(20): 1853–1857.

51. Wang, Y., A. Pan, Q. Zhu, Z. Nie, Y. Zhang, Y. Tang, S. Liang, and G. Cao. 2014. Facile Synthesis of Nanorod-Assembled Multi-shelled Co$_3$O$_4$ Hollow Microspheres for High-Performance Supercapacitors. *J. Power Sources.* 272: 107–112.

52. Wang, X., C. Yan, A. Sumboja, and P. Leen. 2014. High Performance Porous Nickel Cobalt Oxide Nanowires for Asymmetric Supercapacitor. *Nano Energ.* 3: 119–126.

53. Wang, G., X. Shen, J. Horvat, B. Wang, H. Liu, D. Wexler, and J. Yao. 2009. Hydrothermal Synthesis and Optical, Magnetic, and Supercapacitance Properties of Nanoporous Cobalt Oxide Nanorods. *J. Phys. Chem. C.* 113: 4357–4361.

54. Kumar, K., A. Subramania, and K. Balakrishnan. 2014. Preparation of Electrospun Co$_3$O$_4$ Nanofibers as Electrode Material for High Performance Asymmetric Supercapacitors. *Electrochim. Acta.* 149: 152–158.

55. Deng, M.J., C.Z. Song, C.C. Wang, Y.C. Tseng, J.M. Chen, and K.T. Lu. 2015. Low Cost Facile Synthesis of Large-Area Cobalt Hydroxide Nanorods with Remarkable Pseudocapacitance. *ACS Appl. Mater. Interfaces.* 17: 9147–9156.

56. Gu, Z., H. Nan, B. Geng, and X. Zhang. 2012. Construction of Unique Co$_3$O$_4$@CoMoO$_4$ Core/Shell Nanowire Arrays on Ni Foam by Action Exchange Method for High-Performance Supercapacitors. *J. Mater. Chem. A.* 3: 14578–14584.

57. Li, D., F. Yu, Z. Yu, X. Sun, and Y. Li, 2015. Three-Dimensional Flower-Like Co(OH)$_2$ Microspheres of Nanoflakes/Nanorods Assembled on Nickel Foam as Binder-Free Electrodes for High Performance Supercapacitors. *Mater. Lett.* 158: 17–20.

58. He, F., K. Liu, J. Zhong, S. Zhang, Q. Huang, and C. Chen. 2015. One Dimensional Nickel Oxide-Decorated Cobalt Oxide (Co$_3$O$_4$) Composites for High-Performance Supercapacitors. *J. Electroanalytical Chem.* 749: 89–95.

59. Qian, T.M., Q.X. Zhi, H.W. Yin, T. Li, Z.M. Gao, X.Y. Xing, and T.Z. Ren. 2015. Co$_3$O$_4$ Nanorods with Self-Assembled Nanoparticles in Queue for Supercapacitor. *Electrochim. Acta.* 180: 104–111.

60. Li, J., G. Zan, and Q. Wu. 2016. An Ultra-High Performance Anode Material for Supercapacitors: Self-Assembled Long Co$_3$O$_4$ Hollow Tubes Network with Multiple Heteroatoms (C-, N- and S-) Doping. *J. Mater. Chem. A.* 4: 9097–9105.

61. Harilal, M., B. Vidyadharan, I.I. Misnon, G.M. Anilkumar, A. Lowe, J. Ismail, M.B.M. Yusoff, and R. Jose. 2017. One-Dimensional Assembly of Conductive and Capacitive Metal Oxide Electrodes for High Performance Asymmetric Supercapacitors. *ACS Appl. Mater. Interfaces.* 9: 10730–10742.

62. Howli, P., S. Das, S. Sarkar, M. Samanta, K. Panigrahi, N.S. Das, and K.K. Chattopadhyay. 2017. Co_3O_4 Nanowires on Flexible Carbon Fabric as a Binder-Free Electrode for All Solid-State Symmetric Supercapacitor. *ACS Omega.* 2: 4216–4226.

63. Chen, M., Q. Ge, M. Qi, X. Liang, F. Wang, and Q. Chen, 2019. Cobalt Oxides Nanorods Arrays as Advanced Electrode for High Performance Supercapacitor. *Surf. Coat. Tech.* 360: 73–77.

64. Gao, M., W. Wang, Q. Rong, J. Jiang, Y. Zhang, and H. Yu. 2018. Porous ZnO-Coated Co_3O_4 Nanorod as a High-Energy-Density Supercapacitor Material. *ACS Appl. Mater. Interfaces.* 10(27): 23163–23173.

65. Kannana. V, J.H. Choi, H.C. Park, and H.S. Kim. 2018. Ultrahigh Supercapacitance in Cobalt Oxide Nanorod Film Grown by Oblique Angle Deposition Technique. *Curr. Appl. Phys.* 18(11): 1399–1402.

66. Pan, X., F. Ji, Q. Xia, X. Chen, H. Pan, S.N. Khisro, S. Luo, M. Chen, and Y. Zhang. 2018. High-Performance Supercapacitors Based on Superior Co_3O_4 Nanorods Electrode for Integrated Energy Harvesting-Storage System. *Elctrochim. Acta.* 282: 905–912.

67. He, F., K. Liu, J. Zhong, S. Zhang, Q. Huang, and C. Chen. 2015. One Dimensional Nickel Oxide-decorated Cobalt Oxide Composites for High-performance Supercapacitors. *J. Electroanal. Chem.* 749: 89–95.

68. Xiao, Y., A. Dai, X. Zhao, S. Wu, D. Su, X. Wang, and S. Fang. 2019. A Comparative Study of One-Dimensional and Two-Dimensional Porous CoO Nanomaterials for Asymmetric Supercapacitor. *J. Alloy. Compd.* 781: 1006–1012.

69. Xinga, L., Y. Donga, F. Hua, X. Wua, and A. Umarb. 2018. Co_3O_4 Nanowire @ NiO Nanosheet Arrays for High Performance Asymmetric Supercapacitors. *Dalton Trans.* 47: 5687–5694.

70. Wang, X., and Y. Li. 2002. Selected Control Hydrothermal Synthesis of α-and βMnO_2 Single Crystal Nanowires. *J. Am. Chem. Soc.* 124(12): 2880–2881.

71. Kim, J.S., S.S. Shin, H.S. Han, L.S. Oh, D.H. Kim, D.H. Kim, K.S. Kong, J.Y. Kim. 2014. 1-D Structured Flexible Supercapacitor Electrodes with Prominent Electronic/Ionic Transport Capabilities. *ACS Appl. Mater. Interfaces.* 6(1): 268–274.

72. Ataherian, F., K.T. Lee, and N.L. Wu. 2010. Long-Term Chemical Behaviours of Manganese Oxide Aqueous Electrochemical Capacitor Under Reducing Potentials. *Electrochim. Acta.* 55(25): 7429–7435.

73. Bai, X., X. Tong, Y. Gao, W. Zhu, C. Fu, J. Ma, T. Tan, C. Wang, Y. Luo, and H. Sun. 2018. Hierarchical Multidimensional MnO_2 via Hydrothermal Synthesis for High Performance Supercapacitors. *Electrochim. Acta.* 281: 525–533.

74. Chen. S., J. Zhu, Q. Han, Z. Zheng, Y. Yang, and X. Wang. 2009. Shape-Controlled Synthesis of One-Dimensional MnO_2 via a Facile Quick-Precipitation Procedure and its Electrochemical Properties. *Cryst. Grow. Design.* 9: 4356–4361.

75. Davoglio, R.A., G. Cabello, J.F. Marco, and S.R. Biaggio. 2018. Synthesis and Characterization of α-MnO_2 Nanoneedles for Electrochemical Supercapacitors. *Electrochim. Acta.* 261: 428–435.

76. Duay. J., S.A. Sherrill, Z. Gui, E. Gillette, and S.B. Lee.2013. Self-Limiting Electrodeposition of Hierarchical MnO_2 and $M(OH)_2$/MnO_2 Nanofibril/Nanowires: Mechanism and Supercapacitor Properties. *ACS Nano.* 7(2): 1200–1214.

77. Gong, L., X. Liu, and L. Lu. 2012. Synthesis of MnO_2 Nanorods from a ZnO Template and Their Capacitive Performances. *Mater. Lett.* 67: 226–228.

78. Kumara, N., K. Guru Prasada, A. Sena, and T. Maiyalagan. 2018. Enhanced Pseudocapacitance from Finely Ordered Pristine α-MnO_2 Nanorods at Favourably High Current Density using Redox Additive. *Appl. Surf. Sci.* 449: 492–499.

79. Li, Y., H. Fu, Y. Zhang, Z. Wang, and X. Li. 2014. Kirkendall Effect Induced One-Step Fabrication of Tubular Ag/MnOx Nanocomposites for Supercapacitor Application. *J. Phys. Chem.* 118: 6604–6611.

80. Li, F., Y.X. Zhang, M. Huang, Y. Xing, and L.L. Zhang. 2015. Rational Design of Porous MnO_2 Tubular Arrays via Facile and Templated Method for High Performance Supercapacitors. *Electrochim. Acta*. 154: 329–337.

81. Li, N., X. Zhu, C. Zhang, L. Lai, R. Jiang, and J. Zhu. 2017. Controllable Synthesis of Different Microstructured MnO_2 by a Facile Hydrothermal Method for Supercapacitors. *J. Alloy. Compd.* 692: 26–33.

82. Lobinsky, A.A., and V.P. Tolstoy. 2017. Synthesis of γ-MnOOH Nanorods by Successive Ionic Layer Deposition Method and Their Capacitive Performance. *J. Energ. Chem.* 26(3): 336–339.

83. Ma, Z., G. Shao, Y. Fan, G. Wang, J. Song, and D. Shen. 2016. Construction of Hierarchical α-MnO_2 Nanowires@ Ultrathin δ-MnO_2 Nanosheets Core-shell Nanostructure with Excellent Cycling Stability for High-Power Asymmetric Supercapacitor Electrodes. *ACS Appl. Mater. Interfaces.* 8(14): 9050–9058.

84. Maiti, S., A. Pramanik, and S. Mahanty. 2014. Interconnected Network of MnO_2 Nanowires with a "Cocoon" Like Morphology: Redox Couple Mediated Performance Enhancement in Symmetric Aqueous Supercapacitor. *ACS Appl. Mater. Interfaces.* 6(13): 10754–10762.

85. Purushothaman, K.K., M. Cuba, and G. Muralidharan. 2012. Supercapacitor Behavior of α-$MnMoO_4$ Nanorods on Different Electrolytes. *Mater. Res. Bull.* 47: 3348–3351.

86. Qu, Q., P. Zhang, B. Wang, Y. Chen, S. Tian, Y. Wu, and R. Holze. 2009. Electrochemical Performance of MnO_2 Nanorods in Neutral Aqueous Electrolytes as a Cathode for Asymmetric Supercapacitors. *J. Phys. Chem. C.* 113: 14020–14027.

87. Su, D., Z. Tang, J. Xie, Z. Bian, J. Zhang, D. Yang, D. Zhang, J. Wang, Y. Liu, A. Yuan, and Q. Kong. 2018. Co, Mn-LDH Nanoneedle Arrays Grown on Ni Foam for High Performance Supercapacitors. *Appl. Surf. Sci.* 469: 487–494.

88. Sung, D.Y., I.Y. Kim, T.W. Kim, M.S. Song, and S.J. Hwang. 2011. Room Temperature Synthesis Routes to the 2D Nanoplates and 1D Nanowires/Nanorods of Manganese Oxides with Highly Stable Pseudocapacitance Behaviors. *J. Phys. Chem. C.* 115: 13171–13179.

89. Yang, P., Y. Ding, Z. Lin, Z. Chen, Y. Li, P. Qiang, M. Ebrahimi, W. Mai, C.P. Wong, and Z.L. Wang. Low-Cost High-Performance Solid-State Asymmetric Supercapacitors Based on MnO_2 Nanowires and Fe_2O_3 Nanotubes. *Nano Lett.* 14: 731–736.

90. He, H., C.P. Yang, G.L. Zhanga, D.W. Shi, Q.A. Huang, H.B. Xiao, Y. Liub, and R. Xiong. 2016. Supercapacitor of TiO_2 Nanofibers by Electrospinning and KOH Treatment. *Mater. Design.* 106: 74–80.

91. Yang, Y., D. Kim, M. Yang, and P. Schmuki. 2011. Vertically Aligned Mixed V_2O_5–TiO_2 Nanotube Arrays for Supercapacitor Applications. *Chem. Commun.* 47: 7746–7748.

92. Ramadoss, A., and S.J. Kim. 2013. Vertically Aligned TiO_2 Nanorod Arrays for Electrochemical Supercapacitor. *J. Alloy. Compd.* 561: 262–267.

93. Zhou, H., and Y. Zhang. 2017. Electrochemically Self-Doped TiO_2 Nanotube Arrays for Supercapacitors. *J. Phys. Chem. C.* 118: 5626–5636.

94. Choi., H., H. Kim, H.K. Kim, S.H. Lee, and Y.H. Lee. 2016. Improving the Electrochemical Performance of Hybrid Supercapacitor Using Well-Organized Urchin-Like TiO_2 and Activated Carbon. *Electrochim. Acta.* 208: 201–210.

95. Qin, Y., J. Zhang, Y. Wang, X. Shu, C. Yu, J. Cui, H. Zheng, Y. Zhanga, and Y. Wu, 2016. Supercapacitive Performance of Electrochemically Doped TiO_2 Nanotube Arrays Decorated with Cu_2O Nanoparticles. *RSC Adv.* 6: 47669–47675.

96. Cui, H., Y. Chen, S. Lu, S. Zhang, X. Zhu, and Y. Song. 2017. TiO_2 Nanotube Arrays Treated with $(NH_4)_2TiF_6$ Dilute Solution for Better Supercapacitive Performances. *Electrochim. Acta.* 253: 455–462.

97. Barai, H. R., M.M. Rahman, and S. Woo. 2017. Annealing-Free Synthesis of K-doped Mixed-Phase TiO_2 Nanofibers on Ti Foil for Electrochemical Supercapacitor. *Electrochim. Acta.* 253: 563–571.

98. Kim, J., K. Zhu, Y. Yan, C.L. Perkins, and A.J. Frank. 2010. Microstructure and Pseudocapacitive Properties of Electrodes Constructed of Oriented NiO-TiO$_2$ Nanotube Arrays. *Nano Lett.* 10: 4099–4104.

99. Li, J.M, K.H. Chang, and C.C. Hu. 2010. A Novel Vanadium Oxide Deposit for the Cathode of Asymmetric Lithium-Ion Supercapacitor. *Electrochem. Commun.* 12(12): 1800–1803.

100. Rui, X, Z. Lu, H. Yu, D. Yang, H.H. Hng, T.M. Lim, and Q. Yan. 2013. Ultrathin V$_2$O$_5$ Nanosheet Cathode: Realizing Ultrafast Reversible Lithium Storage. *Nanoscale.* 5(2): 556–560.

101. Sathiya, M., A.S. Prakash, K. Ramesha, J.M. Tarascon, and A.K. Shukla. 2011. V$_2$O$_5$-Anchored Carbon Nanotubes for Electrochemical Energy Storage. *J. Am. Chem. Soc.* 133 (40): 16291–16299.

102. Balamuralitharan, B., I.H. Cho, J.S. Bak, and H.J. Kim. 2018. V$_2$O$_5$ Nanorod Electrode Material for Enhanced Electrochemical Properties by Facile Hydrothermal Method for Supercapacitor Applications. *New J. Chem.* 42: 11862–11868.

103. Khoo, E., J.M. Wang, J. Ma, and P.S. Lee. 2010. Electrochemical Energy Storage in a β-Na$_{0.33}$V$_2$O$_5$ Nanobelt Network and Its Application for Supercapacitors. *J. Mater. Chem.* 20: 8368–8374.

104. Mu, J., J. Wang, J. Hao, P. Cao, S. Zhao, W. Zeng, B. Miao, and S. Xu. 2015. Hydrothermal Synthesis and Electrochemical Properties of V$_2$O$_5$ Nanomaterials with Different Dimensions. *Ceram. Int.* 41: 12626–12632.

105. Rudra, S., A.K. Nayak, R. Chakraborty, P.K. Maji, and M. Pradhan. 2018. Synthesis of Au-V$_2$O$_5$ Composite Nanowires Through the Shape Transformation of a Vanadium (III) Metal Complex for High-Performance Solid-State Supercapacitors. *Inorg. Chem. Front.* 5: 1836–1843.

106. Wee, G., H.Z. Soh, Y.L. Cheah, S.G. Mhaisalkar, and M. Srinivasan. 2010. Synthesis and Electrochemical Properties of Electrospun V$_2$O$_5$ Nanofibers as Supercapacitor Electrodes. *J. Mater. Chem.* 20: 6720–6725.

107. Xu, J., F. Zheng, H. Gong, L. Chen, J. Xie, P. Hu, Y. Li, Y. Gong, and Q. Zhen. 2017. V$_2$O$_5$ Nanobelt Arrays with Controllable Morphologies for Enhanced Performance Supercapacitors. *Cryst. Eng. Comm.* 19: 6412–6424.

108. Zhang, Y., J. Zheng, Y. Zhao, T. Hu, Z. Gao, and C. Meng. 2016. Fabrication of V$_2$O$_5$ with Various Morphologies for High-Performance Electrochemical Capacitor. *Appl. Surf. Sci.* 377: 385–393.

109. B. Saravanakumar, K.K. Purushothaman, and G. Muralidharan. 2015. High Performance Supercapacitor Based on Carbon Coated V$_2$O$_5$ Nanorods. *J. Electroanalytical Chem.* 758: 111–116.

110. Cai, D., D. Wang, B. Liu, Y. Wang, Y. Liu, L. Wang, H. Li, H. Huang, Q. Li, and T. Wang. 2013. Comparison of the Electrochemical Performance of NiMoO$_4$ Nanorods and Hierarchical Nanospheres for Supercapacitor Applications. *ACS Appl. Mater. Interfaces* 5: 12905–12910.

111. Fang, L., F. Wang, T. Zhai, Q. Yan, M. Lan, K. Huang, and Q. Jing. 2018. Hierarchical CoMoO$_4$ Nanoneedle Electrodes for Advanced Supercapacitors and Electrocatalytic Oxygen Evolution. *Electrochim. Acta.* 259: 552–558.

112. Lu, X., W. Jia, H. Chai, J. Hu, S. Wang, and Y. Cao. 2019. Solid-State Chemical Fabrication of One-Dimensional Mesoporous β-nickel Molybdate Nanorods as Remarkable Electrode Material for Supercapacitors. *J. Colloid and Interf. Sci.* 534: 322–331.

113. Nti, F., D.A. Anang, and J.I. Han. 2018. Facilely Synthesized NiMoO$_4$/CoMoO$_4$ Nanorods as Electrode Material for High Performance Supercapacitor. *J. Alloy. Compd.* 742: 342–350.

114. Pal, S., and K.K. Chattopadhyay. 2018. Fabrication of Molybdenum Trioxide Nanobelts as High Performance Supercapacitor. *Mater. Today Proce.* 5: 9776–9782.
115. Pham, D.V., R.A. Patil, C. Yang, W.C. Yeh, Y. Liou, and Y.R. Ma. 2018. Impact of the Crystal Phase and 3d-valence Conversion on the Capacitive Performance of One-Dimensional MoO_2, MoO_3, and Magneli-phase Mo_4O_{11} Nanorod-based Pseudocapacitors. *Nano Energ.* 47: 105–114.
116. Rajeswari, J., P.S. Kishore, B. Viswanathan, and T.K. Varadarajan. 2009. One-Dimensional MoO_2 Nanorods for Supercapacitor Applications. *Electrochem. Commun.* 11: 572–576.
117. Xu, C. J. Liao, R. Wang, P. Zou, R. Wang, F. Kang, and C. Yang. 2016. MoO_3@Ni Nanowire Array Hierarchical Anode for High Capacity and Superior Longevity All-Metal-Oxide Asymmetric Supercapacitors. *RSC Advan.* 6: 110112–110119.
118. Nie. G., X. Lu, J. Lei, Z. Jiang, and C. Wang. 2014. Electrospun V_2O_5-doped α-Fe_2O_3 Composite Nanotubes with Tunable Ferromagnetism for High Performance Supercapacitor Electrodes. *J. Mater. Chem. A.* 2: 15495–15501.
119. Rudra, R., A.K. Nayak, S. Koley, R. Chakraborty, P.K. Maji, and M. Pradhan. 2019. Redox-Mediated Shape-Transformation of Fe_3O_4 Nanoflake to Chemically Stable Au-Fe_2O_3 Composite Nanorod for High-Performance Asymmetric Solid-State Supercapacitor Device. *ACS Sustainable Chem. Eng.* 7(1): 724–733.
120. Zhang, S., B. Yin, Z. Wang, and F. Peter. 2016. Super Long-Life All Solid-State Asymmetric Supercapacitor Based on NiO Nanosheets and α-Fe_2O_3 Nanorods. *Chemical Eng. J.* 306: 193–203.
121. Shinde, S.K., D.P. Dubal, G.S. Ghodake, D.Y. Kim, and V.J. Fulari. 2016. Morphological Tuning of CuO Nanostructures by Simple Preparative Parameters in SILAR Method and Their Consequent Effect on Supercapacitors. *Nano Struct. Nano Obj.* 6: 5–13.
122. Zhao, J., X. Shu, Y. Wang, C. Yu, J. Zhang, J. Cui, Y. Qin, H. Zheng, J. Liu, Y. Zhang, and Y. Wu. 2016. Construction of CuO/Cu_2O@CoO Core Shell Nanowire Arrays for High Performance Supercapacitors. *Surf. Coat. Tech.* 299: 15–21.
123. Balasubramaniam, M., and S. Balakumar. 2019. Nanostructuring of Silver Nanoparticles Anchored 1D Zinc Antimonate Electrode Material by Ultrasonication Assisted Chemical Reduction Approach for Supercapacitors. *Mater. Chem. Phys.* 224: 334–348.
124. Li, W., G. He, J. Shao, Q. Liu, K. Xu, J. Hu, and I.P. Parkin. 2015.Urchin-like MnO_2 Capped ZnO Nanorods as High-Rate and High-Stability Pseudocapacitor Electrodes. *Electrochim. Acta.* 186: 1–6.
125. Sun, X.Z.Y., X. Yan, X. Sun, G. Zhang, Q. Zhang, Y. Jiang, W. Gao, and Y. Zhang. 2016. High Carrier Concentration ZnO Nanowire Arrays for Binder-Free Conductive Support of Supercapacitors Electrodes by Al Doping. *J. Colloid Interf. Sci.* 484: 155–161.
126. Shinde, P.A., A.C. Lokhande, N.R. Chodankar, A.M. Patil, J.H. Kim, and C.D. Lokhande. 2017. Temperature Dependent Surface Morphological Modification of Hexagonal WO_3 Thinfilms for High Performance Supercapacitor Application. *Electrochim. Acta.* 224: 397–404.
127. Shinde, P.A., A.C. Lokhande, A.M. Patil, and C.D. Lokhande. 2018. Facile Synthesis of Self-Assembled WO_3 Nanorods for High-Performance Electrochemical Capacitor. *J. Alloy. Compd.* 770: 1130–1137.

10 One-Dimensional Carbon Nanostructures for Supercapacitors

Manas Roy and Mitali Saha

CONTENTS

10.1 INTRODUCTION

A supercapacitor (SC) deals with the power storage device that accumulates electrical charges at the interface created between the associated electrical conductor and ionic conductor. It stores lower amounts of energy, but generates very high power. The important requirements for a supercapacitor are high specific capacitance, low electrode–electrolyte resistance, and long cycling life. Actually, the technology of SC bridges the gap between customary energy storage technologies like parallel plate capacitor and battery by coalescing the components of their corresponding operating mechanisms. There are three electrode materials that are commonly used to make a supercapacitor: activated carbons with large surface area, metal oxides, and conducting organic or inorganic polymers. However, the electrolytes can be aqueous or organic. Supercapacitors are complementary to batteries, but follow a different energy storage mechanism. Batteries store chemical energy by moving charged ions from one electrode to another electrode via an electrolyte where chemical interaction occurs. A supercapacitor, on the other hand, stores an electrical charge physically without any chemical reactions. The efficiency of an SC device exclusively depends on morphology and electrochemical activity/conductivity of the electrode materials.

10.2 CARBON NANOSTRUCTURES AS SUPERCAPACITORS

In 1957, Becker gave the concept of supercapacitor utilizing carbon electrode and aqueous H_2SO_4 as an electrolyte. Later on, NEC (Japan) gave moldable first aqueous electrolyte-based SC for saving the power units in electronics, and it was

227

the initiation of commercialization of the supercapacitors [1]. Carbon nanostructures such as Buckminster fullerenes [2–4], carbon nanotubes (CNTs) [5–9], and graphene [10–14] have a tremendous impact in various applications as they possess higher aspect ratios and surface areas, and outstanding thermal, electrical, optical, and mechanical properties. Two-dimensional carbon nanostructure such as graphene has the benefits of increased surface area and better conductivity; however, it gets restacked easily during the preparation [15]. Three-dimensional carbon nanostructures including activated carbon and template carbon are rich in pores in spite of high surface area, so their specific capacitance becomes limited at high current density [16]. Onion-like carbon possesses limited capacitance (~30 Fg^{-1}) but is fully accessible to ions [17]. In the last two decades, the development in the synthetic techniques of nanodomain carbon in a controlled manner and manipulation of the functionalization and self-assembly techniques in the multifunctional composite synthesis have been appreciably increased making them as the ideal candidates for energy storage devices [18–22]. These carbon nanostructures have been proposed as the electrode materials for SC in near future. Among all, graphene sheets and CNTs play an important role in electronic and energy storage devices nowadays due to good electrical conductivity with highly accessible surface area [5,10,23].

10.3 ONE-DIMENSIONAL CARBON NANOSTRUCTURES AS SUPERCAPACITORS

One can think about the one-dimensional CNTs as films of pure carbon rolled up to give a sheet-like structure (also mentioned in the literature as graphene). These films or sheets can be rolled up evenly or can be twisted after rolling, and so the structure of CNTs actually depends on the diameter of the tubular form or the degree of twisting which usually is defined by two indices labeled as n and m. The tubular structures where n-m is zero or a multiple of 3 possess electrons in conduction bands at room temperature, and thus, they conduct electricity smoothly and are called as metallic nanotubes. Other nanotubes, having a bandgap between 0.5 and 3.5 eV, are semiconductors in nature (Figures 10.1 and 10.2).

The performance of any supercapacitor usually depends on the physiochemical nature of electrode and the applied electrolytic component. In general, the pore size distribution greater than 0.5 nm is suitable, whereas improved surface area-to-volume ratio is required for increased capacitance in case of carbon materials. Therefore, the key factor to achieve high capacitance is maintaining an optimal balance between the mesoporosity and the unsaturated dangling surface orbitals in any nanocarbon material. In the last few years, although activated carbon materials were used due to high surface area, they resulted in limited capacitance owing to reduction of mesoporosity and poor electrolyte accessibility. Recently, CNTs-based nanocomposite materials have been given considerable attention and are frequently investigated as electrode materials in supercapacitors because of their environmental-benign nature and promising electrochemical performances. Chunming Niu et al. prepared flexible CNT sheet electrodes of uniform thickness from catalytically deposited CNTs, and the electrodes showed specific capacitances ranging from 102 and 49 Fg^{-1} at 1 and

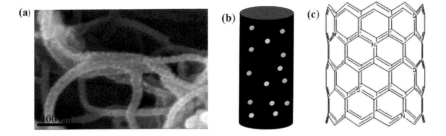

FIGURE 10.1   (a) SEM image of multiwalled CNTs, (b) schematic presentation of N- and S-doped CNTs (white spheres for N and yellow spheres for S), and (c) schematic presentation of zigzag form of N- and S-doped CNTs.

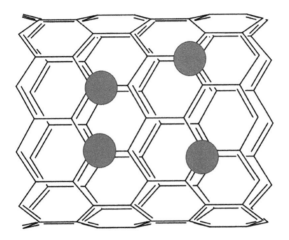

FIGURE 10.2   Schematic presentation of zigzag metal-doped (saffron color) CNTs.

100 Hz, respectively, in the presence of ionic electrolytes such as ~38 wt% H_2SO_4. They suggested that a power density of >8 kW kg^{-1} can be obtained with their CNT electrodes [24]. Frackowiak et al. estimated the capacitance, galvanostatic discharge, and impedance of multiwalled carbon nanotubes (MWCNTs) by cyclic voltammetry at the scan rates ranging from 1 to 10 mV s^{-1}. The specific capacitance values varied from 4 to 135 Fg^{-1}, and the authors ascertained that even with modest specific surface area (≤470 m^{2}g^{-1}), MWCNTs showed better results as supercapacitors compared to the activated carbons [25].

In 2001, Young Hee Lee et al. studied several parameters such as charging time and discharged current density components of the binder materials and hardening temperature of single-walled carbon nanotubes (SWCNTs) which exhibited higher specific capacitance of ~180 Fg^{-1} along with a power density of ~20 kW kg^{-1} in the range of 7–6.5 Wh kg^{-1} at 0.9 V in 7.5 N KOH [26]. In the same year, Frackowiak et al. compared MWCNTs and SWCNTs for supercapacitor applications and found that the former gave higher capacitance than the latter. They also reported pseudocapacitance effects in the presence of metallic particles or by deposition of conducting polypyrrole (PPY) [27]. Kuan-Hong Xue et al. reported the synthesis of MWCNTs and the specific capacitance value of ~365 Fg^{-1} as well as the discharge current density of ~210 mAg^{-1} of the conductor in the presence of 1 mol/L H$_2$SO$_4$ [28]. In 2005, Tadaoki Mitani found that composite of nano-engineered RuO$_2$ on the matrix of CNTs showed increase in capacitance value due to the fact that with the reduction in size, protons were ready to be absorbed by the inner part of RuO$_2$. They suggested that RuO$_2$ is responsible for extending the capacitance of electrode materials constructed using nanodomain carbon [29]. Later on, Chunsheng Du et al. reported a precisely improved specific power density (~30 kW kg^{-1}) for an electrical double-layer capacitor constructed using MWCNT nanocomposites [30].

To get an ultramodern energy storage device, a new hybrid supercapacitor was developed, combined with the advanced lithium-ion batteries and supercapacitor electrode materials [31–33]. Jinghong Li developed a hybrid framework with highly organized CNTs as cathode and a TiO$_2$-B nanowire anode [34] as supercapacitor electrode. Kenji Hata et al. controlled the fabricating process to prepare high-density SWCNTs for flexible heaters and supercapacitor electrodes [35]. Electrochemical performance of ultrathin (~1.0 mm) group of aligned CNT (ACNT) electrodes and entangled CNT (ECNT) electrodes was studied by Hao Zhang et al. by cyclic voltammetry, galvanostatic charge–discharge, and ac impedance. The results indicated that since ACNT electrode consisted of huge porous structure with uniform pore size, it gave superior specific capacitance and lesser equivalent series resistance with enhanced rate capability as compared to ECNT electrode [36]. Similarly, ACNT electrodes having lengths of ~150 μm were deposited on the substrate of metallic alloy by pyrolysis of iron (II) phthalocyanine in ethylene, where the specific capacitance was found to be about 57% (47 Fg^{-1}) at 1000 mV s^{-1} [37].

Gaoping Cao et al. reported ordered growth of manganese oxide nanoflower/CNT composite materials as electrodes having broad surface area and comprehensive conductivity and porous structure. They showed high capacitance (199 Fg^{-1} and 305 F/cm^3), outstanding rate capability (50.8% capacity retention at 77 Ag^{-1}), high-degree cycle life (3%), and capacity loss subsequently after 20,000 charge/discharge cycles [38]. In order to develop high-performance supercapacitors, Wen Lu et al. used the combination of ACNTs and ionic liquids which resulted in an improved overall surface area along with electrolytic accessibility and the capacitance value of the CNT electrode in the range from 35 Fg^{-1} to 440 Fg^{-1} after plasma etching [39]. On similar lines, using the unique properties of vertically organized and plasma-etched CNTs, Wen Lu et al. reported a new class of electrochemical capacitors possessing improved cell voltages (~4 V), energy density (~148 Wh kg^{-1}), and high power density (315 kW kg^{-1}) [40]. X. S. Zhao et al. briefly analyzed the recent progress achieved in performances of various electrode materials based on carbon used for supercapacitors along with significance of electrolytes. They also compared the aqueous and nonaqueous electrolytic solutions used in supercapacitors [41]. George Gruner et al. underlined the efficiency of printed thin-film SWCNT-based supercapacitors. The performances of these electronic materials displayed considerably higher energy and power densities compared to previously reported SWCNT-supported supercapacitor devices [42]. Hanqing Jiang et al. reported stretchable supercapacitors prepared using sinusoidal SWCNT macrofilms, an organic electrolyte, and a polymeric separator. The electrochemical performances were comparable with the supercapacitors which use pristine SWCNTs as electrodes, and interestingly, the performances were found to be unaffected even after applying 30% tensile strain [43].

In 2010, Hui Pan et al. reported in their review article that the capacitance values of CNTs are largely affected by many factors mainly including specific surface area, pore size, size distribution, porous nature, conductivity, and so on. The hybrid supercapacitor needs that either the CNTs should be homogenously protected by the polymers with exactly well-ordered thickness or the oxide nanoparticles must be chemically associated to the walls of CNTs [44]. The lower electrical conductivity of the metal oxide is one of the foremost concerns in pseudocapacitors. To overcome this dispute and for proficient use of MnO_2- and RuO_2-like metal oxides, Jie Liu et al. designed a nanocomposite film using CNT with conducting polymer and evaluated the electrochemical performance. They observed that specific capacitance of the integrated electrode can extend up to 427 Fg^{-1} and the electrode also showed long cycling stability, maintaining 99% of capacitance even after 1000 cycles [45]. Later on, Kenji Hata et al. fabricated a novel composite electrode using SWCNTs and single-walled nanohorns with a meso-/micropore structure which achieved a power rating of 1 MW kg^{-1} surpassing other electrodes [46]. Specific capacitance of 25 Fg^{-1} was obtained when CNT was directly used as a supercapacitor electrode without any catalyst or pre-/posttreatment [47]. A. K. Roy et al. [48] developed novel vertically aligned CNT-graphene structures, tuned their length, and further modified them using functional nanomaterials, like $Ni(OH)_2$, to increase the performance of the supercapacitors and achieve good cycling ability and increased rate capability. CNT networks, acting as common integration platform for photo-supercapacitor and component of symmetric charge storage, have been demonstrated by Madhavi Srinivasan et al. to yield a specific capacitance of 28 Fg^{-1}, and this value was upgraded to 80 Fg^{-1} by joining two organic photovoltaic (OPV) in series. Such assimilation opens up new avenues of flexibility and printability for hybrid prototype charge storage devices [49].

The great challenge in the development of electrochemical capacitors (ECs) is relatively high cost of the carbon nanomaterials and the need for a strategy to design, control, and tailor the synthetic methods for obtaining novel carbon-based composites that can perform more desirable functions for application in ECs, as illustrated in a review article in 2012 [50]. Sishen Xie and co-workers reported a repeated "halving" approach and separated thin films of SWCNTs into ultrathin films having different thicknesses [51]. These as-prepared ultrathin SWCNT films were used to fabricate transparent and flexible SWCNT-based supercapacitors having excellent electrochemical activities. This method improved the utilization of SWCNT films at a faster rate, retaining continuous reticulate structure along with highly uniform SWCNT film. To overcome the operational and cost limitations of thermal CVD, Il-Kwon Oh and co-workers reported 3D carbon nanostructures by growing CNTs on graphene nanosheets under microwave radiation in the presence of an ionic liquid and Pd catalyst. This 3D graphene/CNT/Pd showed extraordinarily high capacitance in 1 M solution of KOH, and the value of capacitance increased up to 46% after 600 cycles, confirming good electrochemical stability of electrode [52]. Usually, supercapacitor electrodes based on pure CNT exhibit low capacitance value, so semiconducting polymers like PPY have drawn much attention nowadays. Liangti Qu et al. tailored the surface property of CNTs and introduced some physical defects into the walls of CNTs which allowed the coating of polymer on both the inner and outer walls of CNTs. As a result, the polypyrrole/d-CNTs exhibited very high capacitance of 587 Fg^{-1} [53]. Huisheng Peng et al. fabricated an Integrated "Energy Wire"-based device by modifying a titanium wire with aligned titania nanotubes followed by twisting aligned fibers of CNT with the modified wire of Ti for simultaneous realization of both energy storage and photoelectric conversion. Interestingly, the device exhibited a high photoelectric conversion as well as storage efficiency of 1.5% [54]. A simplistic protocol has been employed for the designing of low-charge storage device containing composite of CNTs and conducting polymers; PANI was developed by De Chen and co-workers, which enhanced the flexibility of electrode, and as a result, it exhibited high specific capacitance of 705.8 Fg^{-1} and specific energy of 18.9 Wh kg^{-1} in H_2SO_4 (1 M) with high degree of specific power (~11.3 kW kg^{-1}) [55].

The rapid increase in CO_2 emission and global energy consumption, and the overall impact on environment due to the use of conventional energy resources are posing serious human health and environmental problems as well as increasing risks of energy security day by day. By the year 2050, the world will need to increase its energy supply tremendously; therefore, the supercapacitors-based international market is growing rapidly and steadily to meet the stringent requirements of energy crisis. Liming Dai et al. [56] reported in a review article that ACNT films with stretchability, charge mobility, and high transparency are promising electrodes for future stretchable supercapacitors, but more research is required in this area. Later, Huisheng Peng et al. developed MWCNT-PANI-based integrated device which was photo-charged to 0.73 V in 183 s, showing the discharge time of 137 s at the discharge current density of 1.4 mA cm^{-2} [57]. The specific capacitance was found to be 83 Fg^{-1}, and the storage efficiency of energy was 34% with both storage efficiency and energy conversion of 0.79%.

James M. Tour et al. prepared in situ CNT–graphene-based microsupercapacitors on Ni electrodes. Due to exceptional good rate capability of 400 V/s, these G/CNTC microdevices showed a power density of 115 W/cm^3 which can serve as highly compact power sources for the near-future electronic devices [58]. Interestingly, Menghe Miao and co-workers designed and fabricated a flexible thread-like supercapacitor, consisting of two single yarns of CNT along with arrays of polyaniline nanowires, which were about 50 nm in diameter and 400 nm in length. The capacitance of only CNT yarn-based supercapacitor did not reduce even after 800 cycles, but CNT@ PANI composite yarn supercapacitor showed 91% of its original capacitance after the same cycle [59].

10.4   ONE-DIMENSIONAL POROUS CARBON NANOSTRUCTURES AS SUPERCAPACITORS

The electrochemical properties of supercapacitors are dependent on the materials of electrode having suitable morphologies and pore structures. In the recent past, porous carbon materials encapsulated with sulfur were used as cathode material for both supercapacitors and Li-S batteries [60]. Xiaogang Zhang et al. developed N-doped CNTs using tubular-shaped PPY which showed high specific capacitance of 210 Fg^{-1}. Zhengjun Zhang and co-workers [61] found that addition of trace amount of CNTs into reduced graphene oxide (rGO) sheets significantly improved the rate capability and cyclic stability of the electrode than the pure rGO electrode, which increased the specific capacitance of 272 Fg^{-1} in between 0.8 and 0 V. The drawback of using porous carbon materials is the incompatibility of conductive pathways which reduces the electrical conductivity, so these materials always suffer from poor specific capacitances even at high current densities. Hence, Hui Dou et al. fabricated N/P co-doped carbon materials having appropriate pore size distribution which showed high specific capacitance of 280 Fg^{-1} with retention of 94% specific capacitance even after 10,000 cycles [62]. Recently, Zhanhu Guo and co-workers reported an article on the use of various carbon nanostructures including 1D CNTs for energy storage applications [63].

Very recently [64], 1D nanorod-based carbon materials with porous nature were synthesized from Al-based porous coordination polymers using pyrolysis which showed the energy density of 17.6 Wh kg^{-1} at a power density of 272.2 W kg^{-1}. Thus, high degree of ordering of carbon atoms contributed high power and opened up new doors for the construction of porous carbon-based advanced supercapacitors. Shuangxi Liu et al. reported that binary solvent method is efficient to produce electrode materials with high performance and hierarchical porous carbon nanofibers displayed a high capacitance of 251 Fg^{-1} and excellent cycling stability after 5000 cycles [65]. Later, Yusuke Yamauchi and co-workers reported that due to the mechanical instability, hierarchical porous carbon nanostructures are hardly converted into films of flexible nature. So, they designed carbon nanofibers consisting of microporous core and mesoporous shell using electrospinning assembly and demonstrated energy density up to 56.6 Wh kg^{-1} at a power density of 1.76 kW kg^{-1}, and achieved a power density of 113.76 kW kg^{-1} [66]. Amin M. Saleem et al. reviewed the use of 1D and 2D carbon nanomaterials and their nanocomposites as electrode materials for supercapacitors. The CNTs gave high energy and power density after functionalization with oxygen [67]. Although controlled functionalization of graphene and CNTs with tailor-made properties and structures is very difficult and characterization of the active sites in the functionalized CNTs is highly challenging, Liming Dai et al. reviewed the progress of functionalized CNTs and graphene by chemical modification, ball milling, and doping heteroatoms for supercapacitors and batteries. They concluded that the electrochemical activities of the supercapacitors using functionalized CNTs were highly improved due to their self-assembled 3D structures and well-defined mesoporous network [68].

In continuation of the modification of CNT electrodes, Al-Asadi et al. reported high-efficient supercapacitor electrodes giving a maximum specific capacitance of up to 192 Fg^{-1}, with energy of 3.8 Wh kg^{-1} and improved power density of 28 kW kg^{-1}, using zinc oxide nanowires and MWCNT composites. This composite was made by depositing ZnO directly in the matrix of aligned MWCNTs. As compared to pristine MWCNTs, ZnO@MWCNTs have ~12 times superior specific capacitance along with improved power and energy density [69]. Recently, the N-doped CNTs prepared through posttreatment process showed poor electrochemical properties due to low N-doping level but the pyrolysis of polymer precursors containing nitrogen was proved to be more effective to develop N-doped CNTs, as reported by Duck-Joo Yang et al. They prepared N-doped CNF mats by electrospinning the blends of polyacrylonitrile/poly(m-aminophenol) using thermal stabilization and carbonization and achieved the highest specific capacitance of 347.5 Fg^{-1} along with high cycling stability and 90.5% retention of capacitance after 10,000 cycles [69]. Motivated by the combination of vertically aligned structures, Wei Chen and co-workers developed PANI/VA-CNTs-based electrodes, which exhibited high specific capacitance of 403.3 Fg^{-1}, extremely high-distorted CNTs in the presence of HClO$_4$ electrolyte. This composite electrode material possessed high degree of specific capacitance of 314.6 Fg^{-1}, retaining 90.2% after 3000 cycles at 4 Ag^{-1} along with energy density of 98.1 Wh kg^{-1} [70]. Liming Dai summarized the recent developments in high-active supercapacitors using nanomaterials of carbon as well as the design and preparation of structures of electrodes and determination of their charge-storage mechanisms [71]. Following the advanced modification of CNTs, it was felt that electrodes prepared using CNTs usually suffer from severe performance degradation due to restacking during fabrication which hinders the ion accessibility. Hence, Kyunghoon Kim and co-workers concluded that the supercapacitors using the shortened SWCNTs exhibit 7-fold more specific capacitance than SWCNTs-based electrodes and enough good rate capabilities with 100% retention of capacity after 2500 cycles [72]. In view of the urgent demand for flexible energy storage devices, Jintao Zhang and co-workers discussed on the recent research progress in CNTs-based flexible supercapacitors for novel portable and wearable electronics [73].

10.5 CONCLUSION

In the twenty-first century, it is a challenge to achieve high power density and long-term cycling life in the energy storage devices. To overcome this problem, supercapacitors have started to play an important and crucial role in everyday life nowadays, as they have high-rate capability, high power density, and high durability. This chapter describes the improvement of the capacitance properties of 1D carbon nanostructures alone or in modified forms using heteroatoms and functional groups. Besides, considerable attention has been paid to the carbon nanotubes and conducting polymers-based composites, other forms of carbon nanostructures, and transition metal oxides.

REFERENCES

1. Kötz, R., and M. Carlen. 2000. Principles and applications of electrochemical capacitors. *Electrochim. Acta.* 45 (15–16):2483–2498.
2. Kroto, H., J. Heath, O. Brien, S.C. Curl, and R.F. Smalley. 1985. C$_{60}$, Buckminsterfullerene. *Nature* 318 (162):1216.
3. Krätschmer, W., L. D. Lamb, K. Fostiropoulos, and D. R. Huffman. 1990. Solid C60: A new form of carbon. *Nature* 347 (6291):354.
4. Diederich, F., and C. Thilgen. 1996. Covalent fullerene chemistry. *Science* 271 (5247):317–324.
5. Iijima, S. 1991. Helical microtubules of graphitic carbon. *Nature* 354 (6348):56.

6. Thess, A., R. Lee, P. Nikolaev, H. Dai, P. Petit, J. Robert, C. Xu, Y. H. Lee, S. G. Kim, and A. G. Rinzler. 1996. Crystalline ropes of metallic carbon nanotubes. *Science* 273 (5274):483–487.

7. Ebbesen, T., and P. Ajayan. 1992. Large-scale synthesis of carbon nanotubes. *Nature* 358 (6383):220.

8. Wei, B., R. Vajtai, Y. Jung, J. Ward, R. Zhang, G. Ramanath, and P. Ajayan. 2002. Microfabrication technology: Organized assembly of carbon nanotubes. *Nature* 416 (6880):495.

9. Yan, Y., M. B. Chan-Park, and Q. Zhang. 2007. Advances in carbon-nanotube assembly. *Small* 3 (1):24–42.

10. Novoselov, K. S., A. K. Geim, S. V. Morozov, D. Jiang, Y. Zhang, S. V. Dubonos, I. V. Grigorieva, and A. A. Firsov. 2004. Electric field effect in atomically thin carbon films. *Science* 306 (5696):666–669.

11. Geim, A. K., and K. S. Novoselov. 2010. The rise of graphene. *Nat. Mater.* 6:183–191.

12. Sutter, P. 2009. Epitaxial graphene: How silicon leaves the scene. *Nat. Mater.* 8 (3):171.

13. Novoselov, K. S., A. K. Geim, S. Morozov, D. Jiang, M. Katsnelson, I. Grigorieva, S. Dubonos, and A. A. Firsov. 2005. Two-dimensional gas of massless Dirac fermions in graphene. *Nature* 438 (7065):197.

14. Pletikosić, I., M. Kralj, P. Pervan, R. Brako, J. Coraux, A. N'diaye, C. Busse, and T. Michely. 2009. Dirac cones and minigaps for graphene on Ir (111). *Phys. Rev. Lett.* 102 (5):056808.

15. Wang, Y., Q. He, H. Qu, X. Zhang, J. Guo, J. Zhu, G. Zhao, H. A. Colorado, J. Yu, and L. Sun. 2014. Magnetic graphene oxide nanocomposites: Nanoparticles growth mechanism and property analysis. *J. Mater. Chem. C* 2 (44):9478–9488.

16. Jiang, H., P. S. Lee, and C. Li. 2013. 3D carbon based nanostructures for advanced supercapacitors. *Energy Environ. Sci.* 6 (1):41–53.

17. Portet, C., G. Yushin, and Gogotsi, Y. 2007. Electrochemical performance of carbon onions, nanodiamonds, carbon black and multiwalled nanotubes in electrical double layer capacitors. *Carbon* 45 (13):2511–2518.

18. Kimura, H., J. Goto, S. Yasuda, S. Sakurai, M. Yumura, D. N. Futaba, and K. Hata. 2013. Unexpectedly high yield carbon nanotube synthesis from low-activity carbon feedstocks at high concentrations. *ACS Nano* 7 (4):3150–3157.

19. Jiang, C., A. Saha, C. Xiang, C. C. Young, J. M. Tour, M. Pasquali, and A. A. Martí. 2013. Increased solubility, liquid-crystalline phase, and selective functionalization of single-walled carbon nanotube polyelectrolyte dispersions. *ACS Nano* 7 (5):4503–4510.

20. Torres, J. A., and R. B. Kaner. 2014. Graphene synthesis: Graphene closer to fruition. *Nat. Mater.* 13 (4):328.

21. Wang, H., Z. Xu, H. Yi, H. Wei, Z. Guo, and X. Wang. 2014. One-step preparation of single-crystalline Fe_2O_3 particles/graphene composite hydrogels as high performance anode materials for supercapacitors. *Nano Energy* 7:86–96.

22. Habisreutinger, S. N., T. Leijtens, G. E. Eperon, S. D. Stranks, R. J. Nicholas, and H. J. Snaith. 2014. Carbon nanotube/polymer composites as a highly stable hole collection layer in perovskite solar cells. *Nano Lett.* 14 (10):5561–5568.

23. Wang, H., J. T. Robinson, G. Diankov, and H. Dai. 2010. Nanocrystal growth on graphene with various degrees of oxidation. *J. Am. Chem. Soc.* 132 (10):3270–3271.

24. Niu, C., E. K. Sichel, R. Hoch, D. Moy, and H. Tennent. 1997. High power electrochemical capacitors based on carbon nanotube electrodes. *Appl. Phys. Lett.* 70 (11):1480–1482.

25. Frackowiak, E., K. Metenier, V. Bertagna, and F. Beguin. 2000. Supercapacitor electrodes from multiwalled carbon nanotubes. *Appl. Phys. Lett.* 77 (15):2421–2423.

26. An, K. H., W. S. Kim, Y. S. Park, J. M. Moon, D. J. Bae, S. C. Lim, Y. S. Lee, and Y. H. Lee. 2001. Electrochemical properties of high-power supercapacitors using single-walled carbon nanotube electrodes. *Adv. Funct. Mater.* 11 (5):387–392.

27. Frackowiak, E., K. Jurewicz, S. Delpeux, and F. Béguin. 2001. Nanotubular materials for supercapacitors. *J. Power Sources* 97:822–825.

28. Chen, Q.-L., K.-H. Xue, W. Shen, F.-F. Tao, S.-Y. Yin, and W. Xu. 2004. Fabrication and electrochemical properties of carbon nanotube array electrode for supercapacitors. *Electrochim. Acta.* 49 (24):4157–4161.

29. Kim, Y.-T., K. Tadai, and T. Mitani. 2005. Highly dispersed ruthenium oxide nanoparticles on carboxylated carbon nanotubes for supercapacitor electrode materials. *J. Mater. Chem.* 15 (46):4914–4921.

30. Du, C., J. Yeh, and N. Pan. 2005. High power density supercapacitors using locally aligned carbon nanotube electrodes. *Nanotechnology* 16 (4):350.

31. Chen, F., R. Li, M. Hou, L. Liu, R. Wang, and Z. Deng. 2005. Preparation and characterization of ramsdellite $Li_2Ti_3O_7$ as an anode material for asymmetric supercapacitors. *Electrochim. Acta.* 51 (1):61–65.

32. Li, H., L. Cheng, and Y. Xia. 2005. A hybrid electrochemical supercapacitor based on a 5 V Li-ion battery cathode and active carbon. *Solid-State Lett.* 8 (9):A433–A436.

33. Wang, Y.-G., and Y.-Y. Xia. 2005. A new concept hybrid electrochemical surpercapacitor: Carbon/$LiMn_2O_4$ aqueous system. *Electrochem. Commun.* 7 (11):1138–1142.

34. Wang, Q., Z. Wen, and J. Li. 2006. A hybrid supercapacitor fabricated with a carbon nanotube cathode and a TiO2–B nanowire anode. *Adv. Funct. Mater.* 16 (16):2141–2146.

35. Futaba, D. N., K. Hata, T. Yamada, T. Hiraoka, Y. Hayamizu, Y. Kakudate, O. Tanaike, H. Hatori, M. Yumura, and S. Iijima. 2006. Shape-engineerable and highly densely packed single-walled carbon nanotubes and their application as super-capacitor electrodes. *Nat. Mater.* 5 (12):987.

36. Zhang, H., G. Cao, Y. Yang, and Z. Gu. 2008. Comparison between electrochemical properties of aligned carbon nanotube array and entangled carbon nanotube electrodes. *J. Electrochem. Soc.* 155 (2):K19–K22.

37. Gao, L., A. Peng, Z. Y. Wang, H. Zhang, Z. Shi, Z. Gu, G. Cao, and B. Ding. 2008. Growth of aligned carbon nanotube arrays on metallic substrate and its application to supercapacitors. *Solid State Commun.* 146 (9–10):380–383.

38. Zhang, H., G. Cao, Z. Wang, Y. Yang, Z. Shi, and Z. Gu. 2008. Growth of manganese oxide nanoflowers on vertically-aligned carbon nanotube arrays for high-rate electrochemical capacitive energy storage. *Nano Lett.* 8 (9):2664–2668.

39. Lu, W., L. Qu, L. Dai, and K. Henry. 2008. Superior capacitive performance of aligned carbon nanotubes in ionic liquids. *ECS Trans.* 6 (25):257–261.

40. Lu, W., L. Qu, K. Henry, and L. Dai. 2009. High performance electrochemical capacitors from aligned carbon nanotube electrodes and ionic liquid electrolytes. *J. Power Sources.* 189 (2):1270–1277.

41. Zhang, L. L., and X. Zhao. 2009. Carbon-based materials as supercapacitor electrodes. *Chem. Soc. Rev.* 38 (9):2520–2531.

42. Kaempgen, M., C. K. Chan, J. Ma, Y. Cui, and G. Gruner. 2009. Printable thin film supercapacitors using single-walled carbon nanotubes. *Nano Lett.* 9 (5):1872–1876.

43. Yu, C., C. Masarapu, J. Rong, B. Wei, and H. Jiang. 2009. Stretchable supercapacitors based on buckled single-walled carbon-nanotube macrofilms. *Adv. Mater.* 21 (47):4793–4797.

44. Pan, H., J. Li, and Y. Feng. 2010. Carbon nanotubes for supercapacitor. *Nanoscale Res. Lett.* 5 (3):654.

45. Hou, Y., Y. Cheng, T. Hobson, and J. Liu. 2010. Design and synthesis of hierarchical MnO_2 nanospheres/carbon nanotubes/conducting polymer ternary composite for high performance electrochemical electrodes. *Nano Lett.* 10 (7):2727–2733.

46. Izadi-Najafabadi, A., T. Yamada, D. N. Futaba, M. Yudasaka, H. Takagi, H. Hatori, S. Iijima, and K. Hata. 2011. High-power supercapacitor electrodes from single-walled carbon nanohorn/nanotube composite. *ACS Nano.* 5 (2):811–819.

47. Shi, R., L. Jiang, and C. Pan. 2011. A single-step process for preparing supercapacitor electrodes from carbon nanotubes. *Soft Nanosci. Lett.* 1 (1):11.
48. Du, F., D. Yu, L. Dai, S. Ganguli, V. Varshney, and A. Roy. 2011. Preparation of tunable 3D pillared carbon nanotube–graphene networks for high-performance capacitance. *Chem. Mater.* 23 (21):4810–4816.
49. Wee, G., T. Salim, Y. M. Lam, S. G. Mhaisalkar, and M. Srinivasan. 2011. Printable photo-supercapacitor using single-walled carbon nanotubes. *Energy Environ. Sci.* 4 (2):413–416.
50. Li, J., X. Cheng, A. Shashurin, and M. Keidar. 2012. Review of electrochemical capacitors based on carbon nanotubes and graphene. *Graphene.* 1 (1):1.
51. Niu, Z., W. Zhou, J. Chen, G. Feng, H. Li, Y. Hu, W. Ma, H. Dong, J. Li, and S. Xie. 2013. A repeated halving approach to fabricate ultrathin single-walled carbon nanotube films for transparent supercapacitors. *Small.* 9 (4):518–524.
52. Sridhar, V., H.-J. Kim, J.-H. Jung, C. Lee, S. Park, and I.-K. Oh. 2012. Defect-engineered three-dimensional graphene–nanotube–palladium nanostructures with ultrahigh capacitance. *ACS Nano.* 6 (12):10562–10570.
53. Hu, Y., Y. Zhao, Y. Li, H. Li, H. Shao, and L. Qu. 2012. Defective super-long carbon nanotubes and polypyrrole composite for high-performance supercapacitor electrodes. *Electrochim. Acta.* 66:279–286.
54. Chen, T., L. Qiu, Z. Yang, Z. Cai, J. Ren, H. Li, H. Lin, X. Sun, and H. Peng. 2012. An integrated "energy wire" for both photoelectric conversion and energy storage. *Angew. Chem. Int. Ed.* 51 (48):11977–11980.
55. Huang, F., F. Lou, and D. Chen. 2012. Exploring aligned-carbon-nanotubes@ polyaniline arrays on household Al as supercapacitors. *ChemSusChem.* 5 (5):888–895.
56. Chen, T., and L. Dai. 2013. Carbon nanomaterials for high-performance supercapacitors. *Mater. Today.* 16 (7–8):272–280.
57. Yang, Z., L. Li, Y. Luo, R. He, L. Qiu, H. Lin, and H. Peng. 2013. An integrated device for both photoelectric conversion and energy storage based on free-standing and aligned carbon nanotube film. *J. Mater. Chem. A.* 1 (3):954–958.
58. Lin, J., C. Zhang, Z. Yan, Y. Zhu, Z. Peng, R. H. Hauge, D. Natelson, and J. M. Tour. 2012. 3-dimensional graphene carbon nanotube carpet-based microsupercapacitors with high electrochemical performance. *Nano Lett.* 13 (1):72–78.
59. Wang, K., Q. Meng, Y. Zhang, Z. Wei, and M. Miao. 2013. High-performance two-ply yarn supercapacitors based on carbon nanotubes and polyaniline nanowire arrays. *Adv. Mater.* 25 (10):1494–1498.
60. Xu, G., B. Ding, P. Nie, L. Shen, J. Wang, and X. Zhang. 2013. Porous nitrogen-doped carbon nanotubes derived from tubular polypyrrole for energy-storage applications. *Chem. Eur. J.* 19 (37):12306–12312.
61. Cui, X., R. Lv, R. U. R. Sagar, C. Liu, and Z. Zhang. 2015. Reduced graphene oxide/carbon nanotube hybrid film as high performance negative electrode for supercapacitor. *Electrochim. Acta.* 169:342–350.
62. Xu, G., B. Ding, J. Pan, J. Han, P. Nie, Y. Zhu, Q. Sheng, and H. Dou. 2015. Porous nitrogen and phosphorus co-doped carbon nanofiber networks for high performance electrical double layer capacitors. *J. Mater. Chem. A.* 3 (46):23268–23273.
63. Wang, Y., H. Wei, Y. Lu, S. Wei, E. Wujcik, and Z. Guo. 2015. Multifunctional carbon nanostructures for advanced energy storage applications. *Nanomaterials.* 5 (2):755–777.
64. Tan, Y., Q. Gao, J. Xu, and Z. Li. 2016. 1D nanorod-like porous carbon with simultaneous high energy and large power density as a supercapacitor electrode material. *RSC Adv.* 6 (56):51332–51336.
65. Zhang, L., Y. Jiang, L. Wang, C. Zhang, and S. Liu. 2016. Hierarchical porous carbon nanofibers as binder-free electrode for high-performance supercapacitor. *Electrochim. Acta.* 196:189–196.

66. Wang, J., J. Tang, Y. Xu, B. Ding, Z. Chang, Y. Wang, X. Hao, H. Dou, J. H. Kim, and X. Zhang. 2016. Interface miscibility induced double-capillary carbon nanofibers for flexible electric double layer capacitors. *Nano Energy.* 28:232–240.

67. Saleem, A. M., V. Desmaris, and P. Enoksson. 2016. Performance enhancement of carbon nanomaterials for supercapacitors. *J. Nanomater.* 2016. http://dx.doi.org/10.1155/2016/1537269.

68. Wang, B., C. Hu, and L. Dai. 2016. Functionalized carbon nanotubes and graphene-based materials for energy storage. *Chem. Commun.* 52 (100):14350–14360.

69. Al-Asadi, A. S., L. A. Henley, M. Wasala, B. Muchharla, N. Perea-Lopez, V. Carozo, Z. Lin, M. Terrones, K. Mondal, and K. Kordas. 2017. Aligned carbon nanotube/zinc oxide nanowire hybrids as high performance electrodes for supercapacitor applications. *J. Appl. Phys.* 121 (12):124303.

70. Wu, G., P. Tan, D. Wang, Z. Li, L. Peng, Y. Hu, C. Wang, W. Zhu, S. Chen, and W. Chen. 2017. High-performance supercapacitors based on electrochemical-induced vertical-aligned carbon nanotubes and polyaniline nanocomposite electrodes. *Sci. Rep.* 7:43676.

71. Chen, X., R. Paul, and L. Dai. 2017. Carbon-based supercapacitors for efficient energy storage. *Natl. Sci. Rev.* 4 (3):453–489.

72. Kim, T., M. Kim, Y. Park, E. Kim, J. Kim, W. Ryu, H. Jeong, and K. Kim. 2018. Cutting-processed single-wall carbon nanotubes with additional edge sites for supercapacitor electrodes. *Nanomaterials.* 8 (7):464.

73. Li, K., and J. Zhang. 2018. Recent advances in flexible supercapacitors based on carbon nanotubes and graphene. *Sci. Chi. Mater.* 61 (2):210–232.

Index

Note: Page numbers in italic and bold refer to figures and tables, respectively.